Eulerian Numbers

Selected Chapters of Number Theory: Special Numbers

Print ISSN: 2810-9716
Online ISSN: 2810-9724

Series Editor: Elena Deza (*Moscow Pedagogical State University, Russia*)

Published

Vol. 5 *Eulerian Numbers*
by Elena Deza

Vol. 4 *Catalan Numbers*
by Elena Deza

Vol. 3 *Stirling Numbers*
by Elena Deza

Vol. 2 *Perfect and Amicable Numbers*
by Elena Deza

Vol. 1 *Mersenne Numbers and Fermat Numbers*
by Elena Deza

Selected Chapters of Number Theory: Special Numbers – Volume 5

Eulerian Numbers

Elena Deza
Moscow Pedagogical State University, Russia

World Scientific

NEW JERSEY • LONDON • SINGAPORE • BEIJING • SHANGHAI • HONG KONG • TAIPEI • CHENNAI • TOKYO

Published by

World Scientific Publishing Co. Pte. Ltd.

5 Toh Tuck Link, Singapore 596224

USA office: 27 Warren Street, Suite 401-402, Hackensack, NJ 07601

UK office: 57 Shelton Street, Covent Garden, London WC2H 9HE

Library of Congress Cataloging-in-Publication Data
Names: Deza, Elena author
Title: Eulerian numbers / Elena Deza, Moscow Pedagogical State University, Russia.
Description: New Jersey : World Scientific, [2026] |
Series: Selected chapters of number theory: Special numbers, 2810-9716 ; volume 5 |
Includes bibliographical references and index.
Identifiers: LCCN 2025029062 | ISBN 9789819818655 hardcover |
ISBN 9789819818662 ebook | ISBN 9789819818679 ebook other
Subjects: LCSH: Euler's numbers
Classification: LCC QA246 .D483 2026
LC record available at https://lccn.loc.gov/2025029062

British Library Cataloguing-in-Publication Data
A catalogue record for this book is available from the British Library.

For any available supplementary material, please visit
https://www.worldscientific.com/worldscibooks/10.1142/14464#t=suppl

Desk Editors: Nambirajan Karuppiah/Cian Sacker Ooi

Typeset by Stallion Press
Email: enquiries@stallionpress.com

Mathematics is the queen of the Sciences and Number Theory is the queen of Mathematics.

— Gauss

God invented the integers; all else is the work of man.

— Kronecker

I know numbers are beautiful. If they aren't beautiful, nothing is.

— Erdős

Preface

Special numbers, amongst integers, are an important part of number theory, general mathematics, and several applied areas (such as cryptography).

While the "names" of some of them (Fermat numbers, Mersenne numbers, Fibonacci numbers, Catalan numbers, Bernoulli numbers, Stirling numbers, etc.) are familiar to mathematicians, detailed information about them is often scattered throughout the literature on special numbers. There are only a few books which systematically present their history and properties while giving main proofs.

Moreover, though such books exist (see, for example, [Bond93], [CoGu96], [Hogg69], [Line86], [LiNi96], [KLS01], [SlPl95], [Uspe76], and [Voro61]), many new results have appeared during the several decades since their publication; therefore, this material should be updated and reworked.

There are also several classic monographs on number theory ([Apos86], [Dick05], [HaWr79], [MSC96], [Ribe96], [Sier64], etc.) and combinatorics ([Comt74], [FlSe09], [GKP94] [GGL95Û] [Hall70], [Knut97], [LiWi92], [Mazu10], [Rior80], [Stan97], etc.) These are books of fundamental importance. But they treat number theory and combinatorics as a whole, and special numbers receive only limited attention there.

Our first book on special numbers, the monograph *Figurate Numbers* ([DeDe12]), discusses all such problems for polygonal numbers and their different generalisations, giving a full retrospective of the history of the most well-known classes of special integers.

In the current series, *Selected Chapters of Number Theory: Special Numbers*, we consider, following a similar structure, the history and properties of several other famous classes of special numbers. The offered books of the series are *Mersenne and Fermat Numbers*, *Perfect and Amicable Numbers*, *Stirling Numbers*, *Catalan Numbers*, *Eulerian Numbers*, and *Bernoulli Numbers*.

So, the main purpose of the series is to give a complete presentation of the theory of several classes of the classical special numbers, considering much of their properties, facts, and theorems with full proofs.

In particular, we expect:

- to find and organise much of the scattered material;
- present updated material, detailed in a clear and unified manner;
- consider a broad range of well-known as well as hidden connections of a given set of numbers with different mathematical problems;
- draw up a system of multilevel tasks.

The first book of the series, *Mersenne Numbers and Fermat Numbers*, was published in 2021 [Deza21].

The Mersenne numbers, i.e., positive integers of the form $M_n = 2^n - 1$, $n \in \mathbb{N}$, were considered by Euclid (4th century BC) and later by many Arabic and European mathematicians in an attempt to find all perfect numbers. Later, the numbers of the form 2^n-1 were named after Marin Mersenne (1588–1648), a French monk and mathematician, who discussed them in his work *Cogita Physico Mathematics* (1644) and stated some conjectures about the number's occurrence.

A *Fermat number* F_n is a positive integer of the form $F_n = 2^{2^n}+1$, $n \in \mathbb{Z}$, $n \geq 0$.

Fermat numbers were studied by the 17th-century French lawyer and mathematician Pierre de Fermat (1607–1665), who conjectured (1650) that these numbers would always give a prime for $n = 0, 1, 2, \ldots$. In fact, the first five Fermat numbers are primes: $F_0 = 3$, $F_1 = 5$, $F_2 = 17$, $F_3 = 257$, and $F_4 = 65537$. But in 1732,

Leonhard Euler (1707–1783), a Swiss-Russian mathematician, physicist, astronomer, geographer, logician, and engineer, found that F_5 is a composite number. So, Fermat's conjecture that F_n is always prime is clearly false. Moreover, beyond F_4, no further Fermat primes have been found.

Nowadays, the theory of Mersenne and Fermat numbers contains many interesting mathematical facts and theorems, as well as a number of important applications. The mathematical part of this theory is closely connected with classical arithmetic and number theory. It contains much information about the divisibility properties of the Mersenne and Fermat numbers, their connections with other classes of special numbers, etc. Moreover, the Mersenne and Fermat numbers are used in the construction of new large primes and have numerous applications in contemporary cryptography.

The second book of this series, published in 2023 [Deza23], was devoted to perfect and amicable numbers, as well as their numerous analogues and generalisations.

A *perfect number* is a positive integer n which is equal to the sum of its proper divisors: $n = \sum_{d|n, d<n} d$. The first four perfect numbers are 6, 28, 496, and 8128.

Perfect numbers were studied by Pythagoras and his followers, more for their mystical properties than for their number-theoretic properties. The first recorded mathematical result concerning perfect numbers is found in Euclid's *Elements*, written around 300 BC. Euclid proved that all even perfect numbers have the form $2^{k-1}(2^k - 1)$, where $2^k - 1$ is a prime. Two millennia later, Euler proved the converse of Euclid's result by showing that every even perfect number had to be of this form. So, the *Euclid–Euler theorem* states that *an even positive integer is a perfect number if and only if it can be represented as a product,* $2^{k-1}(2^k - 1)$*, where* $2^k - 1$ *is a prime.* In fact, it is a Mersenne prime; therefore, there exists a one-to-one relationship between even perfect numbers and Mersenne primes; each Mersenne prime generates one even perfect number, and vice versa.

As for odd perfect numbers, it is unknown whether any odd perfect numbers exist, though various results have been obtained.

Amicable numbers are two different natural numbers (m, n) related in such a way that the sum of the proper divisors of each is equal to the other number: $n = \sum_{d|m,d<m} d$, and $m = \sum_{d|n,d<n} d$. The smallest pair of amicable numbers $(220, 284)$ was known to the Pythagoreans, who credited them with many mystical properties. A general formula by which some of these numbers could be derived was invented circa 850 by the Iraqi mathematician Thābit ibn Qurra. *Thābit's rule* states that *if* $p = 3 \cdot 2^{n-1} - 1$, $q = 3 \cdot 2^n - 1$, *and* $r = 9 \cdot 2^{2n-1} - 1$ *are primes, then* $A = 2^n pq$ *and* $B = 2^n r$ *are a pair of amicable numbers.* This formula gives the pairs $(220, 284)$ for $n = 2$, $(17296, 18416)$ for $n = 4$, and $(9363584, 9437056)$ for $n = 7$, but no other such pairs are known.

The third book of this series, publishing in 2024 [Deza24], was devoted to the Stirling numbers, as well as their analogues and generalisations. There are two classes of the Stirling numbers. The *Stirling number of the second kind* $S(n, k)$ is the number of partitions of a given n-set into k non-empty subsets, for $n = 1, 2, \ldots$, $k = 1, 2, \ldots, n$. The *Stirling number of the first kind* $s(n, k)$ is defined by the equality $s(n, k) = (-1)^{n+k}|s(n, k)|$, where $|s(n, k)|$ is the number of ways of decomposing a given n-set into k cycles, for $n = 1, 2, \ldots, k = 1, 2, \ldots, n$. Using the Stirling numbers of the second kind, we can express the ordinary powers x^n of x through the falling powers $x^{\underline{k}}$ of x: $x^n = \sum_{k=0}^{n} S(n, k)x^{\underline{k}}$. Conversely, using the Stirling numbers of the first kind, we can express the falling factorials (falling powers) $x^{\underline{n}}$ of x through the ordinary powers x^k of x: $x^{\underline{n}} = \sum_{k=0}^{n} s(n, k)x^k$. These two sets of special numbers have a combinatorial nature. The construction of the Stirling numbers and many of their properties and main applications are closely connected with classical problems in combinatorics.

The fourth book of this series was published in 2025 [Deza25]. It is devoted to the Catalan numbers, as well as their relatives and generalisations. This number set is not as famous as the classes of special numbers listed above but has numerous combinatorial applications, many interesting properties, and different connections with other well-known sets of positive integers. The sequence $C_0, C_1, C_2, \ldots, C_n, \ldots$ of *Catalan numbers* is defined by the following recurrent law: $C_0 = 1$, and $C_n = C_0C_{n-1} + C_1C_{n-2} + \cdots + C_{n-1}C_0$ for $n \geq 1$. However, the Catalan numbers can be obtained using the main diagonal of Pascal's triangle: $C_n = \frac{1}{n+1} \cdot \binom{2n}{n}$, where $\binom{2n}{n} = \frac{(2n)!}{n!n!}$, $n \geq 0$, is a *central binomial coefficient*. There are a number of counting problems related to this numerical sequence. The list of such problems includes *Euler's polygon division problem*, Catalan's problem, Brackets' problem, and the handshake problem, as well as questions of enumeration of several classes of trees. A consideration of different aspects of the theory of Catalan numbers, as well as numerous relatives and generalisations of this class of special numbers (such as Motzkin numbers, Schröder numbers, Delannoy numbers, and Narayana numbers), allows us to give a broad overview of modern problems in enumerative combinatorics.

This fifth book of the series is devoted to the *Eulerian numbers*, as well as their relatives and generalisations.

The three main classes of special numbers represented in the book are defined in terms of different permutations: the *Eulerian numbers $E(n,k)$ of the first order* count the classical permutations of the set $\{1, 2, \ldots, n\}$ with k ascents; the *Eulerian numbers $E^2(n,k)$ of the second order* count the Stirling permutations of the set $\{1, 1, 2, 2, \ldots, n, n\}$ with k ascents; and the *Euler numbers E_n* count the (proper) alternating permutations of the set $\{1, 2, \ldots, n\}$. These number sets have many interesting properties, different combinatorial, arithmetical, and analytical applications, and numerous connections with other well-known sets of numbers, including the Euler zig-zag numbers, Bernoulli numbers, and Stirling numbers. The book describes all these aspects in detail.

In Chapter 1, we consider some important questions which will be taken up again in later chapters. In this chapter, we present the following: main combinatorial configurations and formulas used for their calculation; binomial and multinomial theorems; construction and simple properties of Pascal's triangle; classical combinatorial identities; basic notions of graph theory; main facets of the theory of set partitions; main facets of the theory of recurrent equations and generating functions; and basic notations related to asymptotic formulas. We also present all the necessary definitions, give many examples, and collect the main properties of the considered mathematical objects, mostly without proofs but with a large list of references.

Chapter 2 describes the main aspects of *permutations*. Starting with key definitions, we consider the different representations of permutations and list the various classical types of permutations, including cycles and transpositions. Furthermore, we study basic notions concerning the permutations of totally ordered sets, including ascents, descents, runs, excedances, and inversions; we analyse their properties and reveal important interrelations. At the end of the chapter, we consider the connections of permutations with different classes of special numbers and give a detailed list of the main types of permutations of a given set (or multiset).

Chapter 3 contains a rich review of the mathematical theory of *Eulerian numbers of the first order* $E(n,k)$, with all the necessary proofs and many examples. Starting with the history of the numbers, we consider the combinatorial definition of these numbers in terms of permutations, obtain their recurrent relation, construct the triangular array containing this numerical set (*Euler's triangle*), and collect the most important combinatorial problems related to the Eulerian numbers of the first order. Moreover, we give a closed formula for the numbers $E(n,k)$, construct the generating functions related to the different sequences of the numbers $E(n,k)$, list many properties of this number set, and present a large collection of possible connections of Eulerian numbers of the first order with other classes of special numbers, including the elements of Pascal's triangle, Stirling numbers of the first and second kinds, Bernoulli numbers, factorial numbers, Mersenne numbers, and Catalan numbers.

In Chapter 4, we consider similarly the main aspects of the theory of *Eulerian numbers of the second order* $E^2(n,k)$. The structure of Chapter 4 is the same as that of Chapter 3. Starting with the history of the question, we consider the combinatorial definition of the Eulerian numbers of the second order, obtain the recurrent relation for these numbers, and construct the corresponding triangle, collecting its elementary properties. After we consider some combinatorial problems related to the Eulerian numbers of the second order, we give the closed formula for the numbers $E^2(n,k)$, construct the generating function of this sequence, present a rich collection of possible connections of the Eulerian numbers of the second order with other classes of special numbers, and discuss the Eulerian numbers of the mth order, which are a common generalisation of the two considered classes of the Eulerian numbers.

Chapter 5 contains the classical aspects of the theory of *Euler numbers* E_n, as well as their close relatives, the *Euler zig-zag numbers*, and the *secant* and *tangent numbers*. The structure of this chapter is similar to the previous one, beginning with a combinatorial definition in terms of alternating permutations, followed by the recurrent relation, generation function, closed formula, properties, and connections with other classes of special numbers. However, in the last part of this chapter, we present a list of the other well-known numbers related to the name of Euler. We speak about *Euler's lucky numbers* and the *convenient Euler numbers*, which are connected with the theory of binary quadratic forms. We recall the main properties of the famous *Euler's number* e. We give an overview of widely known facts about the *Euler–Mascheroni constant* $\gamma = \lim_{n\to\infty}\left(\sum_{k=1}^{n}\frac{1}{k} - \log n\right)$.

Chapter 6 collects some remarkable individual numbers related to the theory of Eulerian numbers, as well as the general theory of special numbers.

In Chapter 7, a mini dictionary provides a huge list of all classes of the special numbers mentioned in the text.

Chapter 8, Exercises, contains an exclusive collection of problems connected with the theory of Eulerian numbers and their relatives, with either full solutions or, at least, short proof hints.

The target audience of the book includes university professors and students (especially graduate students) interested in number theory, combinatorics, general algebra, and related fields, as well as school teachers, general readers, and amateur mathematicians.

Specifically, the recurrent and combinatorial constructions of all the considered number sets are accessible to undergraduate students and general readers of mathematics, while some properties are slightly more difficult; this part of the book is more involved and requires some (still basic) mathematical and computational background.

The book is organised in such a way that it can be used as source material for individual scientific work by university students and even, to some extent, school students. In fact, the material in the book was already used by many students of Moscow Pedagogical State University in the preparation of their bachelor's or master's theses.

Notations

- X (Y, Z, A, B, C, S, etc.) – a set; $X = \{x_1, \ldots, x_n\}$ – a *finite set*; $|X| = n$ – the number n of elements in the finite set $X = \{x_1, \ldots, x_n\}$; 2^X – the *power set* of X, i.e., the set of all its subsets; $|2^X| = 2^n$; $V_n = \{1, 2, \ldots, n\}$.
- $X = \{x_1, \ldots, x_1, \ldots, x_n, \ldots, x_n\}$ – a *finite multiset*; $Q_n^m = \{1, \ldots, 1, 2, \ldots, 2, \ldots, n, \ldots, n\}$ – the multiset containing each element $1, 2, \ldots, n$ exactly m times; $Q_n = Q_n^2$.
- $X \times Y = \{(x, y) | x \in X, y \in Y\}$ – the *Cartesian product* of the sets X and Y, i.e., the set of ordered pairs (x, y) of elements $x \in X$ and $y \in Y$.
- $\mathbb{N} = \{1, 2, 3, \ldots\}$ – the set of *positive integers* (or *natural numbers*); $V_n = \{1, 2, 3, \ldots, n\}$ – the set of the first n positive integers.
- $\mathbb{Z} = \{\ldots, -3, -2, -1, 0, 1, 2, 3, \ldots\}$ – the set of *integers*.
- $\mathbb{Q}$ – the set of *rational numbers*.
- $\mathbb{R}$ – the set of *real numbers*.
- $\mathbb{C} = \{a + b \cdot i \,|\, a, b \in \mathbb{R}, i^2 = -1\}$ – the set of *complex numbers*.
- $b|a$ – a non-zero integer b divides an integer a: $a = bc$, where $c \in \mathbb{Z}$.
- $gcd(a_1, \ldots, a_n) = (a_1, \ldots, a_n)$ – the *greatest common divisor* of integers $a_1, \ldots, a_n$, at least one of which is non-equal to 0, i.e., the greatest integer dividing $a_1, \ldots, a_n$. If $gcd(a_1, \ldots, a_n) = 1$, the numbers $a_1, \ldots, a_n$ are called *relatively prime*, or *coprime*; if $gcd(a_i, a_j) = 1$ for any distinct $i, j \in \{1, \ldots, n\}$, the numbers $a_1, \ldots, a_n$ are called *pairwise relatively prime*.

- $lcm(a_1, \ldots, a_n) = [a_1, \ldots, a_n]$ – the *least common multiple* of non-zero integers $a_1, \ldots, a_n$, i.e., the least positive integer divided by $a_1, \ldots, a_n$.
- $rest(a, b)$ – the *reminder* of an integer a after its *integer division* by a positive integer b: $a = b \cdot q + rest(a, b)$, where $q, rest(a, b) \in \mathbb{Z}$ and $0 \leq rest(a, b) < b$.
- $P = \{2, 3, 5, 7, 11, 13, 17, 19, \ldots\}$ – the set of *prime numbers*, i.e., the positive integers having exactly two positive integer divisors; $p, q, p_1, p_2, \ldots, p_k, \ldots, q_1, q_2, \ldots, q_s, \ldots$ – prime numbers.
- $n = p_1^{\alpha_1} \cdot \cdots \cdot p_k^{\alpha_k}$, $p_1 < \cdots < p_k \in P$, $\alpha_1, \ldots, \alpha_k \in \mathbb{N}$ – the *prime factorisation* of a positive integer $n > 1$, i.e., its representation as a product of positive integer powers of different prime numbers $p_1, \ldots, p_k$.
- $S = \{4, 6, 8, 9, 10, 12, 14, 15, 16, 18, \ldots\}$ – the set of *composite numbers*, i.e., the positive integers having more than two positive integer divisors.
- $n = (c_s c_{s-1} \ldots c_1 c_0)_g = c_s g^s + c_{s-1} g^{s-1} + \cdots + c_1 g + c_0$, $g \in \mathbb{N} \backslash \{1\}$, $0 \leq c_i \leq g - 1$, $c_s \neq 0$ – representation of a positive integer n in *base* g.
- $\lfloor x \rfloor$, $x \in \mathbb{R}$ – *floor function*: the largest integer less than or equal to x.
- $\lceil x \rceil$, $x \in \mathbb{R}$ – *ceiling function*: the smallest integer greater than or equal to x.
- $[x]$, $x \in \mathbb{R}$ – *nearest integer function*: the nearest integer to x.
- $||x|| = \min\{\{x\}, 1 - \{x\}\}$, $x \in \mathbb{R}$ – the distance between x and the integer closest to x.
- $a \equiv b(mod\ n)$ – an integer a is *congruent* to an integer b modulo n, $n \in \mathbb{N}$, i.e., $n|(a - b)$.
- π (or α, β, τ, $\ldots$); $\pi = (\pi_1, \pi_2, \ldots, \pi_n) = \pi_1 \pi_2 \cdots \pi_n$ – *permutation* of the set $\{1, 2, \ldots, n\}$, i.e., a bijection π of $\{1, 2, \ldots, n\}$ into itself, $\pi(k) = \pi_k$, $k = 1, 2, \ldots, n$; S_n – *symmetric group* of all permutations on $\{1, 2, \ldots, n\}$.
- $\{a_n\}_{n=0}^{\infty} = \{a_n\}_n$ – *sequence* $a_0, a_1, a_2, \ldots, a_n, \ldots$; $\sum_{n=0}^{\infty} a_n = a_0 + a_1 + a_2 + \cdots + a_n + \cdots$ – the sum of the elements of the sequence $\{a_n\}_n$; $\prod_{n=0}^{\infty} a_n = a_0 \cdot a_1 \cdot a_2 \cdot \cdots \cdot a_n \cdot \cdots$ – the product of the elements of the sequence $\{a_n\}_n$.

- $f(x) = a_0 + a_1x + a_2x^2 + \cdots + a_nx^n + \cdots, |x| < r$ – *generating function* of the sequence $a_0, a_1, a_2, \ldots, a_n, \ldots$.
- $g(x) = a_0 + a_1\frac{x}{1!} + a_2\frac{x^2}{2!} + \cdots + a_n\frac{x^n}{n!} + \cdots, |x| < r$ – *exponential generating function* of the sequence $a_0, a_1, a_2, \ldots, a_n, \ldots$.
- $b_0c_{n+k} + b_1c_{n+k-1} + \cdots + b_nc_k = 0$, $b_0, \ldots, b_n \in \mathbb{R}$, $b_0 \neq 0$ – *linear recurrent equation of nth order* for a sequence $c_0, c_1, c_2, \ldots, c_{n-1}$, $c_n = -\frac{b_1}{b_0}c_{n-1} - \cdots - \frac{b_n}{b_0}c_0$, $c_{n+1} = -\frac{b_1}{b_0}c_n - \cdots - \frac{b_n}{b_0}c_1, \ldots$, $c_{n+k} = -\frac{b_1}{b_0}c_{n+k-1} - \cdots - \frac{b_n}{b_0}c_k, \ldots$ with the *initial values* $c_0, c_1, \ldots, c_{n-1}$.
- $A = ((a_{ij}))$, $1 \leq i, j \leq n$ – square $n \times n$ matrix with real entries a_{ij}. $det\, A$ or $|A|$ – *determinant* of a given $n \times n$ matrix $A = ((a_{ij}))$.
- $\log_a x$, $a, x \in \mathbb{R}$, $a > 0, a \neq 1$, $x > 0$ – *logarithm x with base a*, i.e., a number $y \in \mathbb{R}$ such that $a^x = y$; $\log x = \ln x$ – *natural logarithm* $\log_e x$; $\lg x$ – *decimal logarithm* $\log_{10} x$.
- $n!$ – *factorial* of a non-negative integer n: $n! = 1 \cdot 2 \cdot \cdots \cdot n$, $n \in \mathbb{N}$; $0! = 1$.
- $n\sharp$ – *primorial* of a positive integer n: the product of all primes less than or equal to n.
- $n_\alpha! = n(n-\alpha)(n-2\alpha)\cdots$, $\alpha \in \mathbb{N}$ – *multifactorial* of a positive integer n; $n!!$ – *double factorial* of a positive integer n: $n!! = n \cdot (n-2) \cdot (n-4) \cdot \cdots$.
- $x^{\underline{n}}$, $x \in \mathbb{R}$, $n \in \mathbb{N}$ – *falling power* (or *falling factorial*) of x: $x^{\underline{n}} = x \cdot (x-1) \cdot \cdots \cdot (x-n+1)$; $x^{\underline{0}} = 1$.
- $x^{\overline{n}}$, $x \in \mathbb{R}$, $n \in \mathbb{N}$ – *raising power* (or *raising factorial*) of x: $x^{\overline{n}} = x \cdot (x+1) \cdot \cdots \cdot (x+n-1)$; $x^{\overline{0}} = 1$.
- $\binom{n}{m}$ – *binomial coefficients*; C_n^m – the number of *combinations*, i.e., the ways to choose a non-ordered subset of m distinct elements from a fixed set of n distinct elements; $\binom{n}{m} = C_n^m = \frac{n!}{m!(n-m)!}$, $0 \leq m \leq n$. They form *Pascal's triangle*.
- $\binom{n}{n_1,\, n_2,\, \ldots,\, n_k}$, where $n_1, n_2, \ldots, n_k$ are non-negative integers, such that $n_1 + n_2 + \cdots + n_k = n$ – *multinomial coefficients*; $P_{(n_1, n_2, \ldots, n_k)}$ – the number of *permutations with repetitions, where the first element is used n_1 times, ..., the kth element is used n_k times*, i.e., the ways to construct an ordered multiset of $n = n_1 + n_2 + \cdots + n_k$ elements from a fixed non-ordered multiset of n elements, where the first element is used n_1 times, ..., *the* kth element is used n_k

times; $\binom{n}{n_1,\, n_2,\, \ldots,\, n_k} = P_{(n_1,n_2,\ldots,n_k)} = \frac{n!}{n_1!n_2!\ldots n_k!}$; $\binom{n}{k,l}$ – *threenomial coefficients.*

- P_n – the number of *permutations of n elements*, i.e., the ways to construct an ordered set of n distinct elements from a fixed non-ordered set of n distinct elements: $P_n = n!$, $n \geq 0$.
- A_n^k (or P_k^n, ${}_nP_k$, ${}^nP_k$, $P_{n,k}$, $P(n,k)$) – the number of *partial permutations from n elements by k elements*, i.e., the ways to choose an ordered subset of k distinct elements from a fixed set of n distinct elements: $A_n^k = \frac{n!}{(n-k)!}$, $n, k \geq 0, n \geq k$.
- $\overline{A}_n^k$ – the number of *partial permutations from n elements by k elements with repetitions*, i.e., the ways to choose an ordered multiset of k (not necessary distinct) elements from a fixed set of n distinct elements: $\overline{A}_n^k = n^k$, $n, k \geq 0$.
- $\overline{C}_n^k$ (or $\left(\!\binom{n}{k}\!\right)$) – the number of *combinations from n elements by k elements with repetitions*, i.e., the ways to choose a non-ordered multiset of k (not necessary distinct) elements from a fixed set of n distinct elements: $\overline{C}_n^k = C_{n+k-1}^k = \frac{(n+k-1)!}{k!(n-1)!}$, $n, k \geq 0$.
- $f(x) = O(g(x))$ – *big-O notation*: for a complex-valued function f and a real-valued function g, $g(x) > 0$ for all large enough values of x, there exist a positive real number C and a real number x_0 such that $|f(x)| \leq C \cdot g(x)$ for all $x \geq x_0$.
- $f(x) = o(g(x))$ – *little-o notation*: for a complex-valued function f and a real-valued function g, $g(x) > 0$ for all large enough values of x, $\lim_{x\to\infty} \frac{f(x)}{g(x)} = 0$.
- $f(x) \sim g(x)$: for a complex-valued function f and a real-valued function g, $g(x) > 0$ for all large enough values of x, $\lim_{x\to\infty} \frac{f(x)}{g(x)} = 1$.
- $e = 2.718281828\ldots$ – *Euler's number*: $e = \lim_{n\to\infty}(1 + \frac{1}{n})^n$.
- $\gamma = 0.5772156649\ldots$ – the *Euler–Mascheroni constant*: $\gamma = \lim_{n\to\infty}\left(\sum_{k=1}^{n} \frac{1}{k} - \log n\right)$.
- $\Gamma(z)$ – the *gamma function*: $\Gamma(z) = \int\limits_0^\infty t^{z-1}e^{-t}\,dt, \quad z = x + iy \in \mathbb{C}$, $x > 0$; $\Gamma(n+1) = n!$ for $n \in \mathbb{N}$.
- δ_{nm} – the *Kronecker delta-function*: $\delta_{nm} = 1$ for $n = m$, and $\delta_{nm} = 0$ for $n \neq m$.

About the Author

Elena (Ivanovna) Deza is a full professor at Moscow Pedagogical State University, where she has been since 1988. Her research focuses on analytic number theory, discrete mathematics, the theory of discrete metric spaces, and problems of mathematical education. She has written over 50 research papers, several mathematical textbooks and books on teaching mathematics in Russian, and several books/monographs in English, including *Figurate Numbers* (2012), *Mersenne Numbers and Fermat Numbers* (2021), *Perfect and Amicable Numbers* (2023), and *Stirling Numbers* (2024). Her mathematical teaching publications are mainly published in the following leading Russian journals: *Mathematics in School*, *Kvant*, and the *Herald of the University of Russian Academy of Education.* One of her special courses at the Moscow Pedagogical State and Moscow Independent Universities is "Topics in Special Numbers".

Contents

Chapter 1

Preliminaries

1.1. Combinatorial Configurations

Basic definitions

1.1.1. To eliminate the influence of a particular type of object, we use the general language of set theory: we talk about sets and their subsets, as well as the operations of union, intersection, difference, symmetric difference, Cartesian product, etc., of two or several sets. Usually, we speak about finite sets. For a set A consisting of elements $a_1, a_2, \ldots, a_n$, we write $A = \{a_1, a_2, \ldots, a_n\}$. We call the number n of the elements of the set A the *cardinality* of the set A and denote by $|A| : |A| = n$. In this case, we use for the set A the name *n-set*.

So, the sets $A = \{1, 2, 3\}$, $B = \{a, b, c\}$, and $C = \{\triangle, \diamondsuit, ©\}$ are examples of 3-sets.

Note that $\{1, 2, 3\}$ and $\{3, 1, 2\}$ give the same set A. Thus, the number 123 cannot be considered a 3-set since it differs from the numbers 213, 231, 132, 312, and 321 by the order of its elements.

Sets that take order into account are called *ordered sets*, or *tuples*. An ordered set consisting of elements $a_1, a_2, \ldots, a_n$, is written as $(a_1, a_2, \ldots, a_n)$. So, using a 3-set $A = \{1, 2, 3\}$, we get six ordered sets: $(1, 2, 3), (1, 3, 2), (2, 1, 3), (2, 3, 1), (3.1, 2)$, and $(3, 2, 1)$, which are called *permutations* of the elements of the set A.

1.1.2. A *multiset* (or *bag*, *mset*) is a modification of the concept of a set that, unlike a set, allows multiple instances for each of its elements. The number of instances, given for each element, is called the *multiplicity* of that element in the multiset.

For example, the set $A = \{1, 2, 3\}$ contains only elements 1, 2, and 3, each having multiplicity 1, when $\{1, 2, 3\}$ is seen as a multiset. In the multiset $\{1, 1, 2, 3\}$, the element 1 has multiplicity 2, and the elements 2 and 3 have multiplicities 1. In the multiset $\{1, 1, 2, 2, 3\}$, 1 and 2 both have multiplicity 2, while 3 has multiplicity 1.

As with sets, and in contrast to tuples, order does not matter in discriminating multisets, so $\{1, 1, 2, 3\}$ and $\{1, 2, 1, 3\}$ denote the same multiset.

Multisets that take order into account are called *ordered multisets* (or *strings*).

For example, using the (non-ordered) multiset $A = \{1, 1, 2\}$, we get three ordered multisets: $(1, 1, 2)$, $(1, 2, 1)$, and $(2, 1, 1)$, which are called *permutations with repetitions*.

The *cardinality* of a multiset is the sum of the multiplicities of all its elements. For example, in the multiset $\{1, 1, 2, 2, 2, 3\}$, the multiplicities of the members 1, 2, and 3 are, respectively, 2, 3, and 1; therefore, the cardinality of this multiset is 6.

1.1.3. *Permutations* of m elements (or *m-permutations*) are all kinds of ordered sets that can be obtained from m different elements of a given m-set. The number of permutations of m elements is indicated by the symbol P_m.

For example, $P_3 = 6$ since, using a (non-ordered) 3-set $A = \{1, 2, 3\}$, we can get exactly six permutations (ordered sets): $(1, 2, 3), (1, 3, 2), (2, 1, 3), (2, 3, 1), (3, 1, 2)$, and $(3, 2, 1)$.

Let's try now to get all the subsets of the set $A = \{1, 2, 3\}$, each containing two elements. These subsets are $\{1, 2\}$, $\{1, 3\}$, and $\{2, 3\}$. They are called *combinations from* 3 *elements by* 2 *elements* (or 2*-combinations of a* 3*-set*).

In general, *combinations* from m elements by n elements (or *n-combinations of an m-set*) are all (non-ordered) sets, consisting of n different elements, selected from m different elements of a given

m-set. The number of combinations from m elements by n elements is indicated by the symbol C_m^n.

For example, $C_3^2 = 3$, as there are exactly three 2-combinations of the 3-set $\{1,2,3\}$: $\{1,2\}$, $\{1,3\}$, and $\{2,3\}$.

If we want to obtain from a given 3-set (say, from the set $A = \{1,2,3\}$) all ordered 2-sets, then we get the tuples $(1,2)$, $(2,1)$, $(1,3)$, $(3,1)$, and $(2,3)$, $(3,2)$. They are called *partial permutations from* 3 *elements by* 2 *elements* (or *partial* 2*-permutations of a given* 3*-set*).

Partial permutations from m elements by n elements (or *partial* n*-permutations of a given* m*-set*) are all ordered sets, consisting of n different elements, selected from m different elements of a given m-set. The number of partial n-permutations of a given m-set is indicated by the symbol A_m^n.

So, $A_3^2 = 3$, as there are exactly six partial 2-permutations of the elements $\{1,2,3\}$: $(1,2)$, $(2,1)$, $(1,3)$, $(3,1)$, $(2,3)$, and $(3,2)$.

1.1.4. Similarly, we can define permutations, combinations, and partial permutations with repetitions.

Permutations with repetitions, in which the first element is repeated α_1 times, the second element is repeated α_2 times, ..., the kth element is repeated α_k times (or $(\alpha_1, \ldots, \alpha_k)$*-permutations*), are all ordered multisets, which can be obtained from a non-ordered multiset, in which the first element is repeated α_1 times, the second element is repeated α_2 times, ..., the kth element is repeated α_k times. The number of such permutations with repetitions is indicated by the symbol $P_{(\alpha_1,\ldots,\alpha_k)}$.

For example, permutations with repetitions in which the element 1 repeats two times, and the elements 2 and 3 are each used once, i.e., $(2,1,1)$-permutations of the multiset $\{1,1,2,3\}$, are as follows: $(1,1,2,3)$, $(1,1,3,2)$, $(1,2,1,3)$, $(1,2,3,1)$, $(1,3,1,2)$, $(1,3,2,1)$, $(2,1,1,3)$, $(2,1,3,1)$, $(2,3,1,1)$, $(3,1,1,2)$, $(3,1,2,1)$, and $(3,2,1,1)$. Thus, the number $P_{(2,1,1)}$ of such permutations with repetitions is 12: $P_{(2,1,1)} = 12$. Obviously, $P_{(1,1,1)} = P_3 = 6$. (Compare this result with $P_{(3)} = 1$.)

Combinations with repetitions from m elements by n elements (or n*-combinations with repetitions of a given* m*-set*) are all non-ordered multisets, consisting of n, not necessarily distinct, elements, selected from m different elements of a given m-set. The number of n-combinations with repetitions of a given m-set is indicated by the symbol $\overline{C}_m^n$.

For example, all 2-combinations with repetitions of the 3-set $\{1,2,3\}$ are $\{1,2\}$, $\{1,3\}$, $\{2,3\}$, $\{1,1\}$, $\{2,2\}$, and $\{3,3\}$. Thus, it holds that $\overline{C}_3^2 = 6$.

Partial permutations with repetitions from m elements by n elements (or partial *n-permutations with repetitions of a given m-set*) are all ordered multisets, consisting of n, not necessarily distinct, elements, selected from m different elements of a given m-set. The number of partial n-permutations with repetitions of a given m-set is indicated by the symbol $\overline{A}_m^n$.

For example, all partial 2-permutations with repetitions of the set $\{1,2,3\}$ are $(1,2)$, $(2,1)$, $(1,3)$, $(3,1)$, $(2,3)$, $(3,2)$, $(1,1)$, $(2,2)$, and $(3,3)$. Thus, $\overline{A}_3^2 = 9$.

Addition and multiplication principles

1.1.5. So, the main combinatorial configurations are *combinations*, *partial permutations*, and *permutations*, with or without repetitions.

It is not difficult to show that the number P_m of permutations of m different elements is equal to $m!$, the number A_m^n of partial permutations from m elements by n elements is equal to $\frac{m!}{(m-n)!}$, the number C_m^n of combinations from m elements by n elements is equal to $\frac{m!}{n!\cdot(m-n)!}$, the number $\overline{A}_m^n$ of partial permutations with repetitions from m elements by n elements is equal to m^n, the number $\overline{C}_m^n$ of combinations with repetitions from m elements by n elements is equal to $P_{(m-1,n)} = \frac{(m+n-1)!}{n!\cdot(m-1)!}$, and the number $P_{(\alpha_1,\alpha_2,\ldots,\alpha_k)}$ of permutations with repetitions, in which the ith element is repeated α_i times, $i = 1,\ldots,k$, is equal to $\frac{(\alpha_1+\alpha_2+\ldots+\alpha_k)!}{\alpha_1!\cdot\alpha_2!\cdot\ldots\cdot\alpha_k!}$.

In order to prove these formulas, as well as to solve the majority of simple combinatorial problems, one should use the combinatorial *addition and multiplication principles* (or *rules of sum and product*).

These principles, together with the inclusion–exclusion principle, constitute the basic counting principles, i.e., the fundamental principles of counting. They are often used for different enumerative purposes.

The *rule of sum* is an intuitive principle, stating that if there are A possible outcomes for an event (or ways to do something) and B possible outcomes for another event (or ways to do another thing), and the two events cannot both occur (or the two things cannot both be done), then there are $A+B$ total possible outcomes for the events (or total possible ways to do one of the things). Formally, we use for this conclusion the following well-known fact from set theory: the sum of the cardinalities of two disjoint sets A and B is equal to the cardinality of their union $A \cup B$.

The *rule of product* is another intuitive principle, stating that if there are A ways to do something and B ways to do another thing, then there are $A \cdot B$ ways to do both things. Formally, we use for this conclusion the following well-known fact from set theory: the product of the cardinalities of two sets A and B is equal to the cardinality of their *Cartesian product* $A \times B = \{(a, b) \,| a \in A, b \in B\}$.

1.1.6. Using the multiplication principle, it is easy to prove, for example, the formula

$$A_m^n = \frac{m!}{(m-n)!}.$$

In fact, in order to obtain a partial n-permutation of a given m-set, we should fill n ordered places using m different elements $a_1, \ldots, a_m$. For the first place to be filled we have m possibilities (one can choose any element from $a_1, \ldots, a_m$), for filling the second place we have $m-1$ possibilities (one can select any element from $a_1, \ldots, a_m$, except for the one used in the first step), ..., for filling the nth place we have $m - n + 1$ possibilities. Since we must fill the first *and* the second *and* $\cdots$ *and* the nth places, then, according to the multiplication principle, the total number of possibilities is $m \cdot (m-1) \cdot \cdots \cdot (m - n + 1) = \frac{m!}{(n-m)!}$.

The same or similar reasons can be used for the proofs of the other combinatorial formulas (see, for example, [DeMo10] and [Deza24]).

Theorem. *For the numbers of classical combinatorial configurations, we have the following formulas:*

$$P_m = m!; \quad A_m^n = \frac{m!}{(m-n)!}; \quad C_m^n = \frac{m!}{n! \cdot (m-n)!}; \quad \overline{A}_m^n = m^n;$$

$$P_{(\alpha_1,\alpha_2,\ldots,\alpha_k)} = \frac{(\alpha_1+\alpha_2+\cdots+\alpha_k)!}{\alpha_1! \cdot \alpha_2! \cdot \ldots \cdot \alpha_k!};$$

$$\overline{C}_m^n = P_{(m-1,n)} = \frac{(m+n-1)!}{n! \cdot (m-1)!}.$$

Inclusion–exclusion principle

1.1.7. Many combinatorial problems can be solved using the *inclusion–exclusion principle* (or *inclusion–exclusion method*), a counting technique which generalises the classical method of obtaining the number of elements in the union of two finite sets A and B:

$$|A \cup B| = |A| + |B| - |A \cap B|.$$

The formula expresses the fact that the sum of the cardinalities of two finite sets A and B may become too large since some elements may be counted twice. The double-counted elements are those in the intersection $A \cap B$ of the sets A and B, and the result should be corrected by subtracting the cardinality of the intersection $A \cap B$ of the sets A and B.

The inclusion–exclusion principle, being a generalisation of the two-set case, is perhaps more clearly seen in the case of three sets, which for the sets A, B, and C is given by the formula

$$|A \cup B \cup C| = |A| + |B| + |C| - |A \cap B| - |A \cap C| \\ - |B \cap C| + |A \cap B \cap C|.$$

This formula can be verified by counting how many times each region of the corresponding *Venn diagram* is included on the right-hand side of the formula. In this case, when removing the contributions of over-counted elements, the number of elements in the mutual intersection of the three sets has been subtracted too often and hence must be added back in to get the correct total result.

A generalisation of the results in these examples gives the *inclusion–exclusion principle.*

- *To find the cardinality of the union of n finite sets: include the cardinalities of the sets; exclude the cardinalities of the pairwise intersections; include the cardinalities of the triple-wise intersections; exclude the cardinalities of the quadruple-wise intersections; include the cardinalities of the quintuple-wise intersections; continue until the cardinality of the n-tuple-wise intersection is included (if n is odd) or excluded (if n is even).*

The name "inclusion–exclusion" comes from the idea that the principle is based on over-generous inclusion, followed by compensating exclusion. This concept is attributed to Abraham de Moivre (1718), although it first appears in a paper by Daniel da Silva (1854) and later in a paper by James Joseph Sylvester (1883) (see [LiWi92], [RoTe09], and [Mazu10]). Sometimes the principle is referred to as the *formula of da Silva or Sylvester* due to these publications. The principle can be viewed as an example of the sieve method, which is extensively used in number theory and is sometimes referred to as the *sieve formula.* See, for example, [Stan97] and [Rota64].

1.1.8. Let us illustrate the inclusion–exclusion principle through a discussion of the following problem.

- *At a research institute,* 67 *people work, of which* 47 *speak English,* 35 *speak German, and* 23 *speak both languages; how many people don't speak any of these languages?*

□ To answer the question, we must subtract from the total number of employees the number of employees who speak English and the number of employees who speak German. Since the number of employees speaking both languages has been discarded twice, we must add this number to the resulting expression. Thus, the number of employees who do not know either English or German should be calculated using the formula $67 - 47 - 35 + 23 = 8$. This formula represents the simplest case of the *inclusion–exclusion formula.* □

In general, the following statement holds (see, for example, [DeMo10] and [Deza24]).

Theorem (inclusion–exclusion formula). *Consider a set A, consisting of N different elements. Consider k properties α_1, α_2, …, α_k, which the elements of the set A may or may not possess. Let $N(\alpha_i, \alpha_j^{'})$ be the number of elements of the set A having the property α_i, and not having the property $\alpha_j^{'}$. Then,*

$$\begin{aligned} N(\alpha_1', \alpha_2', \ldots, \alpha_k') = N - N(\alpha_1) - \cdots - N(\alpha_k) + N(\alpha_1, \alpha_2) \\ + \cdots + N(\alpha_{k-1}, \alpha_k) - N(\alpha_1, \alpha_2, \alpha_3) \\ - \cdots - N(\alpha_{k-2}, \alpha_{k-1}, \alpha_k) \\ + \cdots + (-1)^k N(\alpha_1, \alpha_2, \ldots, \alpha_k). \end{aligned}$$

Basic combinatorial identities

There are many useful and beautiful identities connected with combinatorial constructions. For proofs of these identities, a variety of methods are used.

Direct validation

1.1.9. The simplest of these methods is the method of direct algebraic validation. For example, in order to prove the identity

$$C_n^k = C_{n-1}^k + C_{n-1}^{k-1},$$

it is enough to perform the following transformations:

$$\begin{aligned} C_{n-1}^k + C_{n-1}^{k-1} &= \frac{(n-1)!}{k!(n-k-1)!} + \frac{(n-1)!}{(k-1)!(n-k)!} \\ &= \frac{n!}{k!(n-k)!}\left(\frac{n-k}{n} + \frac{k}{n}\right) = \frac{n!}{k!(n-k)!} = C_n^k. \end{aligned}$$

Checking the identity

$$C_n^k = C_n^{n-k}$$

is even easier:

$$C_n^k = \frac{n!}{k!(n-k)!} = \frac{n!}{(n-k)!(n-(n-k))!} = C_n^{n-k}.$$

Combinatorial approach

1.1.10. However, the same identities can be proved solely using a combinatorial approach.

For example, to prove the identity $C_n^k = C_{n-1}^k + C_{n-1}^{k-1}$, it is enough to note that, by fixing one element, for example, a_1, from the set of n different elements $a_1, \ldots, a_n$, we can state that any k-combination of elements $a_1, \ldots, a_n$ either contains this fixed element a_1 or does not contain it. The number of k-combinations of the first type is C_{n-1}^k: one should choose all k elements from the $(n-1)$-set $a_2, \ldots, a_n$ that does not contain a_1. The number of k-combinations of the second type is C_{n-1}^{k-1}: using element a_1 as a member of the considered combination, we should "get" the remaining $k-1$ elements of this combination from the set $a_2, \ldots, a_n$ not containing a_1.

The identity $C_n^k = C_n^{n-k}$ is based on a simple combinatorial observation: choosing k elements from n elements, we left untouched $n-k$ elements (from n elements), i.e., assigning them, for example, to a colleague.

On the other hand, the identity

$$C_n^0 C_m^k + C_n^1 C_m^{k-1} + \cdots + C_n^{k-1} C_m^1 + C_n^k C_m^0 = C_{n+m}^k$$

can be proven if we consider the following combinatorial problem:

- *In how many ways can a team be chosen from n men and m women?*

□ The answer is there are exactly C_{n+m}^k possibilities to choose k people from $n+m$ people. On the other hand, in a given team, there can be zero men, one man, two men, $\ldots, k$ men. So, if in the team there are exactly i men (and, hence, $k-i$ women), we have $C_n^i C_m^{k-i}$ possibilities. So, comparing these two representations, we obtain that

$$C_{n+m}^k = C_n^0 C_m^k + C_n^1 C_m^{k-1} + \cdots + C_n^{k-1} C_m^1 + C_n^k C_m^0 = \sum_{i=0}^{k} C_n^i C_m^{k-i}.$$

□

Binomial theorem approach

1.1.11. Some combinatorial identities can be proved using the binomial theorem (or the multinomial theorem). For example, putting in the binomial formula the values $x = 1$ and $y = 1$, we get the equality $(1+1)^n = C_n^0 + C_n^1 + \cdots + C_n^{n-1} + C_n^n$, or, in other words, the well-known identity

$$C_n^0 + C_n^1 + \cdots + C_n^{n-1} + C_n^n = 2^n.$$

Putting in the binomial formula the values $x = 1$ and $y = -1$, we obtain the identity

$$C_n^0 - C_n^1 + \cdots + (-1)^n C_n^n = 0.$$

Similarly, as

$$(1+a)^n = C_n^0 + C_n^1 a + \cdots + C_n^{n-1} a^{n-1} + C_n^n a^n,$$

we obtain, for example, the identities

$$C_n^0 + 2C_n^1 + 4C_n^2 + \cdots + 2^n C_n^n = 3^n,$$
$$C_n^0 - 3C_n^1 + 9C_n^2 - \cdots + (-3)^n C_n^n = (-2)^n.$$

Special combinatorial transforms

1.1.12. In order to calculate some finite combinatorial sums, special techniques are required. For example, to find the value of the sum $C_n^1 + 2C_n^2 + \cdots + nC_n^n$, one can note that

$$iC_n^i = i\frac{n!}{i!(n-i)!} = n\frac{(n-1)!}{(i-1)!((n-1)-(i-1))!} = nC_{n-1}^{i-1}.$$

Therefore, it holds that

$$C_n^1 + 2C_n^2 + \cdots + nC_n^n = n(C_{n-1}^0 + C_{n-1}^1 + \cdots + C_{n-1}^{n-1}) = n \cdot 2^{n-1}.$$

Exercises

1. Find the values C_5^3, A_6^4, $P_{2,2,3}$, $\overline{A}_2^5$, and $\overline{C}_4^2$.
2. Prove the following identities:
   (a) $C_n^m C_n^k = C_n^k C_{n-k}^{m-k}$;
   (b) $C_n^0 + 2C_n^1 + \cdots + 2^n C_n^n = 3^n$;

(c) $\sum\limits_{i_1+i_2+\cdots+i_m=n} P_{(i_1,i_2,\ldots,i_m)} = m^n$;
(d) $C^k_{n-1} + C^{k-1}_{n-2} + \cdots + C^0_{n-(k+1)} = C^k_n$;
(e) $(C^0_n)^2 + (C^1_n)^2 + \cdots + (C^n_n)^2 = C^n_{2n}$;
(f) $P_{(i-1,j,k)} + P_{(i,j-1,k)} + P_{(i,j,k-1)} = P_{(i,j,k)}$.

3. Find the following sums:

(a) $C^0_n + 2C^1_n + 3C^2_n + \cdots + (n+1)C^n_n$;
(b) $C^1_n - 2C^2_n + 3C^3_n + \cdots + (-1)^{n-1}nC^n_n$;
(c) $C^0_n + \frac{1}{2}C^1_n + \frac{1}{3}C^2_n + \cdots + \frac{1}{n+1}C^n_n$;
(d) $C^0_n - \frac{1}{2}C^1_n + \frac{1}{3}C^2_n + \cdots + (-1)^n\frac{1}{n+1}C^n_n$.

4. Check that the *Vandermond convolution* $C^k_{n+p} = \sum_{m=0}^{M} C^m_n C^{k-m}_p$, where $M = \min\{k, n-1\}$, can be proved using the relationships

$$\begin{aligned} C^k_{n+p} &= C^k_{n+p-1} + C^{k-1}_{n+p-1} \\ &= C^k_{n+p-2} + C^{k-1}_{n+p-2} + C^{k-1}_{n+p-2} + C^{k-3}_{n+p-2} \\ &= C^k_{n+p-2} + 2C^{k-1}_{n+p-2} + C^{k-3}_{n-2} \\ &= C^k_{n-3} + 3C^{k-1}_{n-3} + 3C^{k-2}_{n-3} + C^{k-3}_{n-3} = \cdots. \end{aligned}$$

1.2. Binomial Theorem and Pascal's Triangle

Binomial theorem

1.2.1. The *binomial theorem* (or *binomial expansion*, *Newton's binom*) describes the algebraic expansion of powers of a binomial (for example, the binomial $x+y$ or the binomial $1+x$). According to the theorem, it is possible to expand the polynomial $(x+y)^n$ into a sum, involving terms of the form $B^k_n x^k y^{n-k}$, where the exponents k and $n-k$ are non-negative integers, and the coefficient B^k_n of each term is a specific positive integer, depending on n and k. For example, for $n = 4$, it holds that

$$(x+y)^4 = x^4 + 4x^3y + 6x^2y^2 + 4xy^3 + y^4.$$

Theorem (binomial theorem). *For any $x, y \in \mathbb{R}$ and for any non-negative integer n, the following equality holds:*

$$(x+y)^n = \binom{n}{0}x^n + \binom{n}{1}x^{n-1}y + \binom{n}{2}x^{n-2}y^2 + \cdots + \binom{n}{n-1}xy^{n-1} + \binom{n}{n}y^n,$$

where the summation is taken over all non-negative integers $k, 0 \leq k \leq n$, and

$$\binom{n}{k} = \frac{n!}{k!(n-k)!}.$$

1.2.2. The coefficients $\binom{n}{0}, \binom{n}{1}, \binom{n}{2}, \ldots, \binom{n}{n}$ of this decomposition are called *binomial coefficients*. Combinatorially, the binomial coefficient $\binom{n}{k}$ can be interpreted as the number of ways to choose k elements from an n-element set, i.e., the number C_n^k of k-compositions of a given n-set.

- *Binomial coefficients $\binom{n}{k}$ coincide with the number of compositions C_n^k.*

□ In fact, if we write $(x+y)^n$ as a product

$$(x+y)(x+y)(x+y)\cdots(x+y),$$

then, according to the distributive law, there will be one term in the expansion for each choice of either x or y from each of the binomials of the product. For example, there will only be one term x^n, corresponding to choosing x from each binomial. However, there will be several terms of the form $x^{n-k}y^k$, $k \in \mathbb{N}$, one for each way of choosing exactly k binomials $x+y$ contributing y. Therefore, after combining like terms, the coefficient of $x^{n-k}y^k$ will be equal to the number of ways C_n^k to choose exactly k elements from an n-set. □

1.2.3. Using the summation notation, the *binomial formula* (or *binomial identity*) can be written as

$$(x+y)^n = \sum_{k=0}^{n} \binom{n}{k} x^{n-k}y^k = \sum_{k=0}^{n} \binom{n}{k} x^k y^{n-k}.$$

The final expression follows from the previous one by the symmetry of x and y in the first expression and by the symmetry $\binom{n}{k} = \binom{n}{n-k}$ of the binomial coefficients.

A simple variant of the binomial formula is obtained by substituting 1 for y, so that it involves only a single variable. In this form, the formula is

$$(1+x)^n = \binom{n}{0}x^0 + \binom{n}{1}x^1 + \binom{n}{2}x^2 + \cdots + \binom{n}{n-1}x^{n-1} + \binom{n}{n}x^n,$$

or, equivalently,

$$(1+x)^n = \sum_{k=0}^{n} \binom{n}{k} x^k.$$

More explicitly, we obtain

$$(1+x)^n = 1+nx+\frac{n(n-1)}{2!}x^2+\frac{n(n-1)(n-2)}{3!}x^3+\cdots+nx^{n-1}+x^n.$$

Here are the first few cases of the binomial theorem:

$$\begin{aligned}
(x+y)^0 &= 1, \\
(x+y)^1 &= x+y, \\
(x+y)^2 &= x^2+2xy+y^2, \\
(x+y)^3 &= x^3+3x^2y+3xy^2+y^3, \\
(x+y)^4 &= x^4+4x^3y+6x^2y^2+4xy^3+y^4, \\
(x+y)^5 &= x^5+5x^4y+10x^3y^2+10x^2y^3+5xy^4+y^5, \\
(x+y)^6 &= x^6+6x^5y+15x^4y^2+20x^3y^3+15x^2y^4+6xy^5+y^6, \\
(x+y)^7 &= x^7+7x^6y+21x^5y^2+35x^4y^3+35x^3y^4+21x^2y^5 \\
&\quad +7xy^6+y^7, \\
(x+y)^8 &= x^8+8x^7y+28x^6y^2+56x^5y^3+70x^4y^4+56x^3y^5 \\
&\quad +28x^2y^6+8xy^7+y^8.
\end{aligned}$$

These examples show that the binomial coefficients for varying n and k can be arranged in the form of *Pascal's triangle*.

Multinomial theorem

1.2.4. The binomial theorem is the most important special case of the *multinomial theorem*, which works with powers of sums with more than two terms.

Theorem (multinomial theorem). *For any $x_1, \ldots, x_m \in \mathbb{R}$ and for any non-negative integer n, the following equality holds:*

$$(x_1 + \cdots + x_m)^n = \sum_{k_1+\cdots+k_m=n, k_i \geq 0} \binom{n}{k_1, k_2, \ldots, k_m} x_1^{k_1}, \ldots, x_m^{k_m},$$

where the summation is taken over all sequences of non-negative integers $k_1, \ldots, k_m$ such that the sum of all k_i is n, and

$$\binom{n}{k_1, k_2, \ldots, k_m} = \frac{n!}{k_1! \cdot k_2! \cdot \cdots \cdot k_m!}.$$

1.2.5. The coefficients $\binom{n}{k_1,\ldots,k_m}$ are known as *multinomial coefficients*. Combinatorially, the multinomial coefficient $\binom{n}{k_1,\ldots,k_m}$ counts the number of different ways to partition an n-element set into disjoint subsets of sizes $k_1, \ldots, k_m$, i.e., the number $P_{(k_1,\ldots,k_m)}$ of $(k_1, \ldots, k_m)$-partitions.

- *Multinomial coefficients $\binom{n}{k_1,\ldots,k_m}$ coincide with the numbers $P_{(k_1,\ldots,k_m)}$ of $(k_1, \ldots, k_m)$-partitions.*

□ First, let us open the brackets on the left-hand side of the considered equality:

$$\begin{aligned}(x_1 + \cdots + x_m)^n &= (x_1 + \cdots + x_m) \cdot \cdots \cdot (x_1 + \cdots + x_m) \\ &= x_1 \cdot \cdots \cdot x_1 + \cdots + x_m \cdot \cdots \cdot x_m.\end{aligned}$$

We get a finite number of terms, each of which consists of n factors and each factor taking one of the values $x_1, \ldots, x_m$. By combining similar factors in each of the obtained terms, we get a finite number of terms of the form $x_1^{k_1} \cdots x_m^{k_m}$, where $k_1, \ldots, k_m$ are non-negative integers, such that their sum is equal to n: $k_1 + \cdots + k_m = n, k_1 \geq 0, \ldots, k_m \geq 0$. For a fixed $k_1, \ldots, k_m$, the element $x_1^{k_1} \cdots x_m^{k_m}$ can be obtained from any term, in which x_1 is represented k_1 times, x_2 is represented k_2 times, ..., x_m is represented k_m times. The numbers of such terms is equal to $P_{(k_1,\ldots,k_m)}$. The statement holds. □

It is easy to see that we get the binomial theorem for $k = 2$, noting that, if $k_1 + k_2 = n$, then $k_2 = n - k_1$, and using the equality $P_{(k_1,n-k_1)} = \frac{n!}{k_1!(n-k_1)!} = \binom{n}{k_1} = \binom{n}{n-k_1}$.

Pascal's triangle

1.2.6. *Pascal's triangle* is an arithmetic triangle constructed according to the following law: the top (zeroth) row of the triangle contains only one unity. The next (first) row contains two unities. In the second row, the two extreme elements are again equal to unities, and the internal element is obtained as the sum of the two elements of the first row, located above it, that is, equal to $1 + 1 = 2$. In general, the extreme elements of any row of Pascal's triangle are equal to 1, and any internal element, if it exists, is obtained as the sum of the elements located directly above it in the previous row.

By sequential calculations, we get the following arrangement (sequence A007318 in the OEIS):

$$\begin{array}{ccccccccc} & & & & 1 & & & & \\ & & & 1 & & 1 & & & \\ & & 1 & & 2 & & 1 & & \\ & 1 & & 3 & & 3 & & 1 & \\ 1 & & 4 & & 6 & & 4 & & 1 \\ \vdots & \vdots & \vdots & \vdots & \vdots & \vdots & \vdots & \vdots & \vdots \end{array}$$

Algebraic notation of the law of construction of this rectangular triangle is as follows: if the kth element of the nth row of Pascal's triangle is indicated as T_n^k, then $T_n^k = T_{n-1}^k + T_{n-1}^{k-1}$, and $T_0^0 = T_n^0 = T_n^n = 1$.

1.2.7. It turns out that, for non-negative integers n, the binomial coefficients $\binom{n}{k}$, $k = 0, 1, 2, \ldots$, form the corresponding row of Pascal's triangle.

- *The elements of Pascal's triangle are the binomial coefficients, i.e., for any non-negative integer n, the numbers T_n^k, $k = 0, 1, 2, \ldots, n$,* *are equal to the coefficients* $\binom{n}{k}$ *of the decomposition*

$$(1+x)^n = \binom{n}{0} + \binom{n}{1}x + \cdots + \binom{n}{n}x^n.$$

□ The statement is valid for $n = 0$: we have the equality $(1+x)^0 = 1$, that is, $\binom{0}{0} = T_0^0 = 1$. It is also true for $n = 1$: we have that $(1+x)^1 = 1 + 1 \cdot x$, i.e., $\binom{1}{0} = T_1^0 = 1$, and $\binom{1}{1} = T_1^1 = 1$. In addition,

$$(1+x)^n = (1+x)^{n-1}(1+x) = \left(\binom{n-1}{0} + \binom{n-1}{1}x + \cdots + \binom{n-1}{n-1}x^{n-1}\right)(1+x) = \binom{n-1}{0}$$
$$+ \left(\binom{n-1}{1} + \binom{n-1}{0}\right)x + \left(\binom{n-1}{2} + \binom{n-1}{1}\right)x^2$$
$$+ \cdots + \left(\binom{n-1}{n-1} + \binom{n-1}{n-1}\right)x^{n-1} + \binom{n-1}{n-1}x^n.$$

So, $\binom{n}{0} = \binom{n-1}{0}$, $\binom{n}{n} = \binom{n-1}{n-1}$, and $\binom{n}{k} = \binom{n-1}{k} + \binom{n-1}{k-1}$, $1 \le k \le n-1$. Therefore, the binomial coefficients $\binom{n}{k}$ satisfy the same recurrent relation $\binom{n}{k} = \binom{n-1}{k} + \binom{n-1}{k-1}, 1 \le k \le n-1$, as the elements of Pascal's triangle: $T_n^k = T_{n-1}^k + T_{n-1}^{k-1}, 1 \le k \le n-1$. The initial conditions $\binom{n}{n} = \binom{n-1}{n-1} = \cdots = \binom{1}{1} = \binom{0}{0} = 1$ and $\binom{n}{0} = \binom{n-1}{0} = \cdots = \binom{1}{0} = \binom{0}{0} = 1$ also coincide with the initial conditions of the elements of Pascal's triangle: $T_n^n = T_{n-1}^{n-1} = \cdots = T_1^1 = T_0^0 = 1$ and $T_n^0 = T_{n-1}^0 = \cdots = T_1^0 = T_0^0 = 1$. It holds that $\binom{n}{k} = T_n^k, 0 \le k \le n$. In other words, we have proven the following statement:

$$(1+x)^n = T_n^0 + T_n^1 x + \cdots + T_n^n x^n. \qquad \square$$

As the nth row of Pascal's triangle gives the binomial decomposition's coefficients for $(1+x)^n$, then, in order to obtain this decomposition, it is enough to build the corresponding number of rows of our triangle, or we can simply use already available data. Actually, this is the main practical purpose of Pascal's triangle: it is simply a table of binomial coefficients, which is convenient to use in computational applications and, if necessary, expand to the desired size using an elementary additive algorithm that requires for its implementation only the addition operation.

As binomial coefficients coincide with the number of combinations C_n^k, we get that Pascal's triangle is formed from the numbers C_n^k, $n = 0, 1, 2, \ldots$, $k = 0, 1, 2, \ldots, n$. So, we use the following equalities:

$$T_n^k = \binom{n}{k} = C_n^k, \ n = 0, 1, 2, \ldots, \ \ k = 0, 1, 2, \ldots, n.$$

1.2.8. It is easy to obtain a large list of properties for Pascal's triangle. We consider several of them as follows:

- *Pascal's triangle is symmetrical with respect to its main diagonal, i.e.,*

$$C_n^k = C_n^{n-k}.$$

□ In order to prove this property, it is enough to note that

$$C_n^k = \frac{n!}{k!(n-k)!} = \frac{n!}{(n-k)!(n-(n-k))!} = C_n^{n-k}. \qquad \square$$

- *The sum of the elements of the nth row of Pascal's triangle is equal to 2^n:*

$$C_n^0 + C_n^1 + C_n^2 + \cdots + C_n^n = 2^n.$$

□ Using the binomial decomposition of $(1+x)^n$ for $x = 1$, we get

$$2^n = (1+1)^n = C_n^0 + C_n^1 + C_n^2 + \cdots + C_n^n.$$

- *The alternating sum of the elements of the nth row of Pascal's triangle, $n \geq 1$, is equal to zero:*

$$C_n^0 - C_n^1 + C_n^2 - \cdots + (-1)^n C_n^n = 0.$$

□ To prove this fact, we use the binomial decomposition of $(1+x)^n$, taking $x = -1$:

$$0 = (1-1)^n = C_n^0 - C_n^1 + C_n^2 - \cdots + (-1)^n C_n^n. \qquad \square$$

1.2.9. The diagonals of Pascal's triangle are related to various special numbers.

First of all, we can find many figurate numbers in this triangular construction.

- *Along the descending diagonals of Pascal's triangle, the triangular numbers are arranged, as well as their generalisations to the case of spaces of all dimensions.*

□ Recall that *triangular numbers*, $1, 2, 3, 6, 10, \dots$, correspond to the numbers of balls we can lay out as a triangle; the nth triangular number $S_3(n)$ is calculated using the formula $\frac{n(n+1)}{2}$. It turns out that

$$T_n^2 = C_n^2 = \frac{n(n-1)}{2} = S_3(n-1).$$

So, the second-left (-right) descending diagonal of Pascal's triangle consists of the numbers $C_2^2, C_3^2, C_4^2, \dots, C_n^2, \dots$, that is, the triangular numbers $1, 3, 6, 10, \dots$. Similarly, the third-left (-right) descending diagonal consists of the numbers $C_3^3, C_4^3, C_5^3, \dots, C_n^3, \dots$, that is, the *tetrahedral numbers* $1, 4, 10, 20, \dots$. The numbers $C_4^4, C_5^4, C_6^4, \dots, C_n^4, \dots$ of the fourth diagonal form the sequence $1, 5, 15, 35, \dots$ of simplicial numbers of dimension 4; they are the four-dimensional analogues of the two-dimensional triangular and three-dimensional tetrahedral numbers. □

In general, the numbers $C_k^k, C_{k+1}^k, C_{k+2}^k, \dots, C_n^k, \dots$, forming the kth diagonal, are the k-dimensional analogues of the triangular and tetrahedral numbers. One can say that the sequence $1, 2, 3, 4, \dots$ of the elements $C_1^1, C_2^1, C_3^1, \dots$ of the first diagonal can be considered a one-dimensional analogue of the triangular and tetrahedral numbers, while the unities, forming the sides of the triangle, can be considered their zero-dimensional analogue. From this point of view, we can argue that the following statement holds:

- *Pascal's triangle is formed only from the figurate numbers.*

Exercises

1. In the expansion of $(5a - 2b)^9$, find the coefficient of the term containing a^5.
2. Find the coefficients of the decomposition of the binomial $(1+x)^n$, $n = 5, \dots, 15$, using Pascal's triangle.

3. Verify that the numbers $1, 11, 121, 1331, 14641$, derived from the nth row of Pascal's triangle, $n = 0, 1, 2, 3, 4$, by "gluing" it into the usual decimal number correspond to nth powers of the number 11. Is this statement true for $n = 5, 6, 7, 8$?
4. For small values of n, verify that the following statement is true: *if in the nth row of Pascal's triangle, where n is odd, we add all numbers with indices congruent to* 0 *modulo* 3*, to* 1 *modulo* 3*, and to* 2 *modulo* 3*, i.e., with indices of the from* $3n, 3n+1$*, and* $3n+2$*, then the first two sums will be equal, and the third sum is equal to the previous one minus* 1. Prove this statement.

1.3. Graphs

Definition and examples

1.3.1. A *graph* (or *undirected graph, simple graph*) is a pair $G = (V, E)$, where V is a set whose elements are called *vertices*, and E is a set of unordered pairs $\{u, v\}$ of distinct vertices, whose elements are called *edges* (or *links, lines*): $E \subseteq \{\{u, v\} \mid u, v \in V,\ u \neq v\}$.

Generally, the vertex set V is taken to be finite (which implies that the edge set E is also finite): $V = \{v_1, v_2, \ldots, v_n\}$. Sometimes infinite graphs are considered, but they are usually viewed as a special kind of binary relation because most results on finite graphs either do not extend to the infinite case or need a rather different proof.

A *directed graph* (or *digraph, directed simple graph*) is a graph in which edges have orientations. Formally, a *directed graph* is a pair $D = (V, E)$, where V is a set of vertices (or *nodes, points*) and E a set of *arcs* (or *directed edges, directed links, directed lines, arrows*), which are ordered pairs of distinct vertices: $E \subseteq \{(u, v) \mid u, v \in V,\ u \neq v\}$.

A *multigraph* is a graph which is permitted to have *multiple edges* (or *parallel edges*), that is, edges that have the same end vertices. Thus, two vertices may be connected by more than one edge.

Sometimes, graphs are allowed to contain *loops*, which are edges that join a vertex to itself. To allow loops, the pairs of vertices in E must be allowed to have the same vertex twice. Such generalised graphs are called *graphs with loops* or simply graphs when it is clear from the context that loops are allowed.

A *subgraph* of a graph G is another graph formed from a subset of the vertices and edges of G. A *spanning subgraph* is one that includes all vertices of the graph; an *induced subgraph* is one that includes all the edges whose endpoints belong to the vertex subset.

1.3.2. The vertices u and v of an edge $\{u, v\}$ are called the *edge's endpoints*. The edge is said to *join* u and v and to *be incident* to them. When an edge $\{u, v\}$ exists, the vertices u and v are called *adjacent*.

In any graph, the *degree* $d(v)$ of a vertex v is defined as the number of edges that have v as an endpoint. For graphs that are allowed to contain loops, a loop should be counted as contributing two units to the degree of its endpoint.

A vertex v may not belong to any edge, in which case it is not joined to any other vertex and is called *isolated*, $d(v) = 0$. A vertex v belonging to one edge is called a *leaf* (or *terminal vertex*, *hanging vertex*): $d(v) = 1$. A vertex v belonging to even (odd) number of edges is called *even* (*odd*): in this case, $d(v)$ is even (odd).

1.3.3. The *degree sum formula* states that:

- *the sum of the vertex degrees equals twice the number of edges*:

$$\sum_{v \in V} d(v) = 2|E|.$$

□ The proof is very simple: the sum $\sum_{v \in V} d(v)$ counts each edge $\{u, v\}$ twice. □

The *handshaking lemma* states that:

- *the number of odd vertices in every finite undirected graph is even.*

□ In fact, it holds that

$$2|E| = \sum_{v \in V} d(v) = \sum_{u \in V, d(u)=2k} d(v) + \sum_{v \in V, d(v)=2t+1} d(v).$$

As the values $2|E|$ and $\sum_{u\in V,d(u)=2k} d(v)$ are even, the sum $\sum_{v\in V,d(v)=2t+1} d(v)$ should be even and, hence, should have an even number of summands. □

Types of graphs

1.3.4. There are several important types of graphs.

A *complete graph* is a graph in which each pair of vertices is joined by an edge. In other words, a complete graph contains all possible edges. The complete graph on n vertices is denoted by K_n. It has exactly $\frac{n(n-1)}{2}$ edges.

A *bipartite graph* is a simple graph in which the vertex set can be partitioned into two sets, V_1 and V_2, so that no two vertices in V_1 share a common edge and no two vertices in V_2 share a common edge. In a *complete bipartite graph*, every vertex in V_1 is adjacent to every vertex in V_2, but there are no edges within V_1 or V_2. If $|V_1| = n_1$, and $|V_2| = n_2$, the complete bipartite graph is denoted by K_{n_1,n_2}. It contains n_1n_2 edges.

A *path* (or *simple path, path graph, linear graph*) P_n, $n \geq 1$, is a graph in which the vertices can be listed in an order, $v_1, v_2, \ldots, v_n$, such that its edges are $\{v_1, v_2\}$, $\{v_2, v_3\}$, $\ldots$, $\{v_{n-1}, v_n\}$. The number of edges in P_n is $n-1$. If a path graph occurs as a subgraph of another graph, it is called a *path* in that graph.

A *cycle* (or *simple cycle, cycle graph, circular graph*) C_n, $n \geq 3$, is a graph in which the vertices can be listed in an order, $v_1, v_2, \ldots, v_n$, such that its edges are $\{v_1, v_2\}$, $\{v_2, v_3\}, \ldots, \{v_{n-1}, v_n\}$, $\{v_n, v_1\}$. The number of edges in C_n is n. If a cycle graph occurs as a subgraph of another graph, it is a *cycle* (or *circuit*) in that graph.

In an undirected graph, an unordered pair of vertices u and v is said to be connected if a path leads from u to v. Otherwise, the unordered pair is said to be disconnected.

A *connected graph* is an (undirected) graph in which every two vertices u and v are connected by a path leading from u to v. Otherwise, the graph is called a *disconnected graph.*

It is easy to prove that the number m of edges of a connected graph on n vertices is at least $n - 1$. (See, for example, [DeMo10].)

A *planar graph* is a graph that can be embedded on a plane, i.e., it can be drawn on the plane in such a way that its edges intersect only at their endpoints. In other words, it can be drawn in such a way that no edges cross each other. Such a drawing is called a *plane graph* (or *planar embedding of the graph*). A plane graph can be defined as a planar graph with a mapping from every vertex to a point on a plane and from every edge to a plane curve on that plane, such that the extreme points of each curve are the points mapped from its end vertices and all curves are disjoint except on their extreme points. Every graph that can be drawn on a plane can be drawn on a sphere as well, and vice versa, by means of stereographic projection.

Kuratowski's theorem states (see, for example, [DeMo10]) that:

- *a finite graph is planar if and only if it does not contain a subgraph that is a subdivision of the complete graph* K_5 *or the complete bipartite graph* $K_{3,3}$.

Trees

1.3.5. A *tree* is an (undirected) graph in which any two vertices are connected by exactly one path, or, equivalently, a connected acyclic undirected graph.

A *forest* is an (undirected) graph in which any two vertices are connected by at most one path, or, equivalently, an acyclic undirected graph or a disjoint union of trees.

A *polytree* (or *directed tree, oriented tree, singly connected network*) is a directed acyclic graph whose underlying undirected graph is a tree. In other words, if we replace its directed edges with undirected edges, we obtain an undirected graph that is both connected and acyclic.

Let us list several important properties of trees (see [DeMo10] and [Ore80]):

- Every tree on n vertices has $n - 1$ edges.
- Every tree is a bipartite graph.
- Every tree with countably many vertices is a planar graph.
- Every connected graph G admits a *spanning tree*, which is a tree that contains every vertex of G and whose edges are the edges of G.

- Every finite tree with n vertices, $n > 1$, has at least two terminal vertices (leaves); this minimal number of leaves is characteristic of path graphs; the maximal number, $n - 1$, is attained only by star graphs.

A *rooted tree* is a tree in which one vertex has been designated the *root.*

The edges of a rooted tree can be assigned a natural orientation, either away from or towards the root, in which case the structure becomes a directed rooted tree. When a directed rooted tree has an orientation away from the root, it is called an *arborescence* (or *out-tree*); when it has an orientation towards the root, it is called an *anti-arborescence* (or *in-tree*).

The *tree order* is the partial ordering on the vertices of a tree with $u < v$ if and only if the unique path from the root to v passes through u.

In a rooted tree, the *parent* of a vertex v is the vertex connected to v on the path to the root; every vertex has an unique parent, except the root, which does not have a parent. A *child* of a vertex v is a vertex of which v is the parent. An *ascendant* of a vertex v is any vertex that is either the parent of v or is (recursively) an ascendant of a parent of v. A *descendant* of a vertex v is any vertex that is either a child of v or is (recursively) a descendant of a child of v. A *sibling* to a vertex v is any other vertex on the tree that shares a parent with v. A *leaf* is a vertex with no children. An *internal vertex* is a vertex that is not a leaf.

A *k-ary tree* (for non-negative integers k) is a rooted tree in which each vertex has at most k children. If there are all possible children, we call the tree a *full k-ary tree.* 2-ary trees are often called *binary trees*, while 3-ary trees are sometimes called *ternary trees.*

So, a *full binary tree* (or *proper, plane, strict binary tree*) is a binary tree in which every vertex has either zero or two children. In other words, a full binary tree is either a single vertex (a single vertex as the root) or a tree whose root vertex has two subtrees, both of which are full binary trees.

A rooted tree on n vertices is called *increasing* if its vertices are labelled by numbers $1, 2, \ldots, n$ such that the labels along any path from the root are increasing. If such a rooted tree is a binary tree, we call it an *increasing binary tree* on n vertices.

Rooted trees, often with an additional structure such as an ordering of the neighbours at each vertex, are a key data structure in computer science.

Enumeration of trees

1.3.6. A *labelled tree* (in general, a *labelled graph*) is a tree (a graph) in which each vertex has an unique label.

It is well known (see, for example, [DeMo10]) that:

- *there are* $2^{\frac{n(n-1)}{2}}$ *graphs on* n *labelled vertices.*

Cayley's formula states that:

- *there are* n^{n-2} *trees on* n *labelled vertices.*

A classic proof uses the *Prüfer sequences*, which naturally provide a stronger result:

- *the number of trees with vertices* $1, 2, \ldots, n$ *of degrees* d_1, $d_2, \ldots, d_n$*, respectively, is the multinomial coefficient*
$$\binom{n-2}{d_1 - 1, d_2 - 1, \ldots, d_n - 1}.$$

A more general problem is to count spanning trees in an undirected graph, which is addressed by the *matrix tree theorem.* (Cayley's formula is the special case of spanning trees in a complete graph.)

Counting the number of unlabelled free trees is a harder problem. No closed formula for the number $t(n)$ of trees with n vertices up to graph isomorphism is known. The first few values of $t(n)$ are $1, 1, 1, 1, 2, 3, 6, 11, 23, 47, \ldots$ (sequence A000055 in the OEIS).

Richard Otter (1948, [Otte48]) proved the asymptotic estimate
$$t(n) \sim C\alpha^n n^{-5/2} \quad \text{as } n \to \infty,$$
with $C = 0.534949606\ldots$ and $\alpha = 2.95576528565\ldots$ (sequence A051491 in the OEIS).

This is a consequence of his asymptotic estimate for the number $r(n)$ of unlabelled rooted trees with n vertices:

$$r(n) \sim D\alpha^n n^{-3/2} \quad \text{as } n \to \infty,$$

with $D = 0.43992401257\ldots$ and the same $\alpha = 2.95576528565\ldots$ (see [Knut97] and [FlSe09]).

The first few values of $r(n)$ are $1, 1, 2, 4, 9, 20, 48, 115, 286, 719, \ldots$ (sequence A000081 in the OEIS).

Exercises

1. Find all trees, all binary trees, and all increasing binary trees on n vertices, $n = 1, 2, 3, 4, 5$.
2. Prove that any tree is a bipartite graph. Can a tree be a complete bipartite graph?
3. Find all possible graphs on n vertices, $n = 1, 2, 3, 4, 5$. For each constructed graph, find (if possible) several of its spanning trees.
4. Prove that a tree with n vertices has exactly $n - 1$ edges.

1.4. Partitions

Definitions and examples

1.4.1. A *partition* of a set is a grouping of its elements into non-empty subsets in such a way that every element is included in exactly one subset, i.e., X is a disjoint union of the subsets.

Formally, a family of sets S is a *partition* of X if the following conditions hold:

- $\emptyset \notin S$, the family S does not contain the empty set;
- $\bigcup_{A \in P} A = X$, the union of the sets in S is equal to X;
- $(\forall A, B \in P)\ A \neq B \implies A \cap B = \emptyset$, the intersection of any two distinct sets in S is empty.

The sets in S are called the *parts* (or *blocks*, *cells*) of the partition.

The number of partitions of a given n-set is the nth *Bell number* $B(n)$. Starting with $n = 1$, we get exactly one partition of a 1-set (say, the set $\{1\}$): $\{\{1\}\}$. For the 2-set $\{1, 2\}$, we have two partitions: $\{\{1,2\}\}$ and $\{\{1\},\{2\}\}$. The 3-set $\{1,2,3\}$ has five partitions: $\{\{1\},\{2\},\{3\}\}$, $\{\{1,2\},\{3\}\}$, $\{\{1,3\},\{2\}\}$,$\{\{1\},\{2,3\}\}$, and $\{\{1,2,3\}\}$. So, $B(1) = 1$, $B(2) = 2$, and $B(3) = 5$. In fact, taking into account $B(0) = 1$, we obtain that the sequence of Bell numbers starts with the numbers $1, 1, 2, 5, 15, 52, 203, 877, 4140, \ldots$ (sequence A000110 in the OEIS).

1.4.2. A partition of a set X, together with a total order on the parts of the partition, gives a structure, called an *ordered partition* (or *list of sets*) ([Stan97]). An ordered partition of a finite set may be written as a finite sequence of the sets in the partition: for instance, the three ordered partitions of the set $\{a,b\}$ are $(\{a\},\{b\})$, $(\{b\},\{a\})$, and $(\{a,b\})$.

The number of ordered partitions of a given n-set X is the nth *ordered Bell number* (or *Fubini number*) $OB(n)$. The sequence of ordered Bell numbers, for $n = 0, 1, 2, \ldots$, starts with $1, 1, 3, 13, 75, 541, 4683, 47293, 545835, 7087261, \ldots$ (sequence A000670 in the OEIS).

Partitions of sets and combinatorial configurations

1.4.3. There exist close connections between certain kinds of partitions of a given n-set and the classical combinatorial configurations.

Given positive integers $r_1, r_2, \ldots, r_k$, let the $(r_1, r_2, \ldots, r_k)$-*partition* of a given n-set X be an ordered set $X_1, X_2, \ldots, X_k$ of the subsets of X, such that $X_i \cap X_j = \emptyset$ for $i \neq j$, $X_1 \cup X_2 \cup \cdots \cup X_k = X$, and $|X_1| = r_1$, $|X_2| = r_2, \ldots, |X_k| = r_k$. It is easy to show that the number $\sigma(r_1, r_2, \ldots, r_k)$ of $(r_1, r_2, \ldots, r_k)$-partitions can be calculated using the combinatorial formula for the number $P_{(r_1, r_2, \ldots, r_k)}$ of the corresponding permutations with repetitions.

- *For the number $\sigma(r_1, r_2, \ldots, r_k)$ of $(r_1, r_2, \ldots, r_k)$-partitions, the following identity holds:*

$$\sigma(r_1, r_2, \ldots, r_k) = P_{(r_1, r_2, \ldots, r_k)} = \frac{(r_1 + r_2 + \cdots + r_k)!}{r_1! r_2! \cdots r_k!} = \frac{n!}{r_1! r_2! \cdots r_k!}.$$

□ In fact, the subset $X_1 \subset X$ can be chosen in $C_n^{r_1}$ ways, the subset $X_2 \subset X \backslash X_1$ can be chosen in $C_{n-r_1}^{r_2}$ ways, etc. To choose a subset $X_k \subset X \backslash (X_1 \cup X_2 \cup \cdots \cup X_{k-1})$, we have exactly $C_{n-r_1-r_2-\cdots-r_{k-1}}^{r_k} = C_{r_k}^{r_k} = 1$ ways.

Then, according to the combinatorial rule of product, it holds that

$$\begin{aligned}
&\sigma(r_1, r_2, \ldots, r_k) \\
&\quad = C_n^{r_1} C_{n-r_1}^{r_2} \cdots C_{n-r_1-r_2-\cdots-r_{k-1}}^{r_k} \\
&\quad = \frac{n!}{r_1(n-r_1)} \cdot \frac{(n-r_1)!}{r_2!(n-r_1-r_2)!} \cdot \cdots \cdot \frac{(n-r_1-r_2-\cdots-r_{k-1})!}{r_k!(n-r_1-r_2-\cdots-r_k)!} \\
&\quad = \frac{n!}{r_1! r_2! \cdots r_k! 0!} = \frac{n!}{r_1! r_2! \cdots r_k!}.
\end{aligned}$$

However, the same formula can be obtained as a number $P_{(r_1, r_2, \ldots, r_k)}$ of permutations: r_1 number of 1, r_2 number of 2 , $\ldots, r_k$ number of k, where i denotes the number of the subset X_i to which an element with this index belongs. □

The particular case of $r_1 = r_2 = \cdots = r_k = 1$ gives us the $(1, 1, \ldots, 1)$-partition of an n-set X, with $n = k$. In this case, the number $\sigma(1, 1, \ldots, 1)$ of $(1, 1, \ldots, 1)$-partitions of a given n-set X, i.e., the ordered partitions in which each part contains exactly one element is equal to the number P_n of permutations of n elements:

$$\sigma(1, 1, \ldots, 1) = P_{(1,1,\ldots,1)} = P_n = n!.$$

1.4.4. Going back to the general $(r_1, r_2, \ldots, r_k)$-partitions, we can state that the number $OS(n, k)$ of all ordered partitions of a given n-set into k parts can be represented as

$$OS(n, k) = \sum_{r_1+\cdots+r_k=n} \sigma(r_1, r_2, \ldots, r_k) = \sum_{r_1+\cdots+r_k=n} \frac{n!}{r_1! r_2! \cdots r_k!}.$$

So, the number of all ordered partitions of a given n-set, which is, by definition, the nth *ordered Bell number* (or *Fubini number*)

$OB(n)$, can be represented now as

$$OB(n) = \sum_{k=1}^{n} \sum_{r_1+\cdots+r_k=n} \sigma(r_1, r_2, \ldots, r_k)$$
$$= \sum_{k=1}^{n} \sum_{r_1+\cdots+r_k=n} \frac{n!}{r_1! r_2! \cdots r_k!}.$$

1.4.5. Now, we obtain the number of partitions of an n-set into k non-empty parts such that the order of parts is not important. (By definition, this number is equal to the Stirling number $S(n,k)$ of the second kind.)

- *The number $S(n,k)$ of partitions of an n-set into k non-empty non-ordered parts can be represented as*

$$S(n,k) = \frac{1}{k!} \sum_{r_1+\cdots+r_k=n} \sigma(r_1, r_2, \ldots, r_k)$$
$$= \frac{1}{k!} \sum_{r_1+\cdots+r_k=n} \frac{n!}{r_1! r_2! \cdots r_k!}.$$

□ Each such partition has k non-empty (non-ordered) parts with some cardinalities $r_1, \ldots, r_k$, $r_i \in \mathbb{N}$, $r_1 + \cdots + r_k = n$. Let us choose k positive integers, $r_1, \ldots, r_k$, such that $r_1 + r_2 + \cdots + r_k = n$, and consider the various $(r_1, r_2, \ldots, r_k)$-partitions of the set X. The number $\sigma(r_1, r_2, \ldots, r_k)$ of such partitions is determined by the formula above. Here, since in every such partition the order of parts (i.e., the order of k different elements) is not important, we get that it is equal to

$$\frac{1}{k!} \sum_{r_1+\cdots+r_k=n} \sigma(r_1, r_2, \ldots, r_k) = \frac{1}{k!} \sum_{r_1+\cdots+r_k=n} \frac{n!}{r_1! r_2! \cdots r_k!}.$$

Note that we use for the proof a simple relation: $OS(n,k) = k!S(n,k)$. □

Now, we can obtain the number of partitions of an n-set X into several non-empty parts such that the order of parts is not important. (By definition, this number is equal to the nth Bell number $B(n)$.)

- *The number $B(n)$ of partitions of an n-set into non-empty, non-ordered parts can be represented as*

$$\sum_{k=1}^{n}\frac{1}{k!}\sum_{r_1+\cdots+r_k=n}\sigma(r_1,r_2,\ldots,r_k)=\sum_{k=1}^{n}\frac{1}{k!}\sum_{r_1+\cdots+r_k=n}\frac{n!}{r_1!r_2!\cdots r_k!}.$$

1.4.6. Consider now the so-called *non-crossing partitions.* In fact, consider a finite set that is linearly ordered, or (equivalently, for purposes of this definition) arranged in a cyclic order like the vertices of a regular n-gon. No generality is lost by taking this set to be $V_n = \{1,\ldots,n\}$. So, let us arrange the set $\{1,2,3,\ldots,n\}$ in a cyclic order (for example, placing the numbers $1,2,3,\ldots,n$ in this specified order on a circle). A partition of the set $\{1,2,3,\ldots,n\}$ is called *non-crossing* if no two partition blocks "cross" each other, i.e., if a and b belong to one partition block and x and y belong to another, they are not arranged in the order $axby$.

In other worlds, if we draw an arc based at a and b and another arc based at x and y, then the two arcs cross each other if the order is $axby$ but not if it is $axyb$ or $abxy$. So, a non-crossing partition is a partition without crossing arcs. For example, the partition $\{\{1,2\},\{3,4\}\}$ of the set $\{1,2,3,4\}$ is non-crossing, while the partition $\{\{1,3\},\{2,4\}\}$ is not.

It can be proven (see, for example, [Deza25]) that *the number of non-crossing partitions of an n-set is equal to the nth Catalan number C_n.*

Partitions and Stirling numbers of the second kind

1.4.7. A *Stirling number of the second kind* (or *Stirling set number, Stirling partition number*) $S(n,k)$ is the number of partitions of a given n-set into k non-empty subsets, $n = 1,2,\ldots,k = 1,2,\ldots,n$.

This definition leads to a well-known recurrent relation:

- *For the numbers $S(n,k)$, the following recurrent relation holds:*

$$S(n,k) = kS(n-1,k) + S(n-1,k-1),\ n > 0.$$

□ Let there be an n-set, $n > 0$, and we then find a decomposition of this set into k non-empty parts. In this situation, we can place, on the one hand, the last (nth) element into a new part. There are $S(n-1, k-1)$ of such decompositions, as, constructing one part from the nth element, we should construct additional $k-1$ parts from the first $n-1$ elements of the considered n-set; the number of such decompositions is, by definition, $S(n-1, k-1)$. On the other hand, we can place the last (nth) element into some non-empty subset of the $(n-1)$th set of the first $n-1$ elements. In this case, we have $kS(n-1, k)$ possibilities, as every partition from all $S(n-1, k)$ decompositions of the first $n-1$ elements into k non-empty parts gives k subsets, with which we can combine the last (nth) element. □

Using the conditions $S(0,0) = 1$, $S(n,0) = 0, n > 0$, and $S(n,k) = 0, k > n$, we can obtain all values of the Stirling numbers of the second kind, i.e., we can construct the *triangle of Stirling numbers of the second kind* (sequence A008277 in the OEIS):

$n\backslash k$	0	1	2	3	4	5	6	7	8	9
0	1									
1	0	1								
2	0	1	1							
3	0	1	3	1						
4	0	1	7	6	1					
5	0	1	15	25	10	1				
6	0	1	31	90	65	15	1			
7	0	1	63	301	350	140	21	1		
8	0	1	127	966	1701	1050	266	28	1	
9	0	1	255	3025	7770	6951	2646	462	36	1

A *Bell number* $B(n)$ is defined as the number of all possible partitions of a given n-set into non-empty parts. So, by definition:

- *the sum of elements of the nth row of the triangle of Stirling numbers of the second kind is the nth Bell number $B(n)$, $n \geq 0$.*

1.4.8. The Stirling numbers of the second kind have many interesting properties and important applications. We use some of them in our further consideration. We list these properties in the following. For detailed proofs and additional information see, for example, [Deza24].

- *There exist several special cases for the values of the Stirling numbers of the second kind:*

$$S(n,n)=1, \quad n \geq 0; \quad S(n,1)=1, \quad n \geq 1;$$

$$S(n,n-1)=\binom{n}{2}, \quad n \geq 2.$$

- *The Stirling numbers of the second kind are connected with the Mersenne numbers* $M_n = 2^n - 1$*:*

$$S(n,2)=M_{n-1}, \; n \geq 2.$$

- *The Stirling numbers of the second kind* $S(n,k)$ *satisfy the equality*

$$x^n = S(n,0)x^{\underline{0}} + S(n,1)x^{\underline{1}} + S(n,2)x^{\underline{2}} + \cdots S(n,3)x^{\underline{3}} + \cdots + S(n,n)x^{\underline{n}},$$

i.e., using the Stirling numbers of the second kind, we can express the ordinary powers x^n *of* x *through the falling powers (falling factorials)* $x^{\underline{k}}$ *of* x*.*
- *The generating function of the sequence of Stirling numbers of the second kind* $S(n,k)$, $n=0,1,2,\ldots$, *for a fixed* $k=1,2,\ldots$, *has the form* $\frac{x^k}{(1-x)(1-2x)\cdots(1-kx)}$, *while the exponential generating function of the sequence of Stirling numbers of the first kind* $S(n,k)$, $n=0,1,2,\ldots$, *for a fixed* $k=0,1,2,\ldots$, *has the form* $\frac{(e^x-1)^k}{k!}$.

Enumeration of partitions

1.4.9. In the following table, we present numbers of important ordered partitions of a given n-set:

$\|X\|$	Any	Into k parts	$(r_1, \ldots, r_k)$-partitions	$(n, n-k)$-partitions	$(1, \ldots, 1)$-partitions
0	1	1	1	1	1
1	1	$0, 1$	1	$1, 1$	1
2	3	$0, 1, 2$	$0, 1, 2$	$1, 2, 1$	2
3	13	$0, 1, 6, 6$	$0, 1, 3+3, 6$	$1, 3, 3, 1$	6
4	75	$0, 1, 14, 36, 24$	$0, 1, 4+6+4, 24$	$1, 4, 6, 4, 1$	24
$\vdots$	$\vdots$	$\vdots$	$\vdots$	$\vdots$	$\vdots$
n	$OB(n)$ $= \sum_{k=1}^{n} k!S(n,k)$	$OS(n,k)$ $= k!S(n,k)$	$P_{(r_1,\ldots,r_k)}$	C_n^k	P_n
OEIS	*A000670*	*A019538*	*A019538*	*A007318*	*A000142*

The following table resents the numbers of important non-ordered partitions of a given n-set X:

$\|X\|$	Any	Into k parts	Non-crossing
0	1	0	1
1	1	$0, 1$	1
2	2	$0, 1, 1$	2
3	5	$0, 1, 3, 1$	5
4	15	$0, 1, 7, 6, 1$	14
$\vdots$	$\vdots$	$\vdots$	$\vdots$
n	$B(n) = \sum S(n,k)$	$S(n,k)$	C_n
OEIS	*A000110*	*A008277*	*A000108*

Exercises

1. Find all possible partitions of the sets $\emptyset$, $\{1\}$, $\{1,2\}$, $\{1,2,3\}$, $\{1,2,3,4\}$, $\{1,2,3,4,5\}$. Check the results pertaining to these partitions, as represented above.
2. Find all possible equivalence relations over the sets $\emptyset$, $\{1\}$, $\{1,2\}$, $\{1,2,3\}$, $\{1,2,3,4\}$, $\{1,2,3,4,5\}$. Compare the obtained results with those concerning the non-ordered partitions of the same sets.
3. Find all possible partial orders over the sets $\emptyset$, $\{1\}$, $\{1,2\}$, $\{1,2,3\}$, $\{1,2,3,4\}$, $\{1,2,3,4,5\}$. Compare the obtained results with those concerning the ordered partitions of the same sets.

4. Check the closed formula $S(n,k) = \frac{1}{k!}\sum_{j=0}^{k}(-1)^j C_k^j (k-j)^n$ for $k = 7, 8, 9, 10$. Check that

$$S(n,3) = \frac{1}{6}(3^n - 3\cdot 2^n + 3);\ S(n,4) = \frac{1}{24}(4^n - 4\cdot 3^n + 6\cdot 2^n - 4).$$

1.5. Recurrent Relations

Definitions and examples

1.5.1. For solutions of many problems, we can use the method of moving from a given problem, concerning n objects, to a similar problem, concerning a smaller number ($n-1$, $n-2$, etc.) of objects. Such a method is called *recurrence method* (from Latin *recurrere*, meaning *return*), and we call as *recurrent relations* the relationships that allow us to get information about n objects using information about $n-1$ ($n-2, n-3$, etc.) objects.

A *recurrent relation of order k* for a sequence $f_1, f_2, \dots, f_n, \dots$ is a relationship that allows us to express an element f_{n+k} of the sequence through k previous elements $f_n, f_{n+1}, \dots, f_{n+k-1}$ of the sequence; a *linear recurrent relation of the order k with constant coefficients* is the relation

$$f_{n+k} = a_1 f_{n+k-1} + a_2 f_{n+k-2} + \cdots + a_k f_n,\ a_1, \dots, a_k \in \mathbb{R},$$

where $a_1, \dots, a_k$ are arbitrary constants.

Recurrent relations in combinatorics

1.5.2. Many recurrent relations arise when we solve combinatorial problems. For example, from the theory of combinatorial configurations, it is known that $P_m = mP_{m-1}$, that is, the number of permutations of m elements can be calculated if we know the number of permutations of $m-1$ elements. Thus, knowing the initial condition $P_1 = 1$, one can find $P_2, P_3, \dots, P_n, \dots$ using the recurrence method.

Similarly, the formula $C_n^k = C_{n-1}^k + C_{n-1}^{k-1}$ gives a recurrent relation for the numbers C_n^k of combinations: the number C_n^k of combinations from n elements by k elements can be found by knowing the number of combinations from $n-1$ elements (by k elements and

$k-1$ elements). Thus, using the initial conditions $C_0^0 = 1$, $C_n^0 = 1$, and $C_n^n = 1$, all values of C_n^k can be obtained through sequential calculations:

$$C_0^0 = 1,$$
$$C_1^0 = 1, \quad C_1^1 = 1,$$
$$C_2^0 = 1, \quad C_2^1 = C_1^0 + C_1^1 = 1 + 1 = 2, \quad C_2^2 = 1,$$
$$C_3^0 = 1, \quad C_3^1 = 1 + 2 = 3, \quad C_3^2 = 1 + 2 = 3, \quad C_3^3 = 1,$$
$$\vdots \quad \vdots \quad \vdots$$

Using this algorithm, we get an arithmetic triangle, called *Pascal's triangle*.

Recurrent relations and Fibonacci numbers

1.5.3. One of the most widely known problems connected with recurrent relations is *the rabbit problem*, considered by Leonardo Pisano (c. 1170–c. 1250), also known as Fibonacci – an Italian merchant and one of the most significant Western mathematicians of the Middle Ages – in his 1202 book *Liber Abaci*.

- *A newly born breeding pair of rabbits are put in a field; each breeding pair mates at the age of one month, and at the end of their second month, they always produce another pair of rabbits; and the rabbits never die but continue breeding forever. How many pairs will be there in one year?*

Considering this problem, we obtain a recurrent relation

$$u_n = u_{n-1} + u_{n-2},$$

with initial conditions $u_1 = u_2 = 1$, allowing us to calculate sequentially all elements of the sequence $u_1, u_2, \ldots, u_n, \ldots$, which are called the *Fibonacci numbers*.

It is easy to check that there exist infinitely many sequences satisfying the same recurrent relation $a_n = a_{n-1} + a_{n-2}$. For example, if $a_1 = 3, a_2 = 5$, then we obtain a sequence $3, 5, 8, 13, 21, 34, \ldots$, which is a subsequence of the Fibonacci sequence. If $a_1 = 1, a_2 = 3$, then we get a sequence of *Lucas numbers*: $1, 3, 4, 7, 11, 18, 29, 47, 76, 123, \ldots$ (sequence A000032 in the OEIS). If $a_1 = -1, a_2 = 1$, then the sequence takes the form $-1, 1, 0, 1, 1, 2, 3, 5, \ldots$. If $a_1 = 2, a_2 = 4$, then we get the sequence $2, 4, 6, 10, 16, 26, 42, 68, \ldots$, etc. However, with the initial conditions as $u_1 = 1$ and $u_2 = 1$, the Fibonacci sequence $1, 1, 2, 3, 5, 8, 13, 21, 34, 55, \ldots$ (sequence A000045 in the OEIS), defined by the recurrent ration $u_n = u_{n-1} + u_{n-2}$, is unique.

1.5.4. There exist two properties of Pascal's triangle which can be rewritten in terms of the Fibonacci numbers.

- *The sum of elements of the $(n-1)$th ascending diagonal of Pascal's triangle, $n \geq 1$, is the nth Fibonacci number:*

$$C_{n-1}^0 + C_{n-2}^1 + C_{n-3}^2 + \cdots = u_n.$$

□ First, let us see that *the sum of elements of the $(n+1)$th ascending diagonal of Pascal's triangle, $n \geq 1$, equals the sum of the elements of the two previous ascending diagonals of the triangle.*

Indeed, the sum X_{n+1} of the elements of the $(n+1)$th ascending diagonal satisfies the relation

$$\begin{aligned} X_{n+1} &= C_{n+1}^0 + C_n^1 + C_{n-1}^2 + C_{n-2}^3 + \cdots = C_n^0 + (C_{n-1}^0 + C_{n-1}^1) \\ &\quad + (C_{n-2}^1 + C_{n-2}^2) + (C_{n-3}^2 + C_{n-3}^3) + \cdots \\ &= (C_n^0 + C_{n-1}^1 + C_{n-2}^2 + C_{n-3}^3 + \cdots) \\ &\quad + (C_{n-1}^0 + C_{n-2}^1 + C_{n-3}^2 + C_{n-4}^3 + \cdots) = X_n + X_{n-1}. \end{aligned}$$

Now, the property immediately follows if we note that the sum of the elements of the zeroth ascending diagonal is $u_1 = 1$, and the sum of the elements of the first ascending diagonal is equal to $u_2 = 1$. □

- *The sum of all the elements of Pascal's triangle, placing above the nth ascending diagonal, $n \geq 1$, increased per unity is equal to the $(n+2)$th Fibonacci number:*

$$C_0^0 + C_1^0 + (C_2^0 + C_1^1) + \cdots + (C_{n-1}^0 + C_{n-2}^1 + C_{n-3}^2 + \ldots) + 1 = u_{n+2}.$$

□ First, let us show that *the sum of the elements of the nth ascending diagonal of the Pascal's triangle, $n \geq 2$, is the sum of all the elements of the Pascal's triangle that are located above the $(n-1)$th ascending diagonal increased by one.*

In fact, the sum X_n of the elements of the nth ascending diagonal satisfies the relation

$$\begin{aligned} X_n &= X_{n-2} + X_{n-1} = X_{n-2} + (X_{n-3} + X_{n-2}) \\ &= X_{n-2} + X_{n-3} + (X_{n-4} + X_{n-3}) = \cdots = X_{n-2} + X_{n-3} + X_{n-4} \\ &\quad + \cdots + X_1 + (X_0 + X_1) = X_{n-2} + X_{n-3} + X_{n-4} \\ &\quad + \cdots + X_1 + X_0 + 1. \end{aligned}$$

The property follows now from the well-known identity $u_1 + u_2 + \cdots + u_n = u_{n+2} - 1$. □

Recurrent relations and Bernoulli numbers

1.5.5. Consider another problem that leads to the construction of a recurrent relation; this problem is of interest to us because its solution is closely related to combinatorial constructions, and the obtained sequence forms one of the most important classes of special combinatorial numbers, namely, the sequence of *Bernoulli numbers.*

- *The sequence $1^k, 1^k + 2^k, 1^k + 2^k + 3^k, \ldots$ of the sums S_n, $n = 1, 2, 3, \ldots$, of the kth powers of the first n positive integers satisfies the linear recurrence relation*

  $$S_{n+k+2} - C_{k+2}^1 S_{n+k+1} + C_{k+2}^2 S_{n+k} - \cdots + (-1)^{k+2} C_{k+2}^{k+2} S_n = 0$$

  of the $(k+2)$th order with constant coefficients, where C_{k+2}^m are the numbers of m-combinations of a given $(k+2)$-set.

□ 1. Show that the sequence $a_1 = 1^k, a_2 = 2^k, \ldots, a_n = n^k, \ldots$ satisfies the linear recurrent relation

$$a_{n+k+1} - C_{k+1}^1 a_{n+k} + C_{k+1}^2 a_{n+k-1} - \cdots + (-1)^{k+1} C_{k+1}^{k+1} a_n = 0.$$

For a proof, build a pyramid (i.e., a string) of k cubes, painted in $n+k+1$ different colours. The number N of such pyramids is equal to

$$\overline{A}_{n+k+1}^{k} = (n+k+1)^k = a_{n+k+1}.$$

Consider the following properties of the considered pyramids: property α_1: "there is no colour 1 in the pyramid"; property α_2: "there is no colour 2 in the pyramid"; ...; property α_{k+1}: "there is no colour $k+1$ in the pyramid". Then, the number $N(\alpha_i)$ of pyramids that have the property α_i, is equal to $\overline{A}_{n+k}^{k} = (n+k)^k = a_{n+k}$, the number $N(\alpha_i, \alpha_j)$ of pyramids with properties α_i and α_j is equal to $\overline{A}_{n+k-1}^{k} = (n+k-1)^k = a_{n+k-1}$, ..., the number $N(\alpha_1, \ldots, \alpha_{k+1})$ of pyramids with all the properties $\alpha_1, \ldots, \alpha_{k+1}$, is equal to $\overline{A}_{n}^{k} = n^k = a_n$. Moreover, the number $N(\alpha_1', \ldots, \alpha_{k+1}')$ of pyramids that have no indicated properties is equal to 0: we cannot construct a pyramid from k cubes, containing $k+1$ colours. Therefore, using the *inclusion–exclusion formula*, we get the equality

$$N(\alpha_1', \ldots, \alpha_{k+1}') = N - C_{k+1}^1 N(\alpha_i) + C_{k+1}^2 N(\alpha_i, \alpha_j) - \cdots + (-1)^{k+1} C_{k+1}^{k+1} N(\alpha_1, \ldots, \alpha_{k+1}),$$

or, in other words, the formula

$$a_{n+k+1} - C_{k+1}^1 a_{n+k} + C_{k+1}^2 a_{n+k-1} - \cdots + (-1)^{k+1} a_n = 0.$$

2. As the sequence $a_1 = 1^k, a_2 = 2^k, \ldots, a_n = n^k, \ldots$ satisfies the recurrent relation

$$a_{n+k+1} - C_{k+1}^1 a_{n+k} + C_{k+1}^2 a_{n+k-1} - \cdots + (-1)^{k+1} C_{k+1}^{k+1} a_n = 0,$$

the sequence S_n satisfies the recurrent relation

$$(S_{n+k+1} - S_{n+k}) - C_{k+1}^1 (S_{n+k} - S_{n+k-1}) + \cdots + (-1)^{k+1} C_{k+1}^{k+1} (S_n - S_{n-1}) = 0,$$

or, in other words, the following recurrent relation holds,

$$S_{n+k+1} - (C_{k+1}^0 + C_{k+1}^1) S_{n+k} + (C_{k+1}^1 + C_{k+1}^2) S_{n+k-1} + \cdots = 0.$$

As $C_n^k = C_{n-1}^k + C_{n-1}^{k-1}$, then it holds that

$$S_{n+k+1} - C_{k+2}^1 S_{n+k} + C_{k+2}^2 S_{n+k-1} - \cdots + (-1)^{k+2} C_{k+2}^{k+2} S_n = 0.$$

The statement is proven. □

1.5.6. The *Bernoulli numbers* B_n, $n = 0, 1, 2, \ldots$, are rational numbers which occur frequently in analysis. The kth Bernoulli number B_k can be defined as the coefficient of n in the representation of $0^k + 1^k + 2^k + \cdots + (n-1)^k$ as a polynomial in n. In fact,

$$0^0 + 1^0 + 2^0 + \cdots + (n-1)^0 = n,$$

$$0^1 + 1^1 + 2^1 + \cdots + (n-1)^1 = \frac{n(n-1)}{2} = \frac{1}{2}n^2 - \frac{1}{2}n,$$

$$0^2 + 1^2 + 2^2 + \cdots + (n-1)^2 = \frac{(n-1)n(2n-1)}{6} = \frac{1}{3}n^3 - \frac{1}{2}n^2 + \frac{1}{6}n,$$

etc. So, the sequence of Bernoulli numbers starts with the numbers $1, -\frac{1}{2}, \frac{1}{6}, 0, -\frac{1}{30}, 0, \frac{1}{42}, 0, -\frac{1}{30}, 0, \frac{5}{56}, \ldots$ (sequences A027641 and A027642 in the OEIS).

However, using the above-considered recurrent approach, we can obtain for the sequence B_n, $n = 0, 1, 2, \ldots$, the following recurrent relation:

$$B_n = -\frac{1}{n+1} \sum_{k=1}^{n} \binom{n+1}{k+1} B_{n-k}, \quad \text{with} \quad B_0 = 0.$$

It can be used to arrive at an alternative definition for the Bernoulli numbers. For more information, see, for example, [Deza18].

Recurrent relations and closed formulas

1.5.7. If we have a recurrent relation of the kth order and the initial values $f_1, f_2, \ldots, f_k$ of a sequence, we can write out all members of the considered sequence but cannot compute the next element without the previous ones. Sometimes this is not convenient, and there is often a need for a formula that expresses the nth element f_n of the sequence only as a function of its index n.

A formula, expressing the nth element of a sequence $f_1, f_2, \ldots, f_n, \ldots$ as a function of its index n is called a *closed formula* for the elements of the sequence $f_1, f_2, \ldots, f_n, \ldots$. If the sequence $f_1, f_2, \ldots, f_n, \ldots$ is defined by a recurrent relation, then the closed formula for the elements of the sequence $f_1, f_2, \ldots, f_n, \ldots$ is also called a *solution* of this recurrent relation.

So, the formula $P_n = n!$, giving the sequence 1, 2, 6, 24, 120, $720, \ldots$ of factorials, is a solution of the recurrent relation $P_{n+1} = (n+1)P_n$; it is a closed formula for the number of permutations of n elements.

The formula $C_n^k = \frac{n!}{k!(n-k)!}$ is a closed formula for the number of combinations from n elements by k elements.

The formula $A_n^k = \frac{n!}{(n-k)!}$ is a closed formula for the number of partial permutations from n elements by k elements.

Solutions of recurrent relations

1.5.8. For solutions of recurrent relations, there are no general rules. However, there is a very large class of recurrent relations solved using a common method. These are the *linear recurrent relations with constant coefficients*: $f_{n+k} = a_1 f_{n+k-1} + a_2 f_{n+k-2} + \cdots + a_k f_n$. Their solutions are based on a study of solutions of the equation $x^k = a_1 x^{k-1} + a_2 x^{k-2} + \cdots + a_k$, called the *characteristic equation* of a given recurrent relation $f_{n+k} = a_1 f_{n+k-1} + a_2 f_{n+k-2} + \cdots + a_k f_n$.

For $k = 2$, the theorem concerning the solutions of a linear recurrent relation $f_{n+2} = a_1 f_{n+1} + a_2 f_n$ is as follows ([DeMo10]):

- *If the characteristic equation $x^2 = a_1 x + a_2$ of a linear recurrent relation $f_{n+2} = a_1 f_{n+1} + a_2 f_n$ has two different roots r_1 and r_2, then the general solution of the recurrent relation $f_{n+2} = a_1 f_{n+1} + a_2 f_n$ has the form $f_n = C_1 r_1^{n-1} + C_2 r_2^{n-1}$, where C_1 and C_2 are arbitrary constants; if the equation $x^2 = a_1 x + a_2$ has two coinciding roots $r_1 = r_2 = r$, then the general solution of the recurrent relation $f_{n+2} = a_1 f_{n+1} + a_2 f_n$ has the form $f_n = (C_1 + C_2 n) r^{n-1}$, where C_1 and C_2 are arbitrary constants.*

A solution of a linear recurrent relation $f_{n+k} = a_1 f_{n+k-1} + \cdots + a_k f_n$ of the kth order, $k > 2$, can be found similarly. If all k roots of the characteristic equation $x^k = a_1 x^{k-1} + \cdots + a_k$ of the considered recurrent relation are different, then the general solution has the form

$$f_n = C_1 r_1^{n-1} + C_2 r_2^{n-1} + \cdots + C_k r_k^{n-1}.$$

If we have s coinciding roots, $r_1 = r_2 = \cdots = r_s = r$, then to this value corresponds a contribution $r^{n-1}(C_1 + C_2 n + \cdots + C_s n^{s-1})$ to the general solution.

Consider, for example, a linear recurrent relation $f_{n+4} = 5f_{n+3} - 6f_{n+2} - 4f_{n+1} + 8f_n$ of the fourth order with constant coefficients. The characteristic equation of this relation has the form $x^4 - 5x^3 + 6x^2 + 4x - 8 = 0$. The roots of this equation are $r_1 = -1, r_2 = r_3 = r_4 = 2$. Therefore, the general solution of the considered recurrent relation has the form

$$f_n = C_1(-1)^{n-1} + (C_2 + C_3 n + C_4 n^2)2^{n-1}.$$

1.5.9. The recurrent relation for the Fibonacci numbers has the form $f_n = f_{n-1} + f_{n-2}$. It is a linear recurrent relation of the second order with constant coefficients ($a_1 = a_2 = 1$). The characteristic equation of this relation has the form $x^2 - x - 1 = 0$. There are two distinct roots, $\alpha = \frac{1+\sqrt{5}}{2}$ and $\beta = \frac{1-\sqrt{5}}{2}$. Therefore, the general solution of this recurrent relation has the form $f_n = C_1(\frac{1+\sqrt{5}}{2})^{n-1} + C_2(\frac{1-\sqrt{5}}{2})^{n-1}$. Find now the constants C_1 and C_2 using the initial conditions $f_1 = f_2 = 1$. In order to find them, consider the system

$$\begin{cases} 1 = C_1 + C_2, \\ 1 = C_1 \frac{1+\sqrt{5}}{2} + C_2 \frac{1-\sqrt{5}}{2}. \end{cases}$$

It is easy to check that $C_1 = \frac{1+\sqrt{5}}{2\sqrt{5}}$, $C_2 = 1 - C_1 = -\frac{1-\sqrt{5}}{2\sqrt{5}}$, and the solution takes the form $f_n = \frac{1}{\sqrt{5}}(\frac{1+\sqrt{5}}{2})^n - \frac{1}{\sqrt{5}}(\frac{1-\sqrt{5}}{2})^n$.

So, we get a formula that allows us to compute the nth Fibonacci number u_n as a function of its index n without using any information about the previous elements of the sequence:

$$u_n = \frac{\alpha^n - \beta^n}{\sqrt{5}}, \quad \text{where} \quad \alpha = \frac{1+\sqrt{5}}{2}, \ \beta = \frac{1-\sqrt{5}}{2}.$$

This formula is called *Binet's formula.*

Exercises

1. Prove the following identities:

   (a) $u_1 + u_2 + \cdots + u_n = u_{n+2} - 1$;
   (b) $u_1 + u_3 + \cdots + u_{2n-1} = u_{2n}$;
   (c) $u_2 + u_4 + \cdots + u_{2n} = u_{2n+1} - 1$;
   (d) $u_1^2 + u_2^2 + \cdots + u_n^2 = u_n u_{n+1}$;
   (e) $u_{n+m} = u_{n-1}u_m + u_n u_{m+1}$;
   (f) $u_{2n} = u_{n+1}^2 + u_{n-1}^2$.

2. Prove the following arithmetic properties of the Fibonacci numbers:

   (a) $(u_n, u_m) = u_{(n,m)}$; (b) $m|n \Leftrightarrow u_m|u_n$.

3. Find a recurrent relation for the sequence $1, 3, 6, 10, 15, \ldots$ of *triangular numbers*. Find the closed formula for the sequence of *triangular numbers*.
4. Prove that, for a fixed n, we can use a "one-dimensional" recurrent relation $\binom{n}{k} = \frac{n+1-k}{k}\binom{n}{k-1}$ with the initial condition $\binom{n}{0} = 1$.
5. Find the first 10 elements of the sequence B_n, $n = 0, 1, 2, \ldots$, of Bernoulli numbers using the recurrent approach.

1.6. Generating Functions

Classical generating functions

1.6.1. The method of generating functions is closely connected with the recurrence method; it became widespread in combinatorics in the 19th century (see, for example, [FlSe09] and [Land02]).

A *generating function* of the sequence $a_0, a_1, a_2, \dots, a_n, \dots$ is the formal power series

$$a_0 + a_1x + a_2x^2 + \cdots + a_nx^n + \cdots .$$

If in some area this series converges to the function $f(x)$, then we call the function $f(x)$ the *generating function* of the sequence $a_0, a_1, \dots, a_n, \dots$.

In the following, we consider several important examples of generating functions:

- The function $f(x) = \frac{1}{1-x}$ is the generating function of the sequence $1, 1, \dots, 1, 1, \dots$, since, using the known formula of the sum of an infinite geometric progression, we can say that

$$\frac{1}{1-x} = 1 + 1 \cdot x + 1 \cdot x^2 + \cdots, \quad |x| < 1.$$

- By squaring both sides of the equality of the previous example, we get the equality

$$\begin{aligned}\frac{1}{(1-x)^2} &= (1 + x + x^2 + \cdots)(1 + x + x^2 + \cdots)\\ &= 1 + 2x + 3x^2 + \cdots, \quad |x| < 1;\end{aligned}$$

  thus, the function $f(x) = \frac{1}{(1-x)^2}$ is the generating function for the sequence $1, 2, 3, \dots, n, \dots$ of positive integers.
- Similarly, it holds that

$$\begin{aligned}\frac{1}{(1-x)^3} &= (1 + 2x + 3x^2 + \cdots)(1 + x + x^2 + \cdots)\\ &= 1 + 3x + 6x^2 + \cdots, \quad |x| < 1;\end{aligned}$$

  thus, the function $f(x) = \frac{1}{(1-x)^3}$ is the generating function for the sequence $1, 3, 6, \dots, \frac{n(n+1)}{2}, \dots$ of *triangular numbers*.
- For a fixed integer $n \geq 0$, the function $(1 + x)^n$ is the generating function for the sequence $\binom{n}{k}$, $k = 0, 1, 2, \dots, n$, of *binomial coefficients*:

$$(1 + x)^n = \binom{n}{0} + \binom{n}{1}x + \cdots + \binom{n}{n}x^n, \quad x \in \mathbb{R}.$$

- If we change in the previous example the number $n \in \mathbb{Z}, n \geq 0$, to the number $\alpha \in \mathbb{R}$ and the binomial coefficients $\binom{n}{k} = \frac{n(n-1)\cdots(n-k+1)}{k!}$ to the *generalised binomial coefficients* $\binom{\alpha}{k} = \frac{\alpha(\alpha-1)\cdots(\alpha-k+1)}{k!}$, we get the decomposition

$$(1+x)^\alpha = 1 + \alpha x + \frac{\alpha(\alpha-1)}{1\cdot 2}x^2 + \frac{\alpha(\alpha-1)(\alpha-2)}{1\cdot 2\cdot 3}x^3 + \cdots$$
$$= \binom{\alpha}{0} + \binom{\alpha}{1}x + \cdots + \binom{\alpha}{n}x^n + \cdots, \quad |x| < 1.$$

Therefore, for a given $\alpha \in \mathbb{R}$, the function $f(x) = (1+x)^\alpha$ is the generating function for the infinite sequence $\binom{\alpha}{k} = \frac{\alpha(\alpha-1)\cdots(\alpha-k+1)}{k!}$, $k = 0, 1, 2, \ldots$, of generalised binomial coefficients.

1.6.2. Considering the notion of generating function of a given sequence, the following simple concepts are necessary to study them:

- *A generating function of the sum (or the difference) of two sequences $\{a_n\}_n$ and $\{b_n\}_n$ is equal to the sum (the difference) of the corresponding generating functions: if $f(x) = \sum_{n=0}^{\infty} a_n x^n$, and $g(x) = \sum_{n=0}^{\infty} b_n x^n$, then*

$$f(x) + g(x) = \sum_{n=0}^{\infty}(a_n + b_n)x^n, \;\; f(x) - g(x) = \sum_{n=0}^{\infty}(a_n - b_n)x^n.$$

□ This statement is trivial:

$$f(x) \pm g(x) = \sum_{n=0}^{\infty} a_n x^n \pm \sum_{n=0}^{\infty} b_n x^n = \sum_{n=0}^{\infty}(a_n \pm b_n)x^n. \qquad \square$$

- *The product of two generating functions $f(x) = \sum_{n=0}^{\infty} a_n x^n$ and $g(x) = \sum_{n=0}^{\infty} b_n x^n$ of the sequences $\{a_n\}_n$ and $\{b_n\}_n$ is the generating function of the discrete convolution (the Cauchy product) $c_n = \sum_{k=0}^{n} a_k b_{n-k}$ of these sequences: if $f(x) = \sum_{n=0}^{\infty} a_n x^n$, and $g(x) = \sum_{n=0}^{\infty} b_n x^n$, then*

$$f(x)g(x) = \sum_{n=0}^{\infty} c_n x^n, \;\; c_n = \sum_{k=0}^{n} a_k b_{n-k}.$$

□ This statement can be obtained by direct verification:

$$
\begin{aligned}
f(x)g(x) &= \left(\sum_{n=0}^{\infty} a_n x^n\right)\left(\sum_{k=0}^{\infty} b_k x^k\right) \\
&= (a_0 + a_1x + a_2x^2 + \cdots + a_nx^n + \cdots)(b_0 + b_1x + b_2x^2 \\
&\quad + \cdots + b_nx^n + \cdots) \\
&= a_0b_0 + (a_0b_1 + a_1b_0)x + (a_0b_2 + a_1b_1 + a_2b_0)x^2 \\
&\quad + \cdots + (a_0b_n + a_1b_{n-1} + \cdots + a_{n-1}b_0 + a_nb_0)x^n + \cdots \\
&= \sum_{n=0}^{\infty}\left(\sum_{k=0}^{n} a_k b_{n-k}\right) x^n = \sum_{n=0}^{\infty} c_n x^n. \qquad \square
\end{aligned}
$$

1.6.3. In general, the problem of finding the generating function of a given sequence is a very difficult task. However, there is a class of sequences for which the generating function may be found using a common, relatively simple algorithm. These sequences are given by linear recurrent relations with constant coefficients.

The rule for finding the generating function in this case looks like the following (see [DeMo10]):

- *The generating function of the sequence* $c_0, c_1, \ldots, c_n, \ldots$ *satisfying the linear recurrent relation*

$$b_0c_{m+k} + b_1c_{m+k-1} + \cdots + b_mc_k = 0$$

of the mth order with constant coefficients $b_0, b_1, \ldots, b_m$ *and the initial conditions* $c_0 = C_0, \ldots, c_{m-1} = C_{m-1}$, *where* $C_0, \ldots, C_{m-1}$ *are some constants, has the form* $\frac{f(x)}{\varphi(x)}$, *i.e., it is a ratio of two polynomials* $f(x)$ *and* $\varphi(x)$, *where* $\varphi(x) = b_0 + b_1x + \cdots + b_mx^m$ *and* $f(x) = a_0 + a_1x + \cdots + a_nx^n$, $n < m$; *the coefficients* $a_1, \ldots, a_n$ *of* $f(x)$ *can be found from the following conditions*:

$$
\begin{cases}
a_0 = b_0c_0, \\
a_1 = b_0c_1 + b_1c_0, \\
\vdots \\
a_n = b_0c_0 + b_1c_{n-1} + \cdots + b_nc_0.
\end{cases}
$$

Moreover, it holds that

$$\frac{f(x)}{\varphi(x)} = c_0 + c_1x + \cdots + c_nx^n + \cdots, \quad |x| < \min\{|x_1|, \ldots, |x_m|\},$$

where $x_1, \ldots, x_m$ *are solutions to the equation* $\varphi(x) = 0$.

1.6.4. For example, find the generating function of the Fibonacci sequence $1, 1, 3, 5, 8, \ldots$ Since the elements of this sequence satisfy the recurrent relation $c_{k+2} - c_{k+1} - c_k = 0$, with the initial conditions $c_0 = 1$ and $c_1 = 1$, then, in our notation, $b_0 = 1, b_1 = -1$, and $b_2 = 1$; therefore, it holds that $\varphi(x) = 1 - x - x^2$. By calculating $a_0 = b_0 \cdot c_0 = 1 \cdot 1 = 1, a_1 = b_0 \cdot c_1 + b_1 \cdot c_0 = 1 \cdot 1 - 1 \cdot 1 = 0$, we get that $f(x) \equiv 1$. So, the generating function of the sequence $c_0 = 1, c_1 = 1, c_2 = 2, \ldots, c_n, \ldots$ has the form $\frac{1}{1-x-x^2}$.

The roots of the equation $x^2 + x - 1 = 0$ are $x_1 = \frac{-1+\sqrt{5}}{2}, x_2 = \frac{-1-\sqrt{5}}{2}$, so the decomposition $\frac{1}{1-x-x^2} = c_0 + c_1x + \cdots + c_nx^n + \cdots$, or, in other words, the decomposition $\frac{1}{1-x-x^2} = u_1 + u_2x + \cdots + u_{n+1}x^n + \cdots$ holds for $|x| < \frac{\sqrt{5}-1}{2} = \frac{1}{\alpha}$, where $\alpha = \frac{1+\sqrt{5}}{2}$. Therefore, we get the following statement:

- *The generating function of the sequence of Fibonacci numbers is* $\frac{1}{1-x-x^2}$, *i.e., it holds that*

$$\frac{1}{1-x-x^2} = u_1 + u_2x + u_3x^2 + \cdots + u_{n+1}x^n + \cdots = \frac{1}{1-x-x^2},$$

$$|x| < \frac{1}{\alpha} = \frac{\sqrt{5}-1}{2} \approx 0,6.$$

Exponential generating functions

1.6.5. When we study sequences $\{a_n\}_n$, it is often more convenient to investigate (instead of the properties of the series $\sum_{n=0}^{\infty} a_nx^n$) the behaviour of the series $\sum_{n=0}^{\infty} a_n\frac{x^n}{n!}$.

An *exponential generating function* of the sequence $a_0, a_1, a_2, \ldots,$ $a_n, \ldots$ is a formal power series

$$a_0 + a_1 \frac{x}{1!} + a_2 \frac{x^2}{2!} + \cdots + a_n \frac{x^n}{n!} + \cdots .$$

If in some region this series converges to the function $g(x)$, then we consider the function $g(x)$ the *exponential generating function* of the sequence $a_0, a_1, \ldots, a_n, \ldots$.

Consider some classic examples of exponential generating functions:

- The function e^x is the exponential generating function of the sequence $1, 1, \ldots, 1, 1, \ldots$; it follows from the above definition and the well-known decomposition

$$e^x = 1 + \frac{1}{n!}x + \frac{1}{2!}x^2 + \cdots + \frac{1}{n!}x^n + \cdots ,$$

  which is true for any $x \in \mathbb{R}$.
- Considering the decomposition $(1+x)^\alpha = 1 + \alpha x + \frac{\alpha(\alpha-1)}{1\cdot 2}x^2 + \frac{\alpha(\alpha-1)(\alpha-2)}{1\cdot 2\cdot 3}x^3 + \cdots$, occurring for $|x| < 1$, we can check that for any complex value of α, the function $(1+x)^\alpha$ is the exponential generating function for the sequence $1, \alpha, \alpha(\alpha-1), \ldots,$ $\alpha(\alpha-1)\cdots(\alpha-n+1), \ldots$:

$$(1+x)^\alpha = 1 + \alpha\frac{x}{1!} + \alpha(\alpha-1)\frac{x^2}{2!} + \alpha(\alpha-1)(\alpha-2)\frac{x^3}{3!} + \cdots + \alpha(\alpha-1)(\alpha-2)\cdots(\alpha-n+1)\frac{x^n}{n!} + \cdots ,$$

$$|x| < 1.$$

1.6.6. There are numerous such examples; they can be obtained by simple observation. The notion of exponential generating functions is closely related to the decomposition of functions into the *Taylor series* (or, in fact, the *Maclaurin series*), which is an infinite sum of power functions.

A *Taylor series* (at a point a) for the real- or complex-valued function $f(x)$, which is infinitely differentiable in some neighbourhood of the point a, is a formal power series of the form

$$f(a)+\frac{f'(a)}{1!}(x-a)+\frac{f''(a)}{2!}(x-a)^2+\cdots+\frac{f^{(n)}(a)}{n!}(x-a)^n+\cdots,$$

where $f^{(n)}(a)$ denotes the nth derivative of f, evaluated at the point a. (The derivative of order zero of f is defined to be f itself; $(x-a)^0$ and $0!$ are both defined to be 1.)

If $f(x)$ is equal to the sum of its Taylor series in some area, we write that

$$f(x)=f(a)+\frac{f'(a)}{1!}(x-a)+\frac{f''(a)}{2!}(x-a)^2+\frac{f'''(a)}{3!}(x-a)^3$$
$$+\cdots+\frac{f^{(n)}(a)}{n!}(x-a)^n+\cdots$$

for such values of the argument x. In a more compact notation, this can be written as

$$f(x)=\sum_{n=0}^{\infty}\frac{f^{(n)}(a)}{n!}(x-a)^n.$$

For $a=0$, we obtain a *Maclaurin series* (or *Taylor–Maclaurin series*):

$$f(0)+\frac{f'(0)}{1!}x+\frac{f''(0)}{2!}x^2+\cdots+\frac{f^{(n)}(0)}{n!}x^n+\cdots.$$

Thus, the Maclaurin series for the function $f(x)$ is, by definition, an exponential generating function of the sequence $f(0), f'(0), f''(0), \ldots, f^{(n)}(0), \ldots$ of derivatives of the function $f(x)$ of the orders n, $n=0,1,2,3,\ldots$, evaluated at zero. If the Maclaurin series converges to the function $g(x)$ in some neighbourhood of zero, then we write

$$g(x)=f(0)+\frac{f'(0)}{1!}x+\frac{f''(0)}{2!}x^2+\cdots+\frac{f^{(n)}(0)}{n!}x^n+\cdots$$

and say that the *exponential generating function* of the sequence $f(0), f'(0), f''(0), \ldots, f^{(n)}(0), \ldots$ is a function $g(x)$.

1.6.7. To study exponential generating functions, we can use the following statements:

- *The exponential generating function of the sum (or the difference) of two sequences* $\{a_n\}_n$ *and* $\{b_n\}_n$ *is the sum (the difference) of the corresponding exponential generating functions: if* $f(x) = \sum_{n=0}^{\infty} a_n \frac{x^n}{n!}$ *and* $g(x) = \sum_{n=0}^{\infty} b_n \frac{x^n}{n!}$, *then*

$$f(x) + g(x) = \sum_{n=0}^{\infty} (a_n + b_n)\frac{x^n}{n!}, \quad f(x) - g(x) = \sum_{n=0}^{\infty} (a_n - b_n)\frac{x^n}{n!}.$$

□ This statement is trivial:

$$f(x) \pm g(x) = \sum_{n=0}^{\infty} a_n \frac{x^n}{n!} \pm \sum_{n=0}^{\infty} b_n \frac{x^n}{n!} = \sum_{n=0}^{\infty} (a_n \pm b_n)\frac{x^n}{n!}. \qquad \square$$

- *If* $f(x) = \sum_{n=0}^{\infty} a_n \frac{x^n}{n!}$ *and* $g(x) = \sum_{n=0}^{\infty} b_n \frac{x^n}{n!}$ *are exponential generating functions of the sequences* $\{a_n\}_n$ *and* $\{b_n\}_n$, *then their product is an exponential generating function of the binomial convolution* $c_n = \sum_{k=0}^{n} \binom{n}{k} a_k b_{n-k}$ *of the sequences* $\{a_n\}_n$ *and* $\{b_n\}_n$: $f(x)g(x) = \sum_{n=0}^{\infty} c_n \frac{x^n}{n!}$.

□ This statement can be obtained by direct verification:

$$\begin{aligned} f(x)g(x) &= \left(\sum_{n=0}^{\infty} a_n \frac{x^n}{n!}\right)\left(\sum_{k=0}^{\infty} b_k \frac{x^k}{k!}\right) \\ &= \left(a_0 + a_1\frac{x}{1} + a_2\frac{x^2}{2!} + \cdots + a_n\frac{x^n}{n!} + \cdots\right) \\ &\quad \times \left(b_0 + b_1\frac{x}{1!} + b_2\frac{x^2}{2!} + \cdots + b_n\frac{x^n}{n!} + \cdots\right) \\ &= a_0b_0 + \left(\frac{1!}{0!1!}a_0b_1 + \frac{1!}{1!0!}a_1b_0\right)\frac{x}{1!} \\ &\quad + \left(\frac{2!}{0!2!}a_0b_2 + \frac{2!}{1!1!}a_1b_1 + \frac{2!}{2!0!}a_2b_0\right)\frac{x^2}{2!} \end{aligned}$$

$$+\cdots+\left(\frac{n!}{0!n!}a_0b_n+\frac{n!}{1!(n-1)!}a_1b_{n-1}\right.$$

$$\left.+\cdots+\frac{n!}{(n-1)!1!}a_{n-1}b_0+\frac{n!}{n!0!}a_nb_0\right)\frac{x^n}{n!}$$

$$+\cdots=\sum_{n=0}^{\infty}\left(\sum_{k=0}^{n}\binom{n}{k}a_kb_{n-k}\right)x^n=\sum_{n=0}^{\infty}c_nx^n. \qquad \square$$

1.6.8. In our book, we repeatedly use the decompositions into Maclaurin series of some classic functions. A list of the most famous decompositions of this kind is provided in the following.

Decompositions into Maclaurin series of some classic functions

- The *exponential function* e^x (with base e) has the decomposition

$$\mathrm{e}^x=1+\frac{x}{1!}+\frac{x^2}{2!}+\frac{x^3}{3!}+\cdots=\sum_{n=0}^{\infty}\frac{x^n}{n!},\ x\in\mathbb{C}.$$

- The *natural logarithm* (logarithm with base e) has the decompositions (called the *Mercator series*, or the *Newton–Mercator series*)

$$\log(1+x)=x-\frac{x^2}{2}+\frac{x^3}{3}-\cdots=\sum_{n=0}^{\infty}\frac{(-1)^nx^{n+1}}{(n+1)}=\sum_{n=1}^{\infty}\frac{(-1)^{n-1}x^n}{n},$$

$$-1<x\le 1;$$

$$\log(1-x)=-x-\frac{x^2}{2}-\frac{x^3}{3}-\cdots=-\sum_{n=0}^{\infty}\frac{x^{n+1}}{(n+1)}=-\sum_{n=1}^{\infty}\frac{x^n}{n},$$

$$-1\le x<1.$$

- The *binomial series* is the power series (with $\binom{\alpha}{n}=\frac{\alpha(\alpha-1)\cdots(\alpha-n+1)}{n!}$)

$$(1+x)^\alpha=1+\sum_{n=1}^{\infty}\binom{\alpha}{n}x^n,\ |x|<1,\ \alpha\in\mathbb{C}.$$

- The *geometric series*, a special case of the binomial series with $\alpha = -1$, is

$$\frac{1}{1-x} = 1 + x + x^2 + x^3 + \cdots = \sum_{n=0}^{\infty} x^n, \ |x| < 1.$$

- The *trigonometric function* $\sin x$ has the following Maclaurin series:

$$\sin x = x - \frac{x^3}{3!} + \frac{x^5}{5!} - \cdots = \sum_{n=0}^{\infty} \frac{(-1)^n}{(2n+1)!} x^{2n+1}, x \in \mathbb{C}.$$

- The *trigonometric function* $\cos x$ has the following Maclaurin series:

$$\cos x = 1 - \frac{x^2}{2!} + \frac{x^4}{4!} - \cdots = \sum_{n=0}^{\infty} \frac{(-1)^n}{(2n)!} x^{2n}, x \in \mathbb{C}.$$

1.6.9. In combinatorial problems, the question of convergence of a formal series specifying a generating function of a given sequence is not discussed in detail. In writing out the function to which this series converges and naming it the generating function of the considered sequence, we rely on the fact that the specified convergence is performed in some neighbourhood of zero, but the exact properties of this neighbourhood, as a rule, are not investigated. However, in several classic cases, when information about the convergence of a particular series is well known, we indicate existing boundaries, referring to the corresponding sources.

The reader can investigate the question of convergence of the decompositions presented in this book by themselves ([Fich01]): a *convergence interval* of the power series $\sum_{n=0}^{\infty} a_n x^n$ is a set of values of x satisfying the condition $|x| < r$, and a *convergence radius* r in most cases can be calculated using the *Cauchy–Hadamar formula*:

$$\frac{1}{r} = \overline{\lim_{n \to \infty}} \, |a_n|^{1/n}.$$

Exponential generating functions and special numbers

1.6.10. In computational mathematics, generating functions, especially exponential generating functions, are commonly used for defining number sets.

For example, the Bernoulli numbers B_n, $n = 0, 1, 2, \ldots$, can be defined as coefficients of the decomposition

$$\frac{x}{e^x - 1} = \sum_{n=0}^{\infty} B_n \frac{x^n}{n!}.$$

From this definition, we immediately obtain many useful properties of this number sequence.

For example, as the function $f(x) = \frac{x}{e^x-1} + \frac{x}{2}$ is an even function, i.e., it holds that $f(-x) = f(x)$, we get that *all Bernoulli numbers with odd indices, except $B_1 = -\frac{1}{2}$, are equal to zero.*

From this analytical definition of the Bernoulli numbers, we can obtain many important decompositions, which will be used in our further consideration. We list some of these representations in the following. For more information see, for example, [Deza18].

- *There exist the following decompositions*:

$$\frac{x}{2}\cot\frac{x}{2} = B_0 - \frac{B_2}{2!}x^2 + \frac{B_4}{4!}x^4 - \frac{B_6}{6!}x^6 + \cdots + (-1)^k \frac{B_{2k}}{(2k)!}x^{2k} + \cdots, |x| < 2\pi;$$

$$x\cot x = 1 - \frac{2^2\, B_2}{2!}x^2 + \frac{2^4\, B_4}{4!}x^4 - \frac{2^6\, B_6}{6!}x^6 + \cdots + (-1)^k \frac{2^{2k}\, B_{2k}}{(2k)!}x^{2k} - \cdots, |x| < \pi;$$

$$\tan x = \frac{2^2(2^2-1)B_2}{2!}x - \frac{2^4(2^4-1)B_4}{4!}x^3 + \frac{2^6(2^6-1)B_6}{6!}x^5 + \cdots \frac{2^{2k}(2^{2k}-1)B_{2k}}{(2k)!}x^{2k-1} + \cdots, \quad |x| < \frac{\pi}{2};$$

$$\frac{x}{e^x+1} = \frac{x}{2} - \frac{B_2(2^2-1)}{2!}x^2 - \frac{B_4(2^4-1)}{4!}x^4 - \frac{B_6(2^6-1)}{6!}x^6$$
$$+\cdots, +(-1)^k \frac{B_{2k}(2^{2k}-1)}{(2k)!}x^{2k} + \cdots, \quad |x| < \pi;$$
$$\frac{e^x-1}{e^x+1} = \frac{2(2^2-1)B_2}{2!}x + \frac{2(2^4-1)B_4}{4!}x^3 + \frac{2(2^6-1)B_6}{6!}x^5$$
$$+\cdots + \frac{2(2^{2k}-1)B_{2k}}{(2k)!}x^{2k-1} + \cdots, \quad |x| < \pi.$$

Multidimensional generating functions

1.6.11. In some cases, even the exponential generating function may not be helpful in a considered situation. It happens when the sequence under investigation is "multidimensional", that is, when it contains not one but two or more parameters. An important example of such an array is the array of elements C_n^k of Pascal's triangle.

Of course, any "two-dimensional" set $c_{n,k}$ with non-negative integer indices n, k can be "expanded" into a line, but sometimes this is not convenient. Such sequences can be represented by the *two-dimensional generating function* $\sum_{n,k=0}^{\infty} c_{n,k}x^n y^k$. For example, for the elements of Pascal's triangle, such a function has the form $\sum_{n,k=0}^{\infty} C_n^k x^n y^k$. The following result can be obtained through simple reasoning:

$$\sum_{n,k=0}^{\infty} C_n^k x^k y^n = \sum_{n=0}^{\infty}\left(\sum_{k=0}^{\infty} C_n^k x^k\right) y^n = \sum_{n=0}^{\infty}\left(\sum_{k=0}^{n} C_n^k x^k\right) y^n$$
$$= \sum_{n=0}^{\infty}(1+x)^n y^n = \sum_{n=0}^{\infty}((1+x)y)^n = \frac{1}{1-(1+x)y}.$$

By studying the resulting function $\frac{1}{1-(1+x)y}$, we can investigate the properties of all binomial coefficients. In some situations, they can turn out to be quite useful.

Exercises

1. Let $f(x)$ and $g(x)$ be the generating functions of the sequences $\{a_n\}_n$ and $\{b_n\}_n$, respectively. Prove that $h(x)$ is the generating function of the sequence $\{c_n\}_n$ if:
   - $h(x) = \alpha f(x) + \beta g(x)$, $c_n = \alpha a_n + \beta b_n$;
   - $h(x) = f(x)g(x)$, $c_n = a_0 b_n + a_1 b_{n-1} + \cdots + a_n b_0$;
   - $h(x) = tf(x)$, $c_0 = 0, c_n = a_{n-1}$, $n \geq 1$;
   - $h(x) = f(\alpha x)$, $c_n = \alpha^n a_n$;
   - $h(x) = tf'(x)$, $c_n = na_n$;
   - $h(x) = \frac{f(x)}{1-x}$, $c_n = a_0 + a_1 + \cdots + a_n$.
2. Prove that the generating function of the sequence $1, 2, 3, \ldots, n, \ldots$ of positive integers has the form $\frac{1}{(1-x)^2}$, taking the derivative of the generating function $\frac{1}{1-x}$ of the sequence $1, 1, 1, \ldots$. Find the generating function of the sequence $1^2, 2^2, 3^2, \ldots, n^2, \ldots$ of the squares of positive integers. Find the generating function of the sequence $1^3, 2^3, 3^3, \ldots, n^3, \ldots$ of the cubes of positive integers.
3. Prove that the *square root function* and its inverse have the following decompositions:

$$(1+x)^{\frac{1}{2}} = 1 + \frac{1}{2}x - \frac{1}{8}x^2 + \frac{1}{16}x^3 - \frac{5}{128}x^4 + \frac{7}{256}x^5 - \cdots$$

$$= \sum_{n=0}^{\infty} \frac{(-1)^{n-1}(2n)!}{4^n(n!)^2(2n-1)}x^n,$$

$$(1+x)^{-\frac{1}{2}} = 1 - \frac{1}{2}x + \frac{3}{8}x^2 - \frac{5}{16}x^3 + \frac{35}{128}x^4 - \frac{63}{256}x^5 + \cdots$$

$$= \sum_{n=0}^{\infty} \frac{(-1)^n(2n)!}{4^n(n!)^2}x^n.$$

4. Prove the following decompositions:

   (a) $\tan x = \sum_{n=1}^{\infty} \frac{B_{2n}(-4)^n(1-4^n)}{(2n)!}x^{2n-1} = x + \frac{x^3}{3} + \frac{2x^5}{15} + \cdots$, $|x| < \frac{\pi}{2}$;

   (b) $\sec x = \sum_{n=0}^{\infty} \frac{(-1)^n E_{2n}}{(2n)!}x^{2n} = 1 + \frac{x^2}{2} + \frac{5x^4}{24} + \cdots$, $|x| < \frac{\pi}{2}$;

(c) $\arctan x = \sum_{n=0}^{\infty} \frac{(-1)^n}{2n+1} x^{2n+1} = x - \frac{x^3}{3} + \frac{x^5}{5} - \cdots, \quad |x| \le 1,\ x \ne \pm i.$

All angles are expressed in radians. The numbers B_k appearing in the expansions of $\tan x$ are the *Bernoulli numbers.* The numbers E_k in the expansion of $\sec x$ are the *Euler numbers.*

1.7. Asymptotic Formulas

Infinite sums and generating functions

1.7.1. Generating functions can be used to solve some problems of finding infinite and finite sums.

For example, the generating function of the sequence of Fibonacci numbers has the form $\frac{1}{1-x-x^2}$, and the corresponding decomposition $\frac{1}{1-x-x^2} = u_1 + u_2 x + u_3 x^2 + \cdots$ holds for $|x| < \frac{1}{\alpha}$, $\alpha = \frac{1+\sqrt{5}}{2}$. Taking $x = \frac{1}{2}$, we obtain on the right-hand side of the equality an infinite sum, $u_1 + \frac{u_2}{2} + \frac{u_3}{4} + \cdots + \frac{u_n}{2^{n-1}} + \cdots$, and on the left-hand side, we have the value $\frac{1}{1-\frac{1}{2}-\frac{1}{4}} = \frac{4}{4-2-1} = 4$. Therefore, the following equality holds:

$$u_1 + \frac{u_2}{2} + \frac{u_3}{4} + \cdots + \frac{u_n}{2^{n-1}} + \cdots = \sum_{n=1}^{\infty} \frac{u_n}{2^{n-1}} = 4.$$

The generating function of the sequence $1, 1, \ldots, 1, \ldots$ has the form $\frac{1}{1-x}$, and the decomposition $\frac{1}{1-x} = 1 + x + x^2 + \cdots$ takes place for $|x| < 1$. Taking $x = \frac{1}{2}$, we get the equality $\sum_{n=0}^{\infty} 2^{-n} = 2$. Taking $x = -\frac{1}{2}$, we get the equality $\sum_{n=0}^{\infty} (-2)^{-n} = \frac{2}{3}$. Taking $x = \frac{1}{3}$, we get the equality $\sum_{n=0}^{\infty} 3^{-n} = \frac{3}{2}$.

Big-O and little-o notations

1.7.2. In order to obtain from a given infinite sum the corresponding finite sum, we can usually use the so-called *asymptotic formulas.*

Let $f(x)$ and $g(x)$ be complex-valued functions of a real argument, such that $g(x) > 0$ for all large x, i.e., for all $x > \alpha$, where α is a fixed positive real number. We say that $f(x) = O(g(x))$ if there exists a positive real number x_0 such that for all $x \geq x_0$, the inequality $|f(x)| \leq C \cdot g(x)$ holds, where $C > 0$ is a positive constant.

Consider several simple examples:

- $\sin x = O(1)$ since, for all $x \in \mathbb{R}$, it holds that $|\sin x| \leq 1$, i.e., the inequality $|\sin x| \leq Cg(x)$ holds for $C = 1$ and $g(x) \equiv 1$.
- $\frac{1}{x} = O(1)$ since, for all $x \geq 1$, it holds that $|\frac{1}{x}| = \frac{1}{x} \leq 1$, i.e., the inequality $|\frac{1}{x}| \leq Cg(x)$ holds for $x_0 = 1, C = 1, g(x) \equiv 1$.
- $5x^2 - 2x + 1 = O(x^2)$ since, for all $x \geq 1$, it holds that $|5x^2 - 2x + 1| \leq |5x^2| + |-2x| + 1 = 5|x^2| + 2|x| + 1 \leq 5|x^2| + 2|x^2| + |x^2| = 8|x^2| = 8x^2$, i.e., the inequality $|5x^2 - 2x + 1| \leq Cg(x)$ holds for $x_0 = 1, C = 8, g(x) = x^2$.

Besides the "O()" symbol (*big-O notation*), the "o()" symbol (*little-o notation*) is often used: we say that $f(x) = o(g(x))$ if $\lim_{x\to\infty} \frac{f(x)}{g(x)} = 0$.

In other words, writing $f(x) = O(g(x))$ means that, for large x, the ratio $\frac{|f(x)|}{|g(x)|}$ is bounded from above, whereas the condition $f(x) = o(g(x))$ indicates that, for large x, this ratio is small.

1.7.3. A classic example of using asymptotic formulas involves the study of the behaviour of partial sums $\sum_{n\leq x} \frac{1}{n}$ of the *harmonic series* $\sum_{n=1}^{\infty} \frac{1}{n}$.

- *For the sum* $\sum_{n\leq x} \frac{1}{n}$, *the following asymptotic formula holds:*

$$\sum_{n\leq x} \frac{1}{n} = \log x + O(1).$$

□ It is easy to see that

$$\frac{1}{n+1} \leq \int_{n}^{n+1} \frac{dt}{t} \leq \frac{1}{n};$$

therefore, it holds that

$$\sum_{n=1}^{N} \frac{1}{n+1} \leq \sum_{n=1}^{N} \int_{n}^{n+1} \frac{dt}{t} \leq \sum_{n=1}^{N} \frac{1}{n}.$$

As

$$\sum_{n=1}^{N} \int\limits_{n}^{n+1} \frac{dt}{t} = \int_{1}^{N} \frac{dt}{t} + \int_{N}^{N+1} \frac{dt}{t} = \log N + \int_{N}^{N+1} \frac{dt}{t}$$

and $\sum_{n=1}^{N} \frac{1}{n+1} = \sum_{n=1}^{N} \frac{1}{n} - 1 + \frac{1}{N+1}$, one can rewrite the above inequality as

$$\sum_{n=1}^{N} \frac{1}{n} - 1 + \frac{1}{N+1} \le \log N + \int\limits_{N}^{N+1} \frac{dt}{t} \le \sum_{n=1}^{N} \frac{1}{n}.$$

Therefore, we have

$$\sum_{n=1}^{N} \frac{1}{n} - \log N \le \int_{N}^{N+1} \frac{dt}{t} + 1 - \frac{1}{N+1},$$

and as $\int_{N}^{N+1} \frac{dt}{t} \le \frac{1}{N}$, we get that

$$\sum_{n=1}^{N} \frac{1}{n} - \log N \le \frac{1}{N} + 1 - \frac{1}{N+1} \le 2.$$

On the other hand, similar considerations allow us to obtain an estimation:

$$\sum_{n=1}^{N} \frac{1}{n} - \log N \ge \int_{N}^{N+1} \frac{dt}{t} \ge \frac{1}{N} > 0.$$

So, it holds that

$$\left| \sum_{n=1}^{N} \frac{1}{n} - \log N \right| = \sum_{n=1}^{N} \frac{1}{n} - \log N \le 2 \cdot 1,$$

i.e., we have the asymptotic formula of the form

$$\sum_{n=1}^{N} \frac{1}{n} - \log N = O(1). \qquad \square$$

1.7.4. Using more robust reasons, the previous theorem can be clarified (see [DeMo10]):

- *For the sum* $\sum_{n\le n} \frac{1}{n}$, *there exists an asymptotic formula,*

$$\sum_{n\le x} \frac{1}{n} = \log x + \gamma + O(x^{-1}),$$

where $\gamma = 0,5772215\ldots$ *is the Euler–Mascheroni constant:* $\gamma = \lim_{n\to\infty}\left(\sum_{k=1}^{n}\frac{1}{k} - \log n\right)$.

Asymptotic equivalence

1.7.5. The above results allow us to state that, for large x, the sum $\sum_{n\le x}\frac{1}{n}$ behaves like $\log x$. In fact, since for $x\to\infty$ the value $O(1)$ is bounded by a positive constant, the ratio $\frac{O(1)}{\log x}$ goes to zero, and we have that

$$\lim_{x\to\infty}\frac{\sum_{n\le x}\frac{1}{n}}{\log x} = \lim_{x\to\infty}\frac{\log x + O(1)}{\log x} = \lim_{x\to\infty}\left(1 + O\left(\frac{1}{\log x}\right)\right) = 1.$$

To indicate the asymptotic coincidence of two functions, a special symbol is used.

Let $f(x)$ *and* $g(x)$ *be complex-valued functions of a real argument. If* $\lim_{x\to\infty}\frac{f(x)}{g(x)} = 1$, *we say that the functions* $f(x)$ *and* $g(x)$ *are asymptotically equal, and we write* $f(x)\sim g(x)$.

So, we can write that

$$\sum_{n\le x}\frac{1}{n}\sim\log x.$$

Consider several examples:

- $5x^2 - 2x + 1 \sim 5x^2$, as $\lim_{x\to\infty}\frac{5x^2-2x+1}{5x^2} = \lim_{x\to\infty}(1 - \frac{2x-1}{5x^2}) = 1$.
- $C_n^2 \sim \frac{n^2}{2}$, as $\lim_{n\to\infty}\frac{C_n^2}{n^2/2} = \lim_{n\to\infty}\frac{n(n-1)/2}{n^2/2} = \lim_{n\to\infty}\frac{n^2-n}{n^2} = \lim_{n\to\infty}(1-\frac{1}{n}) = 1$.
- $\sin x \sim x$ for $x\to 0$, as $\lim_{x\to 0}\frac{\sin x}{x} = 1$.

Stirling's formula

1.7.6. For estimations and asymptotic formulas for the majority of combinatorial numbers, particularly for binomial coefficients, we should use *Stirling's formula* (or *Stirling's approximation, de Moivre's formula*) which is a formula for approximative calculations of factorials (see, for example, [GKP94], [Gelf59], and [Kara83]).

It is a good approximation, leading to accurate results even for small values of n. It is named after James Stirling, though a related but less precise result was first stated by de Moivre.

1.7.7. The weak version of the estimation of $n!$ is as follows:

- *for all* $n \geq 2$, *it holds that*

$$\log n! = n \log n + O(n).$$

□ Let us show that

$$n \log n - n < \log n! < n \log n.$$

In fact, the inequality $\log n! < n \log n$ is a consequence of the trivial inequality $n! < n^n$.

On the other hand, a Riemann sum approximation for $\int_1^n \log x dx$ using the right endpoints is $\log 2 + \cdots + \log n = \log n!$, which overestimates; therefore,

$$\log n! > \int_1^n \log x dx = n \log n - n + 1.$$

So, we get the following inequalities:

$$n \log n - n + 1 < \log n! < n \log n,$$

$$-n + 1 < \log n! - n \log n < 0, |\log n! - n \log n| < n - 1 < n.$$

Therefore, we obtain that $\log n! - n \log n = O(n)$, and $\log n! = n \log n + O(n)$. □

Dividing by $n \log n$ both parts of the last equality, we obtain $\frac{\log n!}{n \log n} = 1 + O(\frac{1}{\log n})$. So, it is proven that

$$\log n! \sim n \log n.$$

1.7.8. The simplest variant of Stirling's formula has the form

$$\log n! = n \log n - n + O(\log n).$$

□ As before, a Riemann sum approximation for $\int_1^n \log x dx$ using the right endpoints has the form $\log 2 + \cdots + \log n = \log n!$, and we get

$$\log n! > \int_1^n \log x dx = n \log n - n + 1.$$

On the other hand, a Riemann sum approximation for $\int_1^n \log x dx$ using the left endpoints is $\log 1 + \cdots + \log(n-1) = \log n! - \log n$, and we get

$$\log n! - \log n < \int_1^n \log x dx = n \log n - n + 1,$$

or, equivalently,

$$\log n! < n \log n - n + \log n + 1.$$

So, we get the inequalities

$$n \log n - n + 1 < \log n! < n \log n - n + \log n + 1,$$
$$1 < \log n! - n \log n - n < \log n + 1,$$
$$|\log n! - n \log n - n| < \log n + 1 \leq 2 \log n.$$

Therefore, we obtain that $\log n! - n \log n + n = O(\log n)$, and $\log n! = n \log n - n + O(\log n)$. □

In other words, we have proven that

$$n! = e^{n \log n - n + O(\log n)}.$$

As $e^{n \log n - n} = n^n \cdot e^{-n} = (\frac{n}{e})^n$, we get the equality

$$n! = \left(\frac{n}{e}\right)^n \cdot O(n).$$

1.7.9. The error term $O(\log n)$ in the above asymptotic formula for $\log n!$ can be expressed more precisely as $\frac{1}{2}\log(2\pi n) + O(\frac{1}{n})$, giving the asymptotic formula

$$\log n! = n \log n - n + \tfrac{1}{2}\log(2\pi n) + O(\tfrac{1}{n}).$$

So, a more exact approximation of $n!$ is

$$\lim_{n\to\infty} \frac{n!}{\sqrt{2\pi n}\,\left(\frac{n}{e}\right)^n} = 1,$$

which is equivalent to the classical variant of *Stirling's formula:*

$$n! \sim \sqrt{2\pi n}\left(\frac{n}{e}\right)^n.$$

Exercises

1. Find the following sums:
   - $u_1 + \frac{u_2}{3} + \frac{u_3}{9} + \cdots + \frac{u_n}{3^{n-1}} + \cdots$;
   - $\frac{u_1}{5} + \frac{u_2}{25} + \frac{u_3}{125} + \cdots + \frac{u_n}{5^n} + \cdots$;
   - $5 + \frac{5}{4} + \frac{5}{16} + \cdots + \frac{5}{4^n} + \cdots$;
   - $\frac{3}{8} + \frac{9}{16} + \frac{27}{32} + \cdots + \frac{3^n}{2^{n+2}} + \cdots$.
2. Prove that $e^x = 1 + x + \frac{x^2}{2} + O(x^3)$.
3. Prove that $\sum_{n \le x} \frac{1}{n} = \log x + \gamma + O(\frac{1}{x})$, where γ is the Euler–Mascheroni constant.
4. Check that the relation $f(x) \sim g(x)$ means $f(x) = g(x) + o(g(x))$.

References

[AbSt72], [Abra74], [Adam97], [Aign82], [Ande03], [Andr98], [Apos86], [BaCo87], [Bell34], [BeSl95], [BiBa70], [Bond93], [Bron01], [Brua97], [Buch09], [Cata44], [ChMo75], [Cofm75], [Comt74], [Dave99], [Dede63], [DeZa80], [Deza18], [DeMo10], [Dick05], [Erus00], [Eule48], [Fich01], [FlSe09], [Gelf98], [Gelf59], [GoJa83], [GGL95Û], [GuJa90], [Hagg04], [Hall70], [Hara03], [HaPa77], [GaSa92], [HaWr79], [Hogg69], [Hons91], [IrRo90], [Jablo01], [Jord65], [Kara83], [Klar70], [Knot77], [KoKo77], [Kudr36], [KuAd88], [KPP94], [Land94], [Land02], [Lege79], [LiWi92], [Lips88], [Mark0], [Mazu10], [Mazu10], [Nage51], [Ore48], [Ore80], [Pasc54], [Plou92], [Rior79], [Rior80], [RoTe09], [Rose18], [Sachk77], [Smit84], [Stra16], [Stru87], [Stan97], [Stan86], [Uspe76], [Vile14], [Voro61], [Went99].

Chapter 2

Permutations

2.1. Definitions and Examples

Classical approach

2.1.1. A *permutation* of a set is, in loose terms, an arrangement of its members into a sequence or linear order, or, if the set is already ordered, a rearrangement of its elements. The word "permutation" also refers to the act or process of changing the linear order of an ordered set.

For example, written as tuples, there are six permutations of the set $\{1, 2, 3\}$, namely $(1, 2, 3)$, $(1, 3, 2)$, $(2, 1, 3)$, $(2, 3, 1)$, $(3, 1, 2)$, and $(3, 2, 1)$. These are all the possible orderings of this three-element set.

Permutations are used in almost every branch of mathematics and in many other fields of science. The study of permutations of finite sets is an important topic in the fields of combinatorics and group theory. In computer science, they are used for analysing sorting algorithms; in quantum physics, for describing states of particles; and in biology, for describing RNA sequences.

The number of permutations of n distinct objects is $n!$, which means the product of all positive integers less than or equal to n.

Formally, a permutation of a set S is defined as a bijection from S to itself. That is, it is a function from S to S for which every element occurs exactly once as an image value. This is related to the rearrangement of the elements of S, in which each element s is replaced by the corresponding $f(s)$. For example, the permutation $(3, 1, 2)$, mentioned above, is described by the function α, defined as

$$\alpha(1) = 3, \quad \alpha(2) = 1, \quad \alpha(3) = 2.$$

In mathematical texts, it is customary to denote permutations using lowercase Greek letters. Commonly, either α and β or σ, τ, and π are used. As properties of permutations do not depend on the nature of the set's elements, it is often the permutations of the set $V_n = \{1, 2, \dots, n\}$.

2.1.2. The collection of all permutations of a given set forms a group, called the *symmetric group* of the set.

In particular, all permutations of a set S with n elements form the *symmetric group*, denoted by S_n, where the group operation is the function composition $\circ$. Thus, for the permutations from S_n, the following four group axioms hold:

- *Closure*: if π and σ are in S_n, then so is $\pi \circ \sigma$.
- *Associativity*: for any three permutations $\pi, \sigma, \tau \in S_n$, $(\pi \circ \sigma) \circ \tau = \pi \circ (\sigma \circ \tau)$.
- *Identity*: there is an identity permutation, denoted "id" and defined by $\mathrm{id}(x) = x$ for all $x \in S$; for any $\sigma \in S_n$, $\mathrm{id} \circ \sigma = \sigma \circ \mathrm{id} = \sigma$.
- *Invertibility*: for every permutation $\pi \in S_n$, there exists an inverse permutation $\pi^{-1} \in S_n$ so that $\pi \circ \pi^{-1} = \pi^{-1} \circ \pi = \mathrm{id}$.

In general, a composition of two permutations is not commutative, that is, $\pi \circ \sigma \neq \sigma \circ \pi$.

As a bijection from a set to itself, a permutation is a function that performs a rearrangement of a set and is not an arrangement itself. An older and more elementary viewpoint is that permutations are the arrangements themselves. To distinguish between these two, the identifiers *active* and *passive* are sometimes prefixed to the term permutation, whereas in older terminology, *substitutions* and *permutations* are used.

Other types of permutations

In combinatorial mathematics, other types of permutations are considered, including circular permutations, permutations of multisets, and signed permutations.

2.1.3. Permutations, when considered as arrangements, are sometimes referred to as linearly ordered arrangements. In these arrangements, there is a first element, a second element, and so on. If, however, the objects are arranged in a circular manner, this distinguished ordering no longer exists, that is, there is no "first element" in the arrangement, and any element can be considered the start of the arrangement. The arrangement of objects in a circular manner is called a *circular permutation.* These can be formally defined as equivalence classes of ordinary permutations of the objects for the equivalence relation generated by moving the final element of the linear arrangement to its front.

Two circular permutations are equivalent if one can be rotated into the other (that is, cycled without changing the relative positions of the elements). The four circular permutations $(1,3,2,4)$ $(3,2,4,1)$, $(2,4,1,3)$, and $(4,1,3,2)$ on four letters are considered to be the same.

The circular arrangements are to be read counterclockwise, so $(1,3,2,4)$, and $(1,4,2,3)$ are not equivalent since no rotation can bring one to the other.

- *The number of circular permutations of a set S with n elements is* $(n-1)!$.

□ The result is obvious: exactly n ordinary permutations $(\pi_1, \pi_2, \ldots, \pi_n)$, $(\pi_2, \pi_3, \ldots, \pi_1)$, ..., $(\pi_n, \pi_1, \ldots, \pi_{n-1})$ correspond to a given circular permutation $(\pi_1, \pi_2, \ldots, \pi_n)$. □

2.1.4. A *permutation of a multiset* or, as a rule, of the multiset $Q_n^m = \{1, \ldots, 1, 2, \ldots, 2, \ldots, n, \ldots, n\}$, with the multiplicity of m for each integer, is a one-to-one correspondence, $\pi : Q_n^m \to Q_n^m$.

- *The number of permutations of the set Q_n^m is equal to* $P_{(m,m,\ldots,m)} = \frac{(mn)!}{(m!)^n}$.

□ The result follows from the definition; we simply count all permutations with repetitions of the set $\{1, \ldots, 1, 2, \ldots, 2, \ldots, n, \ldots, n\}$, in which any element i is repeated m times. □

The most widely studied permutations of multisets are the so-called *m-Stirling permutations*, i.e., permutations of the set Q_n^m, such that for each value i appearing in the permutation, any values between the two copies of i are larger than i.

- *The number of m-Stirling permutations is equal to the product* $1 \cdot (m+1)(2m+1)\cdots(m(n-1)+1)$, *i.e., it is represented by the multifactorial* $(m(n-1)+1)!!!\cdots!$. *In particular, the number of 2-Stirling permutations, which are simply called Stirling permutations, is equal to the odd double factorial* $1 \cdot 3 \cdot \cdots \cdot (2n-1) = (2n-1)!!$.

□ The result can be easily derived by induction. For $n = 1$, there exists the only "good" permutation $(1, 1, \ldots, 1)$. Going from $n-1$ to n, we should use any good permutation of the multiset $\{1, \ldots, 1, 2, \ldots, 2, 3, \ldots, 3, \ldots, n-1, \ldots, n-1\}$ and insert in it m elements n, considered as the block $nn\cdots n$. There are $m(n-1)+1$ possible places for $nn\cdots n$: before the first element of the permutation used, between the first and second elements, ..., between the $(m(n-1)-1)$th and $(m(n-1))$th (the last) elements, and after the last element. So, from any Stirling permutation of the $(n-1)$th order, we get exactly $m(n-1)+1$ Stirling permutations of the nth order. The result follows obviously. □

2.1.5. An *indexed permutation* of length n with index set $Z_r = \{0, 1, \ldots, r-1\}$ is an ordinary permutation from the symmetric group S_n, where each letter has been assigned an integer between 0 and $r-1$. The set of all such indexed permutations is denoted by S_n^r. The set S_n^r of all indexed permutations of $\{1, 2, \ldots, n\}$ with the index set Z_r can be considered the Cartesian product of the set S_n of all ordinary permutations of $\{1, 2, \ldots, n\}$ and the set $\{(ind_1, ind_2, \ldots, ind_n) \mid ind_i \in Z_r\}$ of all possible index vectors. For example, the indexed permutation $(3^1, 2^2, 4^0, 1^0, 5^1) \in S_n^3$ can be represented as the product of the permutation $(3, 2, 4, 1, 5) \in S_5$ and the index vector $(1, 2, 0, 0, 1)$ with values from the set $Z_3 = \{0, 1, 2\}$.

With such an approach, it is obvious that:

- *the cardinality of the set S_n^r is equal to $r^n n!$.*

In particular, any ordinary permutation can be considered an indexed permutation with $r = 1$; the index set contains only one element: 0 or, say, 1.

Similarly, any *signed permutation*, formally defined as an ordinary permutation in the symmetric group S_n, where each letter has a sign, $+$ or $-$, can be considered an indexed permutation with $r = 2$; the index set has two elements, $\{0, 1\}$ or, say, $\{-1, 1\}$. The number of signed permutations of the set $\{1, 2, \ldots, n\}$ is equal to $2^n n!$.

Exercises

1. Find all permutations of the set $\{1, 2, \ldots, n\}$ for $n = 1, 2, 3, 4$. Find all permutations of the multiset $\{1, 1, 2, 2, \ldots, n, n\}$ for $n = 1, 2, 3, 4$. Find all permutations of the multiset $\{1, 1, 1, 2, 2, 2, \ldots, n, n, n\}$ for $n = 1, 2, 3, 4$.
2. Find all Stirling permutations of the multiset $\{1, 1, 2, 2, \ldots, n, n\}$ for $n = 1, 2, 3, 4$. Find all 3-Stirling permutations of the multiset $\{1, 1, 1, 2, 2, 2, \ldots, n, n, n\}$ for $n = 1, 2, 3, 4$.
3. Find all signed permutations of the set $\{1, 2, \ldots, n\}$ for $n = 1, 2, 3, 4$.
4. Find all indexed permutations of the set $\{1, 2, \ldots, n\}$ for $n = 1, 2, 3, 4$ if the index set is $\{1, 2, 3\}$.

2.2. Different Representations of Permutations

2.2.1. Since writing permutations element-wise, that is, as piecewise functions, is cumbersome, several notations have been invented to represent them more compactly. Cycle notation is a popular choice for many mathematicians due to its compactness and the fact that it makes a permutation's structure transparent, but other notations are also widely used, especially in application areas.

In Cauchy's *two-line notation*, one lists the elements of S in the first row, and for each, its image is listed below it in the second row. For instance, a particular permutation of the set $S = \{1, 2, 3, 4, 5\}$ can be written as

$$\sigma = \begin{pmatrix} 1 & 2 & 3 & 4 & 5 \\ 2 & 5 & 4 & 3 & 1 \end{pmatrix},$$

which means that σ satisfies $\sigma(1) = 2, \sigma(2) = 5, \sigma(3) = 4, \sigma(4) = 3$, and $\sigma(5) = 1$.

2.2.2. If there is a "natural" order for the elements of S, say $x_1, x_2, \ldots, x_n$, then one uses this for the first row of the two-line notation:

$$\sigma = \begin{pmatrix} x_1 & x_2 & x_3 & \cdots & x_n \\ \sigma(x_1) & \sigma(x_2) & \sigma(x_3) & \cdots & \sigma(x_n) \end{pmatrix}.$$

Under this assumption, one may omit the first row and write the permutation in *one-line notation* as

$$(\sigma(x_1), \ \sigma(x_2), \ \sigma(x_3), \ \ldots, \ \sigma(x_n)),$$

that is, as an ordered arrangement of the elements of S. Care must be taken to distinguish one-line notation from the cycle notation described in the following. In the mathematical literature, a common usage is to omit parentheses for one-line notation while using them for cycle notation. The one-line notation is also called the *word representation* of a permutation. The example above would then be

$$2\,5\,4\,3\,1$$

since the natural order $1\,2\,3\,4\,5$ would be assumed for the first row. (It is typical to use commas to separate these entries only if any of them have two or more digits.) This form is more compact and is common in elementary combinatorics and computer science.

2.2.3. A permutation can be decomposed into one or more disjoint cycles, or *orbits*, which are found by repeatedly tracing the application of the permutation on some elements. For example, the permutation σ with $\sigma(7) = 7$ has a 1-cycle, $(\,7\,)$, while the permutation π

with $\pi(2) = 3$ and $\pi(3) = 2$ has a 2-cycle, $(2\,3)$. In general, a cycle of length k, that is, consisting of k elements, is called a *k-cycle.*

Cycle notation describes the effect of repeatedly applying the permutation on the elements of the set. It expresses the permutation as a product of cycles; since distinct cycles are disjoint, this is referred to as *decomposition into disjoint cycles.*

So, the permutation 2 5 4 3 1 (in one-line notation) could be written as (125)(34) in cycle notation.

While permutations in general do not commute, disjoint cycles do; for example,

$$(1\,2\,5)(3\,4) = (3\,4)(1\,2\,5).$$

In addition, each cycle can be written in different ways by choosing different starting points; for example,

$$(1\,2\,5)(3\,4) = (5\,1\,2)(3\,4) = (2\,5\,1)(4\,3).$$

One may combine these equalities to write the disjoint cycles of a given permutation in many different ways.

1-cycles are often omitted from the cycle notation, provided that the context is clear; for any element x in S not appearing in any cycle, one implicitly assumes $\sigma(x) = x$.

A convenient feature of cycle notation is that the cycle notation of the inverse permutation is given by reversing the order of the elements in the permutation's cycles. For example,

$$[(1\,2\,5)(3\,4)]^{-1} = (5\,2\,1)(4\,3).$$

2.2.4. In some contexts, it is useful to fix a certain order for the elements in the cycles and of the (disjoint) cycles themselves. Thus, the *canonical cycle notation* (or *standard representation* or *standard form*) of a permutation satisfies the following conditions: in each cycle, the largest element is listed first; the cycles are sorted in increasing order of their first elements.

For example, (312)(54)(8)(976) is a permutation in canonical cycle notation. The canonical cycle notation does not omit one-cycles.

Sometimes, one uses the "standard form" terminology but reverses both choices; that is, each cycle lists its least element first, and the cycles are sorted in decreasing order of their least, that is, first elements.

2.2.5. The cycles (including the fixed points) of a permutation σ of a set with n elements partition this set; the lengths of these cycles form an integer partition of the number n, which is called the *cycle type* (or *cycle structure* or *cycle shape*) of σ. For example, the cycle type of the permutation $\beta = (1\,2\,5)(3\,4)(6\,8)(7)$ is $(3, 2, 2, 1)$.

This may also be written in a more compact form as $[1^1 2^2 3^1]$. The general form is $[1^{\alpha_1} 2^{\alpha_2} \cdots n^{\alpha_n}]$, where $\alpha_1, \ldots, \alpha_n$ are the numbers of cycles of respective length, and $n = 1 \cdot \alpha_1 + 2 \cdot \alpha_2 + \cdots + n \cdot \alpha_n$. It is easy to prove the following fact:

- *The number of permutations of the set* $\{1, 2, \ldots, n\}$ *of a given cycle type* $[1^{\alpha_1} 2^{\alpha_2} \cdots n^{\alpha_n}]$ *is*

$$\frac{n!}{1^{\alpha_1} 2^{\alpha_2} \cdots n^{\alpha_n} \alpha_1! \alpha_2! \cdots \alpha_n!}.$$

□ Construct a map between the set S_n of all permutations of $\{1, 2, \ldots, n\}$ and the set $S_{n,\alpha}$ of all permutations from S_n with a given cycle type $[1^{\alpha_1} 2^{\alpha_2} \cdots n^{\alpha_n}]$, $n = 1 \cdot \alpha_1 + 2 \cdot \alpha_2 + \cdots + n \cdot \alpha_n$. For a given permutation $\pi = (\pi_1, \pi_2, \ldots, \pi_n)$ of the set $\{1, 2, \ldots, n\}$, parenthesise the word π so that the first α_1 cycles have length 1, the next α_2 cycles have length 2, and so on. For instance, if $(\alpha_1, \alpha_2, \ldots, \alpha_9) = (1, 2, 0, 1, 0, 0, 0, 0, 0)$, and $\pi = (4, 2, 7, 6, 1, 9, 5, 8, 3)$, then we obtain the decomposition $(4)(27)(61)(9583)$. In general, in this way, we get from a permutation $\pi \in S_n$ a permutation ψ having the disjoint cycle decomposition of type $[1^{\alpha_1} 2^{\alpha_2} \cdots n^{\alpha_n}]$, i.e., belonging to $S_{n,\alpha}$. Hence, we have constructed a map $f : S_n \to S_{n,\alpha}$. It is easy to check that, for a given $\psi \in S_{n,\alpha}$, there are exactly $1^{\alpha_1} 2^{\alpha_2} \cdots n^{\alpha_n} \alpha_1! \alpha_2! \cdots \alpha_n!$ ways to write it in disjoint cycle notation so that the cycle lengths are weakly increasing from left to right: we can order the cycles of length i in $\alpha_i!$ ways, and we can choose the first elements of these cycles in i^{α_i} ways.

Hence, it holds that for each $\psi \in S_{n,\alpha}$, we have $|f^{-1}(\psi)| = 1^{\alpha_1} 2^{\alpha_2} \cdots n^{\alpha_n} \alpha_1! \alpha_2! \cdots \alpha_n!$, and the proof follows since $|S_n| = n!$. □

Exercises

1. Find all permutations of the set $\{1, 2, \ldots, n\}$ for $n = 1, 2, 3, 4$; for any considered permutation, obtain its representation in the Cauchy two-line notation and its word representation.
2. For each permutation of the set $\{1, 2, \ldots, n\}$, $n = 1, 2, 3, 4$, obtain its representation in the form of disjoint cycles and its canonical cycle representation.
3. How many cycle types exist for the set S_n of all permutations of $\{1, 2, \ldots, n\}$, $n = 1, 2, 3, 4$? For any such type $[1^{\alpha_1} 2^{\alpha_2} \cdots n^{\alpha_n}]$, construct the map $f : S_n \to S_{n,\alpha}$, and check that for any permutation $\psi \in S_{n,\alpha}$, there are exactly $1^{\alpha_1} 2^{\alpha_2} \cdots n^{\alpha_n} \alpha_1! \alpha_2! \cdots \alpha_n!$ permutations $\pi \in S_n$, such that $f(\pi) = \psi$.
4. For $n = 1, 2, 3, 4, 5, 6, 7, 8, 9, 10$, find all integer partitions of n, $n = x_1 + x_2 + \cdots + x_n$, $x_i \in \mathbb{Z}$, $x_i \geq 0$. Check that the number $p(n)$ of such partitions (in fact, the number of all non-ordered partitions of n into an arbitrary number of non-empty parts) is equal to 1, 2, 3, 5, 7, 11, 15, 22, 30, 42 for $n = 1, 2, \ldots, 10$.

2.3. Cycles and Transpositions

Enumeration of cycles

2.3.1. Formally, a *cyclic permutation* (or *cycle*) is a permutation of the elements of some set S which maps the elements of some subset X of S to each other in a cyclic fashion while fixing (that is, mapping to themselves) all other elements of S. If X has k elements, the cycle is called a k*-cycle*. The set X is called the *orbit* of the cycle.

For example, given $S = \{1, 2, 3, 4\}$, the permutation $(1, 3, 2, 4)$ that sends 1 to 3, 3 to 2, 2 to 4, and 4 to 1 (so $X = S$) is a 4-cycle, and the permutation $(1, 3, 2)$ that sends 1 to 3, 3 to 2, 2 to 1, and 4 to 4 (so $X = \{1, 2, 3\}$, and 4 is a fixed element) is a 3-cycle.

- *The number of k-cycles in the symmetric group S_n is given, for $1 \leq k \leq n$, by the following equivalent formulas:*

$$\binom{n}{k}(k-1)! = \frac{n(n-1)\cdots(n-k+1)}{k} = \frac{n!}{(n-k)!k}.$$

□ In fact, in order to obtain a k-cycle, we should first choose k elements from n elements (by $\binom{n}{k}$ ways), and then put the chosen k elements into a cyclic order (by $(k-1)!$ ways). □

Derangements

2.3.2. The only 1-cycle is the identity permutation: all points of this permutation are fixed.

In contrast, a permutation with no fixed points is called a *derangement.* The number $!n$ of derangements of a set of size n is known as the *subfactorial* of n (or *derangement number,de Montmort number,* or *rencontres number*).

The first elements of the sequence $!n$, $n = 0, 1, 2, \ldots$, are 1, 0, 1, 2, 9, 44, 265, 1854, 14833, 133496, ... (sequence A000166 in the OEIS).

- *The number $!n$ of derangements of the set $\{1, 2, \ldots, n\}$ can be calculated using the formula*

$$!n = n!\left(1 - \frac{1}{1!} + \frac{1}{2!} - \frac{1}{3!} + \cdots + (-1)^n \frac{1}{n!}\right).$$

□ This fact can be easily established using the inclusion–exclusion method. In fact, if the property α_i, $i = 1, 2, \ldots, n$ of a permutation $\pi \in S_n$ has the form $\pi_i = i$, then $N(\alpha_i) = (n-1)!$, $N(\alpha_i, \alpha_j) = (n-2)!$, and so on. Therefore, we get that

$$\begin{aligned} !n = N(\alpha_1', \ldots, \alpha_n') &= \binom{n}{1} N(\alpha_i) - \binom{n}{2} N(\alpha_i, \alpha_j) \\ &\quad + \cdots + (-1)^n \binom{n}{n} N(\alpha_1, \ldots, \alpha_n) \\ &= n!\left(1 - \frac{1}{1!} + \frac{1}{2!} - \frac{1}{3!} + \cdots + (-1)^n \frac{1}{n!}\right). \end{aligned}$$ □

So, for $n > 0$, the subfactorial $!n$ is equal to $\left[\frac{n!}{e}\right]$, the nearest integer for the value $\frac{n!}{e}$, where $n!$ is the factorial of n, and $e = 2.71828\ldots$ is *Euler's number.* It follows from the well-known decomposition

$$\frac{1}{e} = \lim_{n\to\infty}\left(1 - \frac{1}{1!} + \frac{1}{2!} - \frac{1}{3!} + \cdots + (-1)^n \frac{1}{n!}\right).$$

Moreover, as $n \to \infty$, it holds that $!n \to \frac{n!}{e}$.

Transpositions

2.3.3. 2-cycles are called *transpositions*; such permutations merely exchange two elements, leaving the others fixed.

For example, the permutation $\pi = 1423 = (1)(42)(3)$, which swaps 2 and 4, is a transposition.

The following result is well known (see, for example, [Kost82]):

- *Every permutation of a finite set can be expressed as the product of transpositions.*

So, formally, transpositions are generators for the symmetry group S_n.

□ In fact, when the set being permuted is $\{1, 2, \ldots, n\}$ for some integer n, then any permutation can be expressed as a product of adjacent transpositions $(1, 2), (2, 3), (3, 4)$, and so on. This follows because an arbitrary transposition can be expressed as the product of adjacent transpositions. Concretely, one can express the transposition (k, l), $k < l$ by moving k to l one step at a time, then moving l back to where k was, which interchanges these two elements and makes no other changes:

$$(k, l) = (k, k+1)(k+1, k+2) \cdots (l-1, l)(l-2, l-1) \cdots (k, k+1).$$

Furthermore, the decomposition of a permutation into a product of transpositions can be obtained by writing the permutation as a product of disjoint cycles and then splitting iteratively each of the cycles of length m, $m \geq 3$, into a product of a transposition and a cycle of smaller length:

$$(k_1, k_2, \ldots, k_m) = (k_1, k_2) \cdot (k_2, \ldots, k_m). \qquad \square$$

Although many such expressions for a given permutation may exist, either they all contain an even number of transpositions or they all contain an odd number of transpositions. Thus, all permutations can be classified as *even* or *odd* depending on this number.

Consider, for example, the permutation σ of the set $\{1, 2, 3, 4, 5\}$ defined by $\sigma(1) = 3$, $\sigma(2) = 4$, $\sigma(3) = 5$, $\sigma(4) = 2$, and $\sigma(5) = 1$. In one-line notation, this permutation is denoted as $\sigma = 34521$.

It can be obtained from the identity permutation 12345 by three transpositions: first exchange the numbers 2 and 4, then exchange 3 and 5, and finally exchange 1 and 3. This shows that the given permutation is odd. Following the cycle notation, this could be written as $\sigma = (135)(24) = (13)(35)(24)$.

Cycles and Stirling numbers of the first kind

2.3.4. A *Stirling number of the first kind* $s(n, k)$ is defined by the equality

$$s(n, k) = (-1)^{n+k}|s(n, k)|,$$

where $|s(n, k)|$ is the number of ways of decomposition of a given n-set into k cycles, $n = 1, 2, \ldots; k = 1, 2, \ldots, n$. The numbers $|s(n, k)|$ are called *unsigned Stirling numbers of the first kind.*

Using this definition, we can state that:

- *the number of n-permutations with k cycles is equal to the unsigned Stirling number of the first kind $|s(n, k)|$.*

Moreover, this combinatorial definition allows us to obtain the following recurrent relation:

- *For the Stirling numbers of the first kind, the following recurrent formula holds:*

$$s(n, k) = s(n-1, k-1) - (n-1)s(n-1, k), \quad n > 0, \ k > 0.$$

□ Each representation of n objects in the form of k cycles either places the last object in a separate cycle (in $|s(n-1, k-1)|$ ways) or inserts this object into one of the $|s(n-1, k)|$ cyclic representations of the first $n-1$ objects. In the second case, there exist $n-1$ ways of such insertion. In fact, it is easy to check that there are j ways to put a new element in a j-cycle in order to get a $(j+1)$-cycle. For example, if $j = 3$, then the cycle $[A, B, C]$ leads to the cycles $[A, B, C, D]$, $[A, B, D, C]$, or $[A, D, B, C]$ if we insert a new element D, and no other possibilities exist. Summation over all j gives $n-1$ ways to insert the nth (last) object into a circular partition of the first $n-1$ objects. Thus, it holds that

$$|s(n,k)| = |s(n-1,k-1)| + (n-1)|s(n-1,k)|, \ n > 0, \ k > 0;$$

therefore,

$$s(n,k) = s(n-1,k-1) - (n-1)s(n-1,k), \ n > 0, \ k > 0. \quad \square$$

2.3.5. Using the obtained recurrent relation, we can build the *triangle of Stirling numbers of the first kind* (sequence A048994 in the OEIS). At the vertex of the triangle is the number 1 ($s(0,0) = 1$), the left side is formed by zeros ($s(n,0) = 0$, $n > 0$), and the right side consists of unities ($s(n,n) = 1$, $n \geq 1$).

n/k	0	1	2	3	4	5	6	7	8	9
0	1									
1	0	1								
2	0	−1	1							
3	0	2	−3	1						
4	0	−6	11	−6	1					
5	0	24	−50	35	−10	1				
6	0	−120	274	−225	85	−15	1			
7	0	720	−1764	1624	735	175	−21	1		
8	0	−5040	13268	−13132	6769	−1960	322	−28	1	
9	0	40320	−109584	118124	−67284	22449	−4536	546	−36	1

Examining the sums of elements of the nth row of the constructed triangle, we can see that:

- *the sum of elements of the nth row, $n \geq 2$, of the triangle of Stirling numbers of the first kind is equal to zero; the alternating sum of elements of the nth row, $n \geq 2$, of the triangle of Stirling numbers of the first kind is equal to $(-1)^n n!$.*

$\square$ This statement follows from the recurrent construction used and can be proved by induction. However, it is a simple consequence of the well-known fact that the number of all permutations of an n-set is equal to $n!$. $\square$

2.3.6. For a given permutation $\pi \in S_n$, define its *left-to-right maximum* (or *record* or *outstanding element*) as an element π_i, $i = 1, 2, \ldots, n$, such that $\pi_i > \pi_j$ for every $j < i$.

For example, the permutation $(1,3,2,4)$ has three left-to-right maxima 1, 3, and 4, and the permutation $(1,2,3,4)$ has four left-to-right maxima, 1, 2, 3, and 4, while the permutation $(4,3,2,1)$ has only one left-to-right maximum, 4.

- *The number of n-permutations with exactly k left-to-right maxima is equal to the unsigned Stirling number of the first kind* $|s(n,k)|$.

□ The proof of this fact is based on a very important construction on the set S_n of all permutations of $\{1, 2, \ldots, n\}$, called the *fundamental bijection* (see [Stan97]).

For a given permutation $\pi \in S_n$, write it in standard form and remove all the parentheses. In this way, we obtain some other permutation $\psi \in S_n$, which will be denoted as $b(\pi)$: $\psi = b(\pi)$. For example, for the permutation $\pi = (4, 2, 7, 1, 3, 6, 5) = 4271365$, its standard form is $\pi = (2)(41)(6)(753)$, and we have that $\psi = b(\pi) = (2, 4, 1, 6, 7, 5, 3) = 2416753$. It is easy to see that we can uniquely recover the permutation π from $\psi = b(\pi)$ by inserting a left parenthesis into $\psi = (\psi_1, \psi_2, \ldots, \psi_n) = \psi_1\psi_2\cdots\psi_n$ preceding every left-to-right maximum of ψ and then inserting a right parenthesis where appropriate, that is, before every internal left parenthesis and at the end. In particular, for the permutation 2416753, we should put left parentheses before 2, 4, 6, and 7, obtaining the picture (2(41(6(753, and after that, add right parentheses before (4, (6, and (7 and at the end, obtaining the permutation (2)(41)(6)(753).

Thus, the map $b : S_n \to S_n$, such that $\pi \to b(\pi)$, is a bijection from S_n to itself. It is known as the *fundamental bijection.* The construction used allows us to state that a permutation $\pi \in S_n$ has k cycles if and only if $b(\pi)$ has k left-to-right maxima; therefore, the number $|s(n, k)|$ of elements from S_n, having k cycles, is equal to the number of elements from S_n, having k left-to-right maxima. □

2.3.7. The Stirling numbers of the first kind have many interesting properties and important applications. We will use some of them in our further considerations. We list these properties in the following. For detailed proofs and additional information, see [Deza24].

- *There exist several special cases for the values of the Stirling numbers of the first kind:*

$$s(n,n) = 1,\ n \ge 0;\ s(n,1) = (-1)^{n-1}(n-1)!,\ n \ge 1;$$

$$s(n, n-1) = -\binom{n}{2},\ n \ge 2.$$

- The *Stirling numbers of the first kind are connected with the harmonic numbers* $H_n = 1 + \frac{1}{2} + \cdots + \frac{1}{n}$:
$$s(n+1,2) = (-1)^{n+1} n! H_n.$$
- The *Stirling numbers of the first kind* $s(n,k)$ *satisfy the equality*
$$x^{\underline{n}} = s(n,0) + s(n,1)x + s(n,2)x^2 + \cdots + s(n,n-1)x^{n-1} + s(n,n)x^n,$$
i.e., using the Stirling numbers of the first kind, we can express falling powers (falling factorials) $x^{\underline{n}}$ *of* x *through the ordinary powers* $x^0, x^1, \ldots, x^n$ *of* x.
- *The generating function of the sequence of Stirling numbers of the first kind* $s(n,k)$, $k = 0,1,2,\ldots$, *for a fixed* $n = 0,1,2,\ldots$, *has the form* $x(x-1)\cdots(x-k+1)$, *while the exponential generating function of the sequence of Stirling numbers of the first kind* $s(n,k)$, $n = 0,1,2,\ldots$, *for a fixed* $k = 0,1,2,\ldots$, *has the form* $\frac{(\log(1+x))^k}{k!}$.

Exercises

1. Find the values of the unsigned Stirling numbers of the first kind $|s(n,k)|$, $n = 0,1,2,3,4$, $k = 0,1,2,\ldots,n$, using the definition based on permutations of an n-set with k cycles.
2. Using the recurrent relation for the unsigned Stirling numbers of the first kind $|s(n,k)|$, calculate $|s(n,k)|$, $n = 0,1,2,\ldots,11$; $k = 0,1,2,\ldots,n$.
3. Prove that $s(n,n-2) = \frac{n(n-1)(n-2)(3n-1)}{24}$. Check the result for the values $s(4,2), s(5,3)$, $s(6,4)$, $s(7,5)$, $s(8,6)$, and $s(9,7)$.
4. Prove that $s(n,n-3) = -\frac{n^2(n-1)^2(n-2)(n-3)}{48}$. Check the result for the values $s(5,2)$, $s(6,3)$, $s(7,4)$, $s(8,5)$, and $s(9,6)$.

2.4. Ascents, Descents, Runs, and Excedances

2.4.1. In most applications, the elements of the set S being permuted will be compared with each other. This requires that the set S has a total order, so that any two elements can be compared. The set

$V_n = \{1, 2, \ldots, n\}$ is totally ordered by the usual $\leq$ relation, and so it is the most frequently used set in these applications. Similarly, the set $Q_n^m = \{1, \ldots, 1, 2, \ldots, 2, \ldots, n, \ldots, n\}$ can be totally ordered by the same relation $\leq$.

There are a number of properties that are directly related to the total ordering of the considered set (or multiset) S.

2.4.2. An *ascent* of a permutation σ of the set $\{1, 2, \ldots, n\}$ is any position $i < n$ where the following value is bigger than the current one. That is, if $\sigma = (\sigma_1, \sigma_2, \ldots, \sigma_n)$, then i is an ascent if $\sigma_i < \sigma_{i+1}$.

For example, the permutation $\sigma = 3452167$ has ascents (at positions) 1, 2, 5, and 6.

Similarly, a *descent* is a position $i < n$ with $\sigma_i > \sigma_{i+1}$; if $\sigma = \sigma_1\sigma_2 \cdots \sigma_n$, then i is a descent if $\sigma_i > \sigma_{i+1}$.

So, every position i with $1 \leq i < n$ is either an ascent or a descent of σ.

For a permutation $(\sigma_1, \sigma_2, \ldots, \sigma_n)$ of the set $\{1, 2, \ldots, n\}$, we define the *reversal permutation* (or *mirror permutation*) of σ as the permutation $\sigma^r = (\sigma_n, \sigma_{n-1}, \ldots, \sigma_1)$.

It is obvious that if σ has k ascents, then its reversal σ^r has k descents. So, it holds that:

- *the number of permutations of the set* $\{1, 2, \ldots, n\}$ *with* k *ascents is equal to the number of permutations of the set* $\{1, 2, \ldots, n\}$ *with* k *descents.*

2.4.3. An *ascending run* (or *maximal increasing run*) of a permutation of the set $\{1, 2, \ldots, n\}$ is a non-empty increasing contiguous subsequence of the permutation that cannot be extended at either end; it corresponds to a maximal sequence of successive ascents (the latter may be empty; between two successive descents, there is still an ascending run of length 1).

By contrast, an *increasing subsequence* of a permutation is not necessarily contiguous; it is an increasing sequence of elements obtained from the permutation by omitting the values at some positions.

For example, the permutation $\pi = 2453167$ has the ascending runs 245, 3, and 167, while it has an increasing subsequence 2367.

- *If a permutation has k descents, then it must be the union of $k+1$ ascending runs.*

□ This fact is obvious; it follows from the following definition: any descent is the beginning of the current ascending run. □

The number of permutations of the set $\{1, 2, \ldots, n\}$ with k ascents is, by definition, the *Eulerian number* $E(n, k)$; therefore, this is also the number of permutations of $\{1, 2, \ldots, n\}$ with k descents and with $k+1$ ascending runs (see Chapter 3).

2.4.4. An *excedance* of a permutation $\sigma = (\sigma_1 \sigma_2 \cdots \sigma_n)$ is a position i, $1 \leq i \leq n$, such that $\sigma_i > i$. If the inequality is not strict, that is, $\sigma_i \geq i$, then i is called a *weak excedance.*

- *The number of permutations of the set $\{1, 2, \ldots, n\}$ with k excedances coincides with the number of permutations of the set $\{1, 2, \ldots, n\}$ with k ascents.*

□ For a proof, let us use the fundamental bijection $\pi \to b(\pi)$. Suppose that

$$\pi = (a_1, a_2, \ldots, a_{i_1})(a_{i_1+1}, a_{i_1+2}, \ldots, a_{i_2}) \cdots (a_{i_{k-1}+1}, a_{i_{k-1}+2}, \ldots, a_n)$$

is the representation of a permutation π from S_n in the standard form. Thus, $a_1, a_{i_1}+1, \ldots, a_{i_{k-1}+1}$ are the largest elements of their cycles, and $a_1 < a_{i_1+1} < \cdots < a_{i_{k-1}+1}$.

It follows that if $\pi(a_i) \neq a_{i+1}$, then $a_i < a_{i+1}$. Hence, we have that $a_i < a_{i+1}$, or $i = n$ if and only if $\pi(a_i) \geq a_i$. As, by definition, $b(\pi) = \psi = a_1 a_2 \cdots a_n$, i.e., $\psi_i = a_i$, we can state that

$$|\{i \in \{1, 2, \ldots, n-1\} \mid \psi_i < \psi_{i+1}\}| + 1$$
$$= |\{i \in \{1, 2, \ldots, n\} \mid \pi(i) \geq i\}|.$$

The value $|\{i \in \{1, 2, \ldots, n-1\} \mid \psi_i < \psi_{i+1}\}| + 1$ is the number of ascents of $\psi = b(\pi)$ plus 1 (the position n). The value $|\{i \in \{1, 2, \ldots, n\} \mid \pi(i) \geq i\}|$ is the number of weak excedances of π. So, we have proven that the number of permutations of the set $\{1, 2, \ldots, n\}$ with $k+1$ weak excedances coincides with the number of permutations of the set $\{1, 2, \ldots, n\}$ with k ascents. On the other hand, it is easy to check that a permutation $\pi = (\pi_1, \pi_2, \ldots, \pi_n)$

has $k+1$ weak excedances if and only if the permutation $\sigma = (\sigma_1, \sigma_2, \dots, \sigma_n)$, defined by $\sigma_i = (n+1) - \sigma_{n-i+1}$, has $n-k-1$ excedances: $\pi_{n+1-i} < n+1-i$ (i.e., the position $n+1-i$ is not a weak excedance for π) if and only if $\sigma_i = n+1-\pi_{n+1-i} > i$ (i.e., the position i is an excedance for σ). Moreover, π has $n-1-k$ ascents if and only if its reversal $(\pi_n, \pi_{n-1}, \dots, \pi_1)$ has k ascents. □

We therefore obtain the following result:

- *The number of permutations $\pi \in S_n$ with k excedances, as well as the number of permutations $\pi \in S_n$ with $k+1$ weak excedances, is equal to the Eulerian number $E(n,k)$.*

Exercises

1. Find all permutations of the set $\{1, 2, \dots, n\}$ for $n = 1, 2, 3, 4$; for any considered permutation, obtain the number of its ascents and the number of its descents. Check that the number of permutations of the set $\{1, 2, \dots, n\}$ with k ascents is equal to the number of permutations of the set $\{1, 2, \dots, n\}$ with k descents.
2. For each permutation of the set $\{1, 2, \dots, n\}$, $n = 1, 2, 3, 4$, obtain the number of its ascending runs. Check that the number of permutations of the set $\{1, 2, \dots, n\}$ with k ascents is equal to the number of permutations of the set $\{1, 2, \dots, n\}$ with $k+1$ ascending runs.
3. For each permutation of the set $\{1, 2, \dots, n\}$, $n = 1, 2, 3, 4$, obtain the number of its excedances and weak excedances. Check that the number of permutations of the set $\{1, 2, \dots, n\}$ with k ascents is equal to the number of permutations of the set $\{1, 2, \dots, n\}$ with (i) $k+1$ weak excedances and (ii) with k excedances.
4. For each permutation π of the set $\{1, 2, \dots, n\}$, $n = 1, 2, 3, 4$, obtain the number of its readings, i.e., the number of steps; we must scan the one-line notation of π from left to right to put the numbers $1, 2, \dots, n$ in natural order. Check that the number of permutations of the set $\{1, 2, \dots, n\}$ with k ascents is equal to the number of permutations of the set $\{1, 2, \dots, n\}$ with $k+1$ readings.

2.5. Inversions

2.5.1. An inversion of a permutation σ is a pair (i, j) of positions where the entries of a permutation are in the opposite order: $i < j$, and $\sigma_i > \sigma_j$.

For example, the permutation $\sigma = 23154$ has three inversions: $(1, 3), (2, 3)$, and $(4, 5)$, for the pairs of entries $(2, 1)$, $(3, 1)$, and $(5, 4)$.

So, a descent is simply an inversion at two adjacent positions.

Sometimes, an inversion is defined as the pair of values (σ_i, σ_j) whose order is reversed; this makes no difference for the number of inversions, and this pair (reversed) is also an inversion in the above sense for the inverse permutation σ^{-1}. The number of inversions is an important measure for the degree to which the entries of a permutation are out of order; it is the same for σ and for σ^{-1}.

It is always possible to bring a permutation with k inversions into order (that is, transform it into the identity permutation) by successively applying (right-multiplication by) adjacent transpositions, and this requires a sequence of k such operations.

- *The number of permutations of n elements with k inversions is expressed by a Mahonian number, which is the coefficient $[x^k]\prod_{m=1}^{n}\sum_{i=0}^{m-1} x^i$ of x^k in the expansion of the product*

$$\prod_{m=1}^{n}\sum_{i=0}^{m-1} x^i = 1\,(1+x)\left(1+x+x^2\right)\cdots\left(1+x+x^2+\cdots+x^{n-1}\right).$$

The triangle of Mahonian numbers starts with elements 1, 1, 1, 1, 2, 2, 1, 1, 3, 5, 6, 5, 3, 1, 1, ... (sequence A008302 in the OEIS).

The product $1\cdot(1+q)\cdots(1+q+\cdots+q^{n-2})\cdot(1+q+\cdots+q^{n-1})$ is also known as the *q-factorial* $(n)_q!$: if we denote $1+q+\cdots+q^{k-1} = \frac{q^k-1}{q-1}$ by $(k)_q$, we get that

$$(n)_q! = (1)_q(2)_q\cdots(n)_q.$$

The corresponding expansion appears in the *Necklace problem.*

□ In order to prove this fact, let us associate with each permutation $\pi \in S_n$ the string $(a_1, a_2, \ldots, a_n)$, such that for any element $i \in \{1, 2, \ldots, n\}$, the value a_i is the number of entries j of π to the left of i, satisfying $j > i$. For example, if $\pi = 3421$, then we get that $a_4 = 0$, $a_3 = 0$, $a_2 = 2$, and $a_1 = 3$; therefore, it holds that $(a_1, a_2, a_3, a_4) = (3, 2, 0, 0)$. The string $(a_1, a_2, \ldots, a_n)$ is called the *inversion table* of the permutation π and denoted by $I(\pi)$. By the construction, it holds that $0 \le a_i \le n - i$, and the number $inv(\pi)$ of inversions in π is equal to $a_1 + a_2 + \cdots + a_n$. So, the generating function $\sum_{\pi \in S_n} x^{inv(\pi)}$ is represented as

$$\begin{aligned}\sum_{\pi \in S_n} x^{inv(\pi)} &= \sum_{a_1=0}^{n-1} \sum_{a_2=0}^{n-2} \cdots \sum_{a_n=0}^{0} x^{a_1+a_2+\cdots+a_n} \\ &= \left(\sum_{a_1=0}^{n-1} x^{a_1}\right)\left(\sum_{a_2=0}^{n-2} x^{a_2}\right)\cdots\left(\sum_{a_n=0}^{0} x^{a_n}\right) \\ &= 1\,(1+x)\left(1+x+x^2\right)\cdots\left(1+x+x^2+\cdots+x^{n-1}\right) \\ &= \prod_{m=1}^{n} \sum_{i=0}^{m-1} x^i = \prod_{m=1}^{n} (m)_x = (n)_x!. \qquad \square\end{aligned}$$

2.5.2. An interesting statistic, associated with permutations of the set $\{1, 2, \ldots, n\}$ and closely connected with descents and inversions, is the *major index*. For a given permutation $\pi \in S_n$, its *major index* $maj(\pi)$ is defined as the sum of all descents of π:

$$maj(\pi) = \sum_{1 \le i < n:\ \pi_i > \pi_{i+1}} i.$$

It can be proven that the statistics *inv* and *maj* are equidistributed.

- *For any k, $0 \le k \le \frac{n(n-1)}{2}$, The number of permutations of the set $\{1, 2, \ldots, n\}$ having k inversions is equal to the number of permutations of the set $\{1, 2, \ldots, n\}$ having the major index k:*

$$|\{\pi \in S_n \mid inv(\pi) = k\}| = |\{\sigma \in S_n \mid maj(\sigma) = k\}.$$

$\square$ In terms of generating functions, the result takes the form

$$\sum_{\pi \in S_n} x^{inv(\pi)} = \sum_{\sigma \in S_n} x^{maj(\sigma)}.$$

In order to obtain this expression, let us construct a bijection, ϕ: $S_n \to S_n$, using the following algorithm.

For a given permutation $\pi \in S_n$, $\pi = (\pi_1, \pi_2, \ldots, \pi_n)$, construct a sequence $\gamma_1, \gamma_2, \ldots, \gamma_n$ of strings, such that every string γ_i is a permutation of elements, $\{\pi_1, \pi_2, \ldots, \pi_i\}$ (see [Stan97]). First, let $\gamma_1 = \pi_1$. Assume that γ_k has been defined for $1 \leq k < n$, and construct γ_{k+1} following a simple rule: if the last letter of γ_k (which turns out to be π_k) is greater (respectively, smaller) than π_{k+1}, then split γ_k after each letter greater (respectively, smaller) than π_{k+1}, obtaining several compartments; cyclically shift each compartment of one unit to the right, and place π_{k+1} at the end of the obtained string; define γ_{k+1} as the resulting string with π_{k+1} at the end. If all strings $\gamma_1, \gamma_2, \ldots, \gamma_n$ are constructed, set $\phi(\pi) = \gamma_n$.

For example, if $\pi = 324651$, then $\gamma_1 = 3$, $\gamma_2 = 32$, $\gamma_3 = 324$, $\gamma_4 = 3246$, $\gamma_5 = 63245$, $\gamma_6 = 632451$, and $\phi(\pi) = 632451$. Note that $maj(\pi) == 10$, and $inv(\phi(\pi)) == 10$. It is easy to prove now that, in general, the function ϕ is a bijection, transforming maj into inv, i.e.,

$$maj(\pi) = inv(\phi(\pi)).$$

In fact, we can show by induction that, for a string $\eta_k = (\pi_1, \pi_2, \ldots, \pi_k)$, it holds that $inv(\gamma_k) = maj(\eta_k)$. It is obvious for $k = 1$: $inv(\gamma_1) = maj(\eta_1) = 0$. Assume that $inv(\gamma_k) = maj(\eta_k)$ for some $k < n$.

Suppose that the last letter π_k of γ_k is greater than π_{k+1}. Therefore, k is a descent of π, and we need to show that $inv(\gamma_{k+1}) = k + inv(\gamma_k)$. The last letter of any compartment V_j of γ_k is the largest letter of the compartment. Hence, when we cyclically shift this compartment, we create $|V_j| - 1$ new inversions. Each compartment contains exactly one letter larger than π_{k+1}, so when we put π_{k+1} at the end of γ_k, the number of new appearing inversions is equal to the number m of compartments. So, altogether, we have created $\sum_{j=1}^{m}(|V_j| - 1) + m = \sum_{j=1}^{m} |V_j| = k$ new inversions. The proof for the case of $\pi_k < \pi_{k+1}$ is similar. For more details and additional examples, see [Stan97]. □

Exercises

1. Find all permutations of the set $\{1, 2, \ldots, n\}$ for $n = 1, 2, 3, 4$; for any considered permutation, obtain the number of its ascents, the number of its descents, the number of its inversions, and its

major index. Check that the number of permutations of the set $\{1, 2, \ldots, n\}$ with k inversions is equal to the number of permutations of the set $\{1, 2, \ldots, n\}$ with major index equal to k.

2. For each permutation π of the set $\{1, 2, \ldots, n\}$, $n = 1, 2, 3, 4$, check that the number of inversions of π coincides with the number of permutations of π^{-1}. Prove this fact for any permutation $\pi \in S_n$.
3. For each permutation π of the set $\{1, 2, \ldots, n\}$, $n = 1, 2, 3, 4$, obtain the permutation $\phi(\pi)$, described above. Check that $maj(\pi) = inv(\phi(\pi))$.
4. Give several examples of permutations from S_n, $n = 10, 11, 12$. For each constructed permutation π, obtain the permutation $\phi(\pi)$, described above. Check that $maj(\pi) = inv(\phi(\pi))$.

2.6. Alternating Permutations

2.6.1. An *alternating permutation* (or *zigzag permutation*) of the set $\{1, 2, 3, \ldots, n\}$ is an arrangement of those numbers so that each entry is alternately greater or less than the preceding entry.

Formally, a permutation $\sigma = \sigma_1 \ldots \sigma_n$ is said to be *alternating* if its entries alternately rise and descend. Thus, each entry other than the first and the last should be either larger or smaller than both of its neighbours. The *up-down* permutations are such permutations for which the conditions $\sigma_1 < \sigma_2 > \sigma_3 < \cdots$ hold, while the *down-up* permutations satisfy the conditions $\sigma_1 > \sigma_2 < \sigma_3 > \cdots$. Many authors use the term "alternating" to refer only to the *up-down* permutations.

For example, the five alternating (up-down) permutations of $\{1, 2, 3, 4\}$ are 1324 (as $1 < 3 > 2 < 4$); 1423 (as $1 < 4 > 2 < 3$); 2314 (as $2 < 3 > 1 < 4$); 2413 (as $2 < 4 > 1 < 3$); 3412 (as $3 < 4 > 1 < 2$). There is a simple one-to-one correspondence between the down-up and up-down permutations: replacing each entry σ_i with $n + 1 - \sigma_i$ reverses the relative order of the entries.

By convention, in any naming scheme, the unique permutations of length 0 (the permutation of the empty set) and 1 (the permutation consisting of a single entry, 1) are taken to be alternating.

This type of permutation was first studied by Désiré André in the 19th century (see [MSY96] and [Andr79]).

2.6.2. The numbers A_n of *up-down* permutations on $\{1, 2, \ldots, n\}$ are known as *Euler zigzag numbers* (or *up/down numbers*). The first few values of A_n are 1, 1, 1, 2, 5, 16, 61, 272, 1385, 7936, 50521, ... (sequence A000111 in the OEIS).

The numbers Z_n count the permutations of $\{1, \ldots, n\}$, that are either up-down or down-up (or both, for $n < 2$). So, $Z_n = 2A_n$ for $n \geq 2$; the first few values of Z_n are 1, 1, 2, 4, 10, 32, 122, 544, 2770, 15872, 101042, ... (sequence A001250 in the OEIS).

These numbers satisfy a simple recurrence, similar to that of the Catalan numbers: by splitting the set of alternating permutations (both down-up and up-down) of the set $\{1, 2, 3, \ldots, n, n+1\}$ according to the position k of the largest entry $n+1$, one can show that

$$Z_n = 2A_{n+1} = \sum_{k=0}^{n} \binom{n}{k} A_k A_{n-k}, \quad n \geq 1.$$

This recurrence allows us to obtain the exponential generating function $A(x) = \sum_{n=0}^{\infty} A_n \frac{x^n}{n!}$ for the sequence A_n, $n = 0, 1, 2, \ldots$. In fact, it holds that

$$A(x) = \tan\left(\frac{\pi}{4} + \frac{x}{2}\right) = \sec x + \tan x,$$

i.e., $A(x)$ is the sum of the secant and tangent functions. This result is known as *André's theorem* (see [Stan10]).

2.6.3. Since the secant function is even and the tangent is odd, we obtain two additional classes of numbers.

The numbers A_{2n} with even indices are called *secant numbers* (or *zig numbers*); they are the numerators in the Maclaurin series of $\sec x$. The first few values are 1, 1, 5, 61, 1385, 50521, 2702765, 199360981, 19391512145, 2404879675441, ... (sequence A000364 in the OEIS). The secant numbers are related to the *Euler numbers* E_n, which are defined by the Taylor series expansion

$$\frac{1}{\cosh t} = \frac{2}{e^t + e^{-t}} = \sum_{n=0}^{\infty} \frac{E_n}{n!} \cdot t^n.$$

The sequence E_n, $n = 0, 1, 2, \ldots$ starts with the numbers 1, 0, -1, 0, 5, 0, -61, 0, 1385, 0, ... (sequence A122045 in the OEIS), and it is easy to prove that Euler numbers with odd indices are equal to zero, while $E_{2n} = (-1)^n A_{2n}$. So, it holds that

$$A_{2n} = |E_{2n}|.$$

The numbers A_{2n+1} with odd indices are called *tangent numbers* (or *zag numbers*). The first few values are 1, 2, 16, 272, 7936, 353792, 22368256, 1903757312, 209865342976, 29088885112832, ... (sequence A000182 in the OEIS). The tangent numbers are closely related to *Bernoulli numbers.* The relation is given by the formula

$$B_{2n} = (-1)^{n-1} \frac{2n}{4^{2n} - 2^{2n}} A_{2n-1}, \quad n \geq 1.$$

The relationships of the Euler zigzag numbers with the Euler numbers and the Bernoulli numbers can be used to prove the following identity, connecting the Euler zigzag numbers A_n with the Stirling numbers $S(n,k)$ of the second kind (see [Mend07]):

$$A_n = -\frac{4^n}{a_n} \sum_{k=1}^{n} \frac{(-1)^k \, S(n,k)}{k+1} \left(\frac{3}{4}\right)^{\overline{k}}, \quad \text{where}$$

$$a_n = \begin{cases} (-1)^{\frac{n-1}{2}} (1 + 2^{-n}) & \text{if } n \text{ is odd}, \\ (-1)^{\frac{n}{2}} & \text{if } n \text{ is even}. \end{cases}$$

Exercises

1. Find all permutations of the set $\{1, 2, \ldots, n\}$ for $n = 1, 2, 3, 4$; list all obtained alternating (up-down) permutations, all obtained proper alternating (up-down) permutations, and all obtained down-up permutations. What can we say about the number of descents (ascents) in such permutations? Check that the first five Euler numbers E_n are 1, 0, −1, and 5. Check that the first five Euler zigzag numbers A_n are 1, 1, 1, 2, and 5.
2. Prove that the number of up-down permutations of the set $\{1, 2, \ldots, n\}$ is equal to the number of down-up permutations of the set $\{1, 2, \ldots, n\}$.
3. Find the first five Euler numbers E_n using their exponential generating function $\frac{2}{e^x + e^{-x}}$.

4. Find the first five secant numbers A_{2n} using their exponential generating function $\sec x$. Check that $A_{2n} = |E_{2n}|$.
5. Find the first five tangent numbers A_{2n+1} using their exponential generating function $\tan x$.

2.7. Enumeration of Permutations

2.7.1. In the following table, we represent the numbers of the main types of permutations of a given n-set.

$\|X\|$	Any	Circular	k-cycles	Derangements	With k cycles	With k ascents	Alternating
0	1	1	1	1	1	1	1
1	1	1	0, 1	0	0, 1	1, 0	1
2	2	1	0, 2, 1	1	0, 1, 1	1, 1, 0	1
3	6	2	0, 3, 3, 2	2	0, 2, 3, 1	1, 4, 1, 0	2
4	24	6	0, 4, 6, 8, 6	9	0, 6, 11, 6, 1	1, 11, 11, 1, 0	5
⋮	⋮	⋮	⋮	⋮	⋮	⋮	
n	$n!$	$(n-1)!$	$C_{n-1}^{k}(k-1)!$	$[\frac{n}{e}]$	$\|s(n,k)\|$	$E(n,k)$	A_n
OEIS	A000142	A000142	A111492	A000166	A008275	A008292	A000111

In fact, we have obtained the following results:

- The number of all permutations of the set $\{1, 2, \ldots, n\}$ is $n!$.
- The number of circular permutations of the set $\{1, 2, \ldots, n\}$ is $(n-1)!$.
- The number of k-cycles for the set $\{1, 2, \ldots, n\}$ is $\binom{n}{k}(k-1)! = \frac{n(n-1)\cdots(n-k+1)}{k} = \frac{n!}{(n-k)!k}$.
  The number $!n$ of derangements of the set $\{1, 2, \ldots, n\}$, $n > 0$, is equal to $\left[\frac{n!}{e}\right]$.
- The number of permutations of the set $\{1, 2, \ldots, n\}$ with k disjoint cycles (equivalently, the number of permutations of the set $\{1, 2, \ldots, n\}$ with exactly k left-to-right maxima) is the unsigned Stirling number of the first kind $|s(n,k)|$.
- The number of permutations of the set $\{1, 2, \ldots, n\}$ with k ascents (equivalently, the number of permutations of the set $\{1, 2, \ldots, n\}$ with k descents, with k excedances, and with $k+1$ ascending runs) is the Eulerian number $E(n,k)$.
- The number of permutations of the set $\{1, 2, \ldots, n\}$ with k inversions (equivalently, the number of permutations of the

set $\{1, 2, \ldots, n\}$ with major index k) is the Mahonian number $inv(n, k)$.

- The number of alternating (up-down) permutations of the set $\{1, 2, \ldots, n\}$ is the nth Euler zigzag number A_n; the number of all up-down and down-up permutations is Z_n: $Z_n = 2A_n$, $n \geq 2$.

2.7.2. Adding in the consideration of more general constructions, including permutations of multisets and indexed permutations, we can list several additional results.

- The number of permutations of a multiset, $\{1, \ldots, 1, 2, \ldots, 2, \ldots, n, \ldots, n\}$, in which any element is used m times is equal to $P_{m,m,\ldots,m} = \frac{(m+m+\cdots+m)!}{m!m!\cdots m!} = \frac{(mn)!}{(m!)^n}$; in particular, the number of permutations of a multiset, $\{1, 1, 2, 2, \ldots, n, n\}$, in which any element is used twice is equal to $P_{2,2,\ldots,2} = \frac{(2n)!}{(2)^n}$.
- The number of m-Stirling permutations of a multiset, $\{1, \ldots, 1, 2, \ldots, 2, \ldots, n, \ldots, n\}$, in which any element is used m times is equal to the multifactorial $1 \cdot (m+1)(2m+1) \cdots ((n-1)m+1) = ((n-1)m+1)!!! \cdots !$; in particular, the number of Stirling permutations of a multiset, $\{1, 1, 2, 2, \ldots, n, n\}$, in which any element is used twice is equal to $1 \cdot 3 \cdot \cdots \cdot (2n-1) = (2n-1)!!$.
- The number of indexed permutations of the set $\{1, 2, \ldots, n\}$ with index set $Z_r = \{0, 1, \ldots, r-1\}$ is equal to $r^n n!$; in particular, the number of signed permutations of the set $\{1, 2, \ldots, n\}$ is equal to $2^n n!$.

2.7.3. Moreover, there are some additional statistics. For example, the number of permutations of the set $\{1, 2, \ldots, n\}$ that avoid the pattern 213 is equal to the nth Catalan number C_n, $n \geq 0$. Here, C_n can be obtained using Pascal's triangle: $C_n = \frac{1}{n+1}\binom{2n}{n}$.

This approach leads to a large spectrum of problems concerned with the representations of different classes of permutations using special graphs, primarily using several types of trees.

It is widely known that Catalan numbers appear quite often at counting the number of some classes of trees (see, for example, [Deza25]).

In particular, the English mathematician Arthur Cayley (1821–1895) proved the following fact:

- *The number of plane binary trees with n internal vertices is equal to the nth Catalan number C_n, $n \geq 0$.*

So, we can state that plane binary trees with n internal vertices count the permutations of the set $\{1, 2, \ldots, n\}$ that avoid the pattern 213.

2.7.4. However, there exist at least two constructions of special types of trees which count all permutations on the set $\{1, 2, \ldots, n\}$. Both constructions use the so-called *increasing binary trees*, i.e., rooted binary trees with n vertices, labelled by the numbers $1, 2, \ldots, n$, such that the labels along any path from the root are increasing.

The first method gives a bijection between the set S_n of all permutations of the set $\{1, 2, \ldots, n\}$ and the set of all increasing binary trees on n vertices.

The increasing binary tree $T(\pi)$, corresponding to a permutation π, can be constructed inductively. In fact, represent π as $\pi = \rho 1 \delta$, where ρ and δ are permutations of some sets A_1 and A_2, forming a partition (A_1, A_2) of the set $\{2, 3, \ldots, n\}$; choose 1 as the root of $T(\pi)$, and use the trees $T(\rho)$ and $T(\delta)$ as left and right subtrees of $T(\pi)$; the tree $T(\rho)$ (as well as $T(\delta)$) is constructed using the same algorithm by choosing as the root of $T(\rho)$ as the least element of ρ; $T(\emptyset) = \emptyset$. To obtain π from $T(\pi)$, we can simply read the numbers of vertices in the symmetric order, i.e., first the labels of the left subtree (in symmetric order, recursively), then the label of the root, and then the labels of the right subtree.

By the construction, the tree $T(\pi)$, corresponding to the permutation $\pi = (\pi_1, \pi_2, \ldots, \pi_n)$, has the following properties (for convenience, consider the permutation π as the string $(0, \pi_1, \pi_2, \ldots, \pi_n, 0)$, i.e., the set $\pi_0 = \pi_{n+1} = 0$):

- a vertex i of $T(\pi)$ has precisely the successor below right if $\pi_{i-1} < \pi_i < \pi_{i+1}$ (element i is a *double ascent*);
- a vertex i of $T(\pi)$ has precisely the successor below left if $\pi_{i-1} > \pi_i > \pi_{i+1}$ (element i is a *double descent*);

- a vertex i of $T(\pi)$ has precisely the successors below left and below right if $\pi_{i-1} > \pi_i < \pi_{i+1}$ (element i is a *valley*);
- a vertex i of $T(\pi)$ has no successors if $\pi_{i-1} < \pi_i > \pi_{i+1}$ (element i is a *peak*).

So, we can state that:

- *there are exactly $n!$ rooted binary trees on n vertices;*
- *the number of increasing binary trees on n vertices with exactly k vertices, having left successors, is equal to the number of permutations of the set $\{1, 2, \dots, n\}$, having exactly k descents;*
- *the number of complete (i.e., every vertex is either an endpoint or has two successors) increasing binary trees on $2n + 1$ vertices is equal to the number of alternating permutations of the set $\{1, 2, \dots, 2n + 1\}$.*

2.7.5. A second way to represent a permutation of the set $\{1, 2, \dots, n\}$ by a tree uses the bijection between the set S_n of all n-permutations and the set of all *increasing trees* on $n + 1$ vertices, labelled $0, 1, 2, \dots, n$.

In fact, given $\pi = (\pi_1, \pi_2, \dots, \pi_n) \in S_n$, construct an (unordered) tree $T(\pi)$ with vertices $0, 1, \dots, n$ by defining vertex i to be the successor of the rightmost element j of π, which precedes i and is less than i. If there is no such element j, then let i be the successor of the root 0.

It is easy to see that the successors of the root 0 are simply the *left-to-right minima* (*retreating elements*) of π, i.e., the elements π_i, such that $\pi_i < \pi_j$ for all $j < i$. Moreover, the leaves of the tree $T(\pi)$ are simply the elements π_i, for which $\pi_i > \pi_{i+1}$ (including $\pi_n > \pi_{n+1} = 0$).

So, we can state that:

- *the number of unordered increasing trees on $n + 1$ vertices is $n!$;*
- *the number of such trees with k leaves is the number of n-permutations with $k - 1$ descents;*
- *the number of such trees for which the root has k successors is the unsigned Stirling number of the first kind $|s(n, k)|$.*

Exercises

1. Find all permutations of the set $\{1, 2, \ldots, n\}$ for $n = 1, 2, 3, 4$. For any constructed permutation, give its characteristic: the number of its ascents, descents, disjoint cycles, etc.
2. For any permutation of the set $\{1, 2, \ldots, n\}$, $n = 1, 2, 3, 4$, construct the increasing binary tree on n vertices $1, 2, \ldots, n$, corresponding to this permutation. Which properties of the permutation can be obtained from the properties of the constructed binary tree?
3. For any permutation of the set $\{1, 2, \ldots, n\}$, $n = 1, 2, 3, 4$, construct the increasing binary tree on $n + 1$ vertices $0, 1, 2, \ldots, n$, corresponding to this permutation. Which properties of the permutation can be obtained from the properties of the constructed binary tree?
4. Construct all rooted binary trees with n internal vertices, $n = 1, 2, 3, 4, 5$. Check that the number of such trees is equal to the nth Catalan number $C_n = \frac{1}{n+1}\binom{2n}{n}$.

References

[AbSt72], [Aign82], [Ande03], [Andr79], [Apos86], [BaCo87], [BeSl95], [Bhar00], [BiBa70], [BuGr94], [Cata44], [ChMo75], [ClKi08], [Cofm75], [Comt74], [CoGu96], [DeZa80], [Deza17], [Deza18], [Dick05], [Gard88], [Gelf98], [Goul85], [GoJa83], [GGL95Û], [GuJa90], [Hagg04], [Hall70], [Hara03], [KoKo77], [Kost82], [MSY96], [Pudw24], [RoTe09], [Sachk77], [SlPl95], [Sloa25], [Sier64], [Stan10], [Stan15], [Stei94], [Stra16], [Vell08], [Weis25], [Wiki25].

Chapter 3

Eulerian Numbers of the First Order

3.1. Construction of Eulerian Numbers of the First Order

A history of the question

3.1.1. Leonard Euler (1707–1783) was a Swiss, German, and Russian mathematician, physicist, astronomer, logician, geographer, and engineer, who made fundamental contributions to the development of these sciences.

He founded the studies of graph theory and topology and made influential discoveries in many other branches of mathematics, such as analytic number theory, complex analysis, and infinitesimal calculus. He also introduced much of the modern mathematical terminology and notation, including the notion of a mathematical function.

Euler authored more than 850 works (including two dozen fundamental monographs) on mathematical analysis, differential geometry, number theory, approximate calculations, celestial mechanics, mathematical physics, optics, ballistics, theory of music, etc. He served as an academician at St. Petersburg, Berlin, Turin, Lisbon, and Basel Academies of Sciences and was a foreign member of the Paris Academy of Sciences.

Euler has been called a "universal genius", who "was fully equipped with almost unlimited powers of imagination, intellectual gifts and extraordinary memory".

3.1.2. There are many mathematical and physical objects named in honour of Euler. The most famous among them include *Euler's totient theorem* on modular exponentiation in number theory, generalising Fermat's little theorem; *Euler's polyhedral formula* for planar graphs or polyhedra, establishing a relationship between the numbers of their vertices, edges, and faces; a special case of the *Euler characteristic* in topology; *Euler's totient function* in number theory, $\varphi(n)$, giving the number of positive integers not exceeding n and coprime with it; *Euler's formula* $e^{i\pi} + 1 = 0$, linking the five fundamental mathematical constants. For a long list of objects named after Euler, see [Wiki25].

3.1.3. A separate list can be compiled of the numbers associated with the name of this great scientist.

- The *Euler–Mascheroni constant* $\gamma = \lim_{N\to\infty} \sum_{n=1}^{N} n^{-1} - \log N$ is the limit of the difference between the partial sum of the harmonic series and the natural logarithm of the number of terms used.
- The number e, the base of natural logarithm, is often called *Euler's number.*
- The *lucky numbers of Euler* are positive integers n such that the polynomial $x^2 - x + n$ takes only prime values for all $x = 0, 1, \ldots, n-1$.
- *Euler's convenient numbers* are the positive integers D such that any integer, having the only proper representation by the binary quadratic form $x^2 \pm Dy^2$ (i.e., a presentation in which x is relatively prime to Dy^2) is a prime power or twice a prime power.
- The *Euler numbers* and *Euler zig-zag numbers* are closely related to the counting of alternating permutations, etc.

This list also includes the *Eulerian numbers of the first and second orders*, which are the central objects of our consideration.

3.1.4. The Eulerian numbers (of the first order) were originally defined by Euler in a non-combinatorial way.

These numbers appeared in Euler's textbook on differential calculus [Eule55] in the context of solving the differential equation for the exponential generating function $\frac{x-1}{x-e^{y(x-1)}}$. Euler obtained these numbers by expanding the series $\frac{x-1}{x-e^{y(x-1)}} = \sum_{n,k=0}^{\infty} E(n,k)x^k \frac{y^n}{n!}$. It turns out that the *Carlitz identity*

$$\sum_{k=0}^{n} E(n,k)x^k = (1-x)^{n+1} \sum_{k=0}^{\infty} (k+1)^n x^k$$

was already known to Euler. We refer to it as the Carlitz identity because of a generalisation of the identity obtained by Leonard Carlitz in 1975, [Carl75].

The main motivation for Euler seemed to have been obtaining a formula for the alternating sums of powers $1^n - 2^n + 3^n - 4^n + \cdots$ in a manner analogous to what Jacques Bernoulli had done for the unsigned sums of powers.

The connection with the Bernoulli numbers motivated the work of Julius Worpitzky in 1883, [Worp83], and Ferdinand Georg Frobenius in 1910, [Frob10]. *Worpitzky's identity*

$$x^n = \sum_{k=0}^{n} E(n,k) \binom{x+k}{n}$$

dates from the 1883 year paper by Worpitzky, though Don Knuth attributes the identity to an 1867 publication by the Chinese mathematician Li Shan-Lan and remarks that special cases of $n \leq 5$ were known to a Japanese mathematician of the 17th century, Yoshisuke Matsunaga.

The investigation into the descent set and the number of descents of a permutation of a set or a multiset was begun by Percy Alexander MacMahon ([MacM08] and [MacM16]). He was also the first person to consider the excedance of a permutation and showed the equidistribution of the number of descents with the number of excedances. However, MacMahon apparently did not realise that the number of permutations of the set $\{1, 2, \ldots, n\}$ with k descents (as well as with k ascents) is an Eulerian number.

The first statement connecting the Eulerian numbers with descents was made by Carlitz and John Riordan in 1953, [CaRi53].

Carlitz and Riordan showed that the numbers defined by Euler using special series count the numbers of permutations of a fixed size with a given number of descents. They began to study the Eulerian numbers as combinatorial quantities, following the works of Simon Newcomb and MacMahon. In fact, a simple way for the construction of the Eulerian numbers is to obtain the answer to *Simon Newcomb's problem*:

> *... a deck of cards of arbitrary specification is dealt out into a single pile so long as cards are in rising order, with like cards counted as rising, and a new pile is started whenever a non-rising card appears; with all possible arrangements of the deck, in how many ways do k piles appear?*

If there are no ties among the cards, then we can consider the deck of cards as a permutation, and the stacks correspond to maximal increasing runs in the permutation. The Eulerian numbers count the number of permutations of fixed size with a given number of increasing runs.

3.1.5. Such a simple concept has a deep and rich theory, with connections to a vast number of other combinatorial problems. The Eulerian numbers provide connections between ordinary powers, consecutive binomial coefficients, the Stirling numbers of the first and second kinds, the Bernoulli numbers, etc. The Eulerian numbers and their generalisations arise naturally in areas such as partially ordered sets, hyperplane arrangements, Coxeter groups, simplicial complexes, and convex polytopes. They are used in number theory, statistics, control theory, etc., mostly in connection with interesting examples.

In addition to the already mentioned works on the Eulerian numbers, there exist the monograph by Dominique Foata and Marcel-Paul Schützenberger, *Théorie Géométrique des Polynômes Eulériens* [FoSh70a], published in 1970, as well as the book by T. Kyle Petersen, *Eulerian Numbers* [Pete15], published in 2015. Today, the number of articles on the Eulerian numbers and their generalisations is numerous. We cite many works in this chapter.

In the literature, many different notations for the Eulerian numbers are used. The most common ones are $E(n,k)$, $\left\langle {n \atop k} \right\rangle$, $A(n,k)$, $A_{n,k}$, $T_{n,k}$.

In this book, we use for the Eulerian numbers of the first order the notation $E(n,k)$; it is the number of permutations of the set $\{1, 2, \ldots, n\}$ which have exactly k ascents.

Combinatorial definition of Eulerian numbers of the first order

3.1.6. An *Eulerian number of the first order* $E(n,k)$ is the number of permutations of the set $\{1, 2, \ldots, n\}$ with k *ascents*, i.e., such permutations $\pi = (\pi_1, \pi_2, \ldots, \pi_n)$, for which there are exactly k indices j, $1 \le j \le n-1$, with $\pi_j < \pi_{j+1}$.

For example, if $\pi = (3, 1, 2, 5, 6, 4, 7)$, then there are ascents in position 2 (since $1 < 2$), in position 3 (since $2 < 5$), in position 4 (since $5 < 6$), and in position 6 (since $4 < 7$).

Similarly, the permutation $\pi = (3, 1, 2, 6, 5, 4, 7)$ has two *descents*, i.e., indices j, $1 \le j \le n-1$, with $\pi_j > \pi_{j+1}$; in fact, there are descents in position 2 (since $3 > 1$) and in position 5 (since $6 > 4$).

The permutation $(n, n-1, \ldots, 2, 1)$ is the only permutation of the set $\{1, 2, \ldots, n\}$ with no ascents (and with the maximal number $n-1$ of descents), while the permutation $(1, 2, \ldots, n-1, n)$ has the maximal number $n-1$ of ascents (and no descents).

3.1.7. For $n = 1$, there exists exactly one permutation (1), and it has exactly 0 ascents, i.e., by definition, $E(1,0) = 1$.

For $n = 2$, there are two permutations. The permutation $(2, 1)$ has no ascents; the permutation $(1, 2)$ has one ascent, and we get $E(2,0) = 1$ and $E(2,1) = 1$.

For $n = 3$, we have six permutations. The permutation $(3, 2, 1)$ has no ascents. The permutation $(1, 2, 3)$ has two ascents. Four other permutations $(1, 3, 2)$, $(2, 1, 3)$, $(2, 3, 1)$, and $(3, 1, 2)$ have one ascent each. Therefore, we get $E(3,0) = 1$, $E(3,1) = 4$, and $E(3,2) = 1$.

For the fourth-order permutations, the situation becomes more complicated, although the consideration will be similar. Exactly one permutation $(4,3,2,1)$ has no ascents. Exactly one permutation $(1,2,3,4)$ has three ascents. The permutations π having exactly two ascents must satisfy one of the following three inequalities: $\pi_1 < \pi_2 > \pi_3 < \pi_4$, $\pi_1 < \pi_2 < \pi_3 > \pi_4$, or $\pi_1 > \pi_2 < \pi_3 < \pi_4$. We have exactly 11 such permutations: $(1,3,2,4)$, $(1,4,2,3)$, $(2,3,1,4)$, $(2,4,1,3)$, $(3,4,1,2)$, $(1,2,4,3)$, $(1,3,4,2)$, $(2,3,4,1)$, $(2,1,3,4)$, $(3,1,2,4)$, and $(4,1,2,3)$. All other permutations have exactly one ascent; the number of such permutations is $24 - 2 - 11 = 11$. So, we get that $E(4,0) = 1$, $E(4,1) = 11$, $E(4,2) = 11$, and $E(4,3) = 1$.

The coincidence of two numbers $E(4,1) = E(4,2) = 11$ is not an accident. Considering the *reversal* (or *mirrored image*) $\pi^r = (\pi_4, \pi_3, \pi_2, \pi_1)$ of any permutation $\pi = (\pi_1, \pi_2, \pi_3, \pi_4)$ with two ascents, we get a permutation with two descents and, therefore, with one ascent, and vice versa. Thus, using reversals of the fourth-order permutations with exactly two ascents (complete list of which is presented above), we will obtain all fourth-order permutations with exactly one ascent; they are $(4,2,3,1)$, $(3,2,4,1)$, $(4,1,3,2)$, $(3,1,4,1)$, $(2,1,4,3)$, $(3,4,2,1)$, $(2,4,3,1)$, $(1,4,3,2)$, $(4,3,1,2)$, $(4,2,1,3)$, and $(3,2,1,4)$.

3.1.8. Our discussion allows us to draw some preliminary conclusions. First, for a given positive integer n, there is the only permutation without ascents, namely, the permutation $(n, n-1, \ldots, 2, 1)$. There is also a single permutation which has $n-1$ ascents; it is the permutation $(1, 2, \ldots, n-1, n)$. Thus, we have the following relations:

$$E(n,0) = 1, \quad E(n,n-1) = 1, \quad n \geq 1.$$

It is clear that a permutation of the nth order cannot have n (or more) ascents; therefore, using the value $E(0,0) = 1$, we get the following relations:

$$E(0,0) = 1, \quad E(n,n) = 0, \quad E(n,n+k) = 0, \quad n \geq 1, k \geq 1.$$

The reversal of a permutation of the nth order with m ascents is a permutation of the nth order with m descents and, therefore, with $n-1-m$ ascents. Thus, it holds that

$$E(n,k) = E(n,n-k-1), \quad n \geq 1, 0 \leq k \leq n-1.$$

3.1.9. We constructed the Eulerian numbers of the first order by counting ascents in a given permutation of the nth order. However, there exist other possibilities to count the Eulerian numbers of the first order. In fact, we can also use descents, excedances, maximal increasing runs, and readings.

- *For the Eulerian numbers of the first order* $E(n, k)$, $n \geq 0$, $0 \leq k \leq n-1$, *the following statements hold:*
- $E(n, k)$ *is the number of permutations of the set* $\{1, 2, \ldots, n\}$ *with* k *descents, i.e., such permutations* $\pi = (\pi_1, \pi_2, \ldots, \pi_n)$, *for which there are exactly* k *indices* j, $1 \leq j \leq n-1$, *with* $\pi_j > \pi_{j+1}$;
- $E(n, k)$ *is the number of permutations of the set* $\{1, 2, \ldots, n\}$ *with* k *excedances, i.e., such permutations* $\pi = (\pi_1, \pi_2, \ldots, \pi_n)$, *for which there are exactly* k *indices* j, $1 \leq j \leq n-1$, *with* $\pi_j > j$;
- $E(n, k)$ *is the number of permutations of the set* $\{1, 2, \ldots, n\}$ *with* $k+1$ *maximal increasing runs, i.e., such permutations* $\pi = (\pi_1, \pi_2, \ldots, \pi_n)$, *for which there are exactly* $k+1$ *subsequences,* $\pi_i < \pi_{i+1} < \cdots < \pi_{i+m}$, *such that* $\pi_{i-1} > \pi_i$ *if* π_{i-1} *exists and* $\pi_{n+m+1} < \pi_{n+m}$ *if* π_{n+m+1} *exists;*
- $E(n, k)$ *is the number of permutations of the set* $\{1, 2, \ldots, n\}$ *with* $k+1$ *readings, i.e., such permutations* $\pi = (\pi_1, \pi_2, \ldots, \pi_n)$, *for which there are exactly* $k+1$ *steps we must scan the one-line notation of* π *from left to right to put the numbers* $1, 2, \ldots, n$ *in natural order.*

□ For $n = 2$, the permutation $(1, 2)$ has one ascent, zero descents, zero excedances, one run (exactly, 12), and one reading (the first step gives both numbers 1 and 2 in natural order), while the permutation $(2, 1$ has zero ascents, one descent, one excedance ($\pi_1 > 1$), two runs (exactly, 2 and 1), and two readings (the first step gives 1, and the second step gives 2). So, we have exactly one permutation of the set $\{1, 2\}$ with zero ascents, as well as with zero descents, zero excedances, one run and one reading: $E(2, 0) = 1$. Similarly, we have exactly one permutation of the set $\{1, 2\}$ with one ascent, as well as with one descent, one excedance, two runs and two readings: $E(2, 1) = 1$.

For $n = 3$, the permutation $(1, 2, 3)$ has two ascents, zero descents, zero excedances, one run (exactly, 123), and one reading (the first step gives all three numbers 1, 2, and 3 in natural order). The permutation $(1,3,2)$ has one ascent, one descent, one excedance, two runs (exactly, 13 and 2), and two readings (first step gives 1 and 2, and the second step gives 3). The permutation $(2,1,3)$ has one ascent, one descent, one excedance, two runs (exactly, 2 and 13), and two readings (first step gives 1, and the second step gives 2 and 3). The permutation $(2,3,1)$ has one ascent, one descent, two excedances, two runs (exactly, 23 and 1), and two readings (first step gives 1, and the second step gives 2 and 3). The permutation $(3,1,2)$ has one ascent, one descent, two runs (exactly, 3 and 12), and two readings (the first step gives 1 and 2, and the second step gives 3). The permutation $(3,2,1)$ has zero ascents, two descents, three runs (exactly, 3, 2 and 1), and three readings (first step gives 1, the second step gives 2, and the third step gives 3). So, we have exactly one permutation of the set $\{1,2,3\}$ with zero ascents, as well as with zero descents, zero excedances, one run and one reading: $E(3,0)=1$. We have exactly four permutations of the set $\{1,2,3\}$ with one ascent, as well as with one descent, one excedance, two runs, and two readings: $E(3,1)=4$. At last, we have exactly one permutation of the set $\{1,2,3\}$ with two ascents, as well as with two descents, two excedances, three runs, and three readings: $E(3,2)=1$.

In general, it is easy to prove that the number of permutations of the nth order with k ascents is equal to the number of permutations of the nth order with k descents, as well as with k excedances, $k+1$ maximal increasing runs, and $k+1$ readings. The proofs of these statements can be found in Chapter 2. □

Recurrent relation for Eulerian numbers of the first order

3.1.10. Now, we can obtain a Pascal's-like recurrent relation for the construction of the Eulerian numbers of the first order.

- *Recurrent relation for the Eulerian numbers of the first order* $E(n,k)$, $n \geq 1$, $1 \leq k \leq n-1$, *has the form*

$$E(n,k) = (k+1)E(n-1,k) + (n-k)E(n-1,k-1), \quad E(n-1,0) = 1.$$

□ In fact, each permutation $\rho = (\rho_1, \rho_2, \ldots, \rho_{n-1})$ of the elements $\{1, 2, 3, \ldots, n-1\}$ leads to n permutations of the elements $\{1, 2, 3, \ldots, n-1, n\}$ if we insert the new nth element in all possible ways: between two elements ρ_i, ρ_{i+1}, $i = 1, 2, 3, \ldots, n-1$ ($n-2$ positions); before the first element ρ_1 (one position); and after the last element ρ_{n-1} (one position).

Inserting n in the jth position, we get the permutation $\pi = (\rho_1, \rho_2, \ldots, \rho_{j-1}, n, \rho_j, \ldots, \rho_{n-1})$ of the nth order. On the other hand, any permutation of the nth order can be obtained in this way from a permutation of the $(n-1)$th order.

The number of ascents in the permutation π is equal to the number of ascents in the permutation ρ if $j = 1$ or if $\rho_{j-1} < \rho_j$. The number of ascents in the permutation π is greater than the number of ascents in the permutation ρ if $j = n-1$ or if $\rho_{j-1} > \rho_j$. Therefore, in order to obtain a permutation π of the nth order with k ascents, we can:

- use any permutation ρ of the $(n-1)$th order with k ascents, putting element n in the first position, or between two elements, forming an ascent, altogether giving $k+1$ possibilities;
- use any permutation ρ of the $(n-1)$th order with $k-1$ ascents (and, hence, with $(n-2)-(k-1) = n-k-1$ descents), putting element n in the last position, or between two elements, forming an descent, altogether giving $n-k$ possibilities.

In other words, there are $(k+1)E(n-1, k)$ ways to construct a permutation π of the nth order with k ascents from the permutations ρ of the $(n-1)$th order with k ascents and $(n-k)E(n-1, k-1)$ ways to construct a permutation π of the nth order with k ascents from the permutations ρ of the $(n-1)$th order with $k-1$ ascents. The sum of all possibilities gives the recurrent formula

$$E(n, k) = (k+1)E(n-1, k) + (n-k)E(n-1, k-1). \qquad \square$$

For example, if $n = 3$, $k = 2$, we have

$$\begin{array}{rcl} & \to & (1,4,2,3) \\ (1,2,3) & \to & (1,2,4,3) \\ & \to & (1,2,3,4) \end{array}$$

$$(1,3,2) \begin{array}{ll} \to & (1,3,4,2) \\ \to & (1,3,2,4) \end{array} \quad (2,1,3) \begin{array}{ll} \to & (2,1,4,3) \\ \to & (2,1,3,4) \end{array}$$

$$(2,3,1) \begin{array}{ll} \to & (2,3,4,1) \\ \to & (2,3,1,4) \end{array} \quad (3,1,2) \begin{array}{ll} \to & (3,4,1,2) \\ \to & (3,1,2,4) \end{array}.$$

So, $E(4,2) = 3E(3,2) + 2E(3,1) = 3 \cdot 1 + 2 \cdot 4 = 11$.

3.1.11. Using the initial conditions $E(0,0) = 1$ and $E(n,0) = 0$ and noting that $E(n,n) = 0$ for $n > 0$, we can obtain all the values of the Eulerian numbers of the first order.

So, starting from $E(1,0) = 1$ and $E(1,1) = 0$, we obtain that

$$E(2,1) = 2 \cdot E(1,1) + 1 \cdot E(1,0) = 2 \cdot 0 + 1 \cdot 1 = 1.$$

Starting from $E(2,0) = 1$ and $E(2,1) = 1$, we obtain that

$$E(3,1) = 2 \cdot E(2,1) + 2E(2,0) = 2 \cdot 1 + 2 \cdot 1 = 4.$$

Similarly, as $E(2,1) = 1$ and $E(2,2) = 0$, we obtain that

$$E(3,2) = 3 \cdot E(2,2) + 1 \cdot E(2,1) = 3 \cdot 0 + 1 \cdot 1 = 1.$$

Starting from $E(3,0) = 1$ and $E(3,1) = 4$, we obtain that

$$E(4,1) = 2 \cdot E(3,1) + 3E(3,0) = 2 \cdot 4 + 3 \cdot 1 = 11.$$

Similarly, as $E(3,1) = 4$ and $E(3,2) = 1$, we get that

$$E(4,2) = 3 \cdot E(3,2) + 2 \cdot E(3,1) = 3 \cdot 1 + 2 \cdot 4 = 11.$$

As $E(3,2) = 1$ and $E(3,3) = 0$, we obtain that

$$E(4,3) = 4 \cdot E(3,3) + 1 \cdot E(3,2) = 4 \cdot 0 + 1 \cdot 1 = 1.$$

Triangle of Eulerian numbers of the first order

3.1.12. The recurrent relation

$$E(n,k) = (k+1)E(n-1,k) + (n-k)E(n-1,k-1), n, k \in \mathbb{N}, k \leq n-1,$$

for calculating the Eulerian numbers of the first order and the conditions $E(0,0) = 1$, $E(n,0) = 1$, and $E(n,n) = 0$, $n \in \mathbb{N}$, allow us to construct the *triangle of Eulerian numbers of the first order,* which is commonly known as *Euler's triangle* (see sequence A008292 in the OEIS). It can be considered as an edge-weighted version of Pascal's triangle (sequence A007318 in the OEIS) and shares some common characteristics with this arithmetic construction.

Formally, we can define Euler's triangle as an arithmetic triangle, consisting of the numbers $E(n,k)$, $n,k \in \mathbb{Z}, n \geq 0, 0 \leq k \leq n$, constructed according the following law: the top (zero) row of the triangle contains only one unity:

$$E(0,0) = 1.$$

The next (first) row contains one (left) unity and one (right) zero:

$$E(1,0) = 1, E(1,1) = 0.$$

In the second row, the two extreme elements are again equal to unity (extreme left) and zero (extreme right):

$$E(2,0) = 1, E(2,2) = 0.$$

The internal element of the second row is obtained as the sum of the two elements of the first row, located above it, up and up-left, with weights 2 and 1, correspondingly:

$$E(2,1) = 2 \cdot E(1,1) + 1 \cdot E(1,0) = 2 \cdot 0 + 1 \cdot 1 = 1.$$

In general, the extreme-left elements of the nth row of Euler's triangle, $n \in \mathbb{N}$, are equal to 1:

$$E(n,0) = 1, n \in \mathbb{N}.$$

The extreme-right elements of the nth row of Euler's triangle, $n \in \mathbb{N}$, are equal to 0:

$$E(n,n) = 0, n \in \mathbb{N}.$$

Any internal element $E(n,k)$ is obtained as the sum of two elements located in the previous row above it (up and up-left), taken with weights: the weight for the element $E(n-1,k)$, located in the same column k as the calculated element, is equal to the number $k+1$ of the next column; the weight for the element $E(n-1,k-1)$, located up-left, i.e., in the previous column $k-1$, is equal to the difference $n-k$ between the numbers $n-1$ of the row and $k-1$ of the column, where this element is located:

$$E(n,k) = (k+1)\cdot E(n-1,k) + (n-k)\cdot E(n-1,k-1).$$

So, one can say that Euler's triangle represents a sort of a generalisation of Pascal's triangle, where the recurrent relation used is the weighted sum of the two neighbours of the element being calculated, and the weights are simple functions of the corresponding row's and column's numbers.

By sequential calculations, we get, for $0 \le n \le 10$, $0 \le k \le n$, the following construction:

$n\backslash k$	0	1	2	3	4	5	6	7	8	9	10
0	1										
1	1	0									
2	1	1	0								
3	1	4	1	0							
4	1	11	11	1	0						
5	1	26	66	26	1	0					
6	1	57	302	302	57	1	0				
7	1	120	1191	2416	1191	120	1	0			
8	1	247	4293	15619	15619	4293	247	1	0		
9	1	502	14608	88234	156190	88234	14608	502	1	0	
10	1	1013	47840	455192	1310354	1310354	455192	47840	1013	1	0

This version of the triangle contains the Eulerian numbers $E(n,k)$, with $n,k \in \mathbb{Z}, n \ge 0, 0 \le k \le n$. This approach corresponds to the classical Euler's definition and coincides with the approach presented in [GKP94]. This version, read by rows, gives the sequence $1,1,0,1,1,0,1,4,1,0,\ldots$ (sequence A173018 in the OEIS).

3.1.13. Due to the property

$$E(n,k) = E(n, n-1-k), n \geq 1, 0 \leq k \leq n-1,$$

the constructed triangle has a special symmetry: the elements of each row which are located at the same distance from the two outermost unities are equal. So, if we delete the vertex $E(0,0) = 1$ and the zeros $E(n,n) = 0$, $n \geq 1$, forming the right-hand side of the triangle, we can replace the remaining elements in the form of an isosceles triangle, i.e., we get a quite symmetrical arrangement:

$$\begin{array}{ccccccccccccc}
 & & & & & & 1 & & & & & & \\
 & & & & & 1 & & 1 & & & & & \\
 & & & & 1 & & 4 & & 1 & & & & \\
 & & & 1 & & 11 & & 11 & & 1 & & & \\
 & & 1 & & 26 & & 66 & & 26 & & 1 & & \\
 & 1 & & 57 & & 302 & & 302 & & 57 & & 1 & \\
1 & & 120 & & 1191 & & 2416 & & 1191 & & 120 & & 1 \\
\vdots & & \vdots & & \vdots & & \vdots & & \vdots & & \vdots & & \vdots
\end{array}$$

This version of Euler's triangle, read by rows, gives the sequence $1, 1, 1, 1, 4, 1, 1, 11, 11, 1, \ldots$ (sequence A008292 in the OEIS). This version also contains the Eulerian numbers $E(n,k)$ with $n, k \in \mathbb{Z}, n \geq 1, 0 \leq k \leq n-1$. In fact, in this sequence, the numbers $A(n,k) = E(n, k-1)$, with $n, k \in \mathbb{Z}, n \geq 1, 1 \leq k \leq n$, are given (see [Rior80] and [Comt74], etc.). The sequence A008292 is considered the main entry for the Eulerian numbers in the OEIS.

3.1.14. The triangle of Eulerian numbers of the first order, represented by the sequence $1, 0, 1, 0, 1, 1, 0, 1, 4, 1, \ldots$ (sequence A123125 in the OEIS), also does not quite agree with the usual version. In fact, each row of the triangle from A123125 is the reverse of the corresponding row of the triangle from A173018:

$n\backslash k$	0	1	2	3	4	5	6	7
0	1							
1	0	1						
2	0	1	1					
3	0	1	4	1				
4	0	1	11	11	1			
5	0	1	26	66	26	1		
6	0	1	57	302	302	57	1	
7	0	1	120	1191	2416	1191	120	1

The reader can obtain the complete description of the structure and properties of these modifications of the triangle of Eulerian numbers of the first order by themselves.

3.1.15. The small special cases of $E(n,1)$, $E(n,2)$, and $E(n,3)$, $n \in \mathbb{N}$, are given by the sequences $0, 1, 4, 11, 26, 57, 120, 247, 502, 1013, \ldots$ (sequence A000295 in the OEIS), $0, 0, 1, 11, 66, 302, 1191, 4293, 14608, \ldots$ (sequence A000460 in the OEIS), and $0, 0, 0, 1, 26, 302, 2416, 15619, 88234, \ldots$ (sequence A000498 in the OEIS). These sequences can be described by simple recurrences.

- *For the numbers $E(n,1)$, $n = 1, 2, \ldots$, the following recurrent relation holds*:

$$E(n+1,1) = 2E(n,1) + n, \quad E((1,1) = 0;$$

for the numbers $E(n,2)$, $n = 2, 3, \ldots$, the following recurrent relation holds:

$$E(n+1,2) = 3E(n,2) + 2^n(n-1) - n^2 + 1, \quad E(2,2) = 0;$$

for the numbers $E(n,3)$, $n = 3, 4, \ldots$, the following recurrent relation holds:

$$E(n+1,3) = 4E(n,2) + 3^n(n-2) - 2^n(n-2)(n+1) + \frac{(n-2)n(n+1)}{2}, \quad E(3,3) = 0.$$

□ For example, considering the sequence of numbers $E(n,1)$, $n = 1, 2, \ldots$, we obtain that $E(1,1) = 0$ and

$$E(n+1,1) = 2E(n,1) + nE(n,0).$$

As $E(n,0) = 0$ for any positive integer n, we get the recurrent relation

$$E(n+1,1) = 2E(n,1) + n, \quad n \geq 1,$$

with the initial condition $E(1,1) = 0$.

Now, we can easily prove by induction the formula

$$E(n,1) = 2^n - n - 1.$$

In fact, it is true for $n = 1$ and $n = 2$:

$$E(1,1) = 0 = 2^1 - 1 - 1, \quad \text{and} \quad E(2,1) = 1 = 2^2 - 2 - 1.$$

Assuming that $E(n,1) = 2^n - n - 1$ and using the recurrent relation $E(n+1,1) = 2E(n,1) + n$, we get that

$$E(n+1,1) = 2(2^n - n - 1) + n = 2^{n+1} - (n+1) - 1. \qquad \square$$

3.1.16. Each odd row (i.e., $(2n+1)$th row, $n = 0, 1, 2, \ldots$) of Euler's triangle has a central element, which is in fact the number $E(2n+1, n)$. The numbers $E(2n+1, n)$ are called the *central Eulerian numbers.* These numbers form the main diagonal of the isosceles triangle of Eulerian numbers of the first order. The first few elements of this sequence are 1, 4, 66, 2416, 156190, 15724248, 2275172004, 447538817472, 114890380658550, 37307713155613000, ... (sequence A025585 in the OEIS).

All these numbers are even. Moreover, they are maximal elements of the corresponding row. The first property is very simple. The second property we will consider later (see also [LeNi92]).

- *The numbers $E(2n+1, n)$, $n = 0, 1, 2, \ldots$, are even.*

□ Using the recurrent relation for the Eulerian numbers of the first order and the property $E(n,k) = E(n, n-1-k)$, we get that

$$\begin{aligned} E(2n+1, n) &= (n+1)E(2n, n) + (n+1)E(2n, n-1) \\ &= (n+1)E(2n, n) + (n+1)E(2n, 2n-1-(n-1)) \\ &= (n+1)E(2n, n) + (n+1)E(2n, n) \\ &= 2(n+1)E(2n, n). \qquad \square \end{aligned}$$

For an even row (i.e., $(2n)$th row, $n = 0, 1, 2, \ldots$), we have, starting from the second row, two central elements: $E(2n, n-1) = E(2n, n)$. The numbers $E(2n, n)$, $n = 0, 1, 2, \ldots$, give the number of permutations of the set $\{1, 2, \ldots, 2n\}$ with exactly n ascents. This sequence starts with the elements 1, 1, 11, 302, 15619, 1310354, 162512286, 27971176092, 6382798925475, 1865385657780650, ... (sequence A180056 in the OEIS). In can be proven (see, for example, [LeNi92]) that the numbers $E(2n, n-1) = E(2n, n)$ are the maximal elements of the corresponding row.

3.1.17. The sum of the elements of the nth row of Euler's triangle gives the number of all permutations of the nth order, which is equal to $n!$. Therefore, we get the following property:

- *The sum of the elements of the nth row of Euler's triangle is equal to $n!$:*

$$E(n, 0) + E(n, 1) + E(n, 2) + \cdots + E(n, n) = n!.$$

Since Euler's triangle has a certain symmetry, we can obtain useful information about the behaviour of the alternating sum of the elements of the its nth row. This sum behaves differently depending on the parity of the number n. For even n, the situation is very simple: the corresponding sum is equal to zero.

- *The alternating sum of the elements of any even row of Euler's triangle, starting with the second row, is equal to zero:*

$$\begin{aligned} E(2n, 0) - E(2n, 1) + \cdots + E(2n, 2n-2) - E(2n, 2n-1) \\ + E(2n, 2n-1) + E(2n, 2n) = 0. \end{aligned}$$

□ As $E(2n,k) = E(2n, 2n-1-k)$ and $E(n,n) = 0$ for $n \geq 1$, we get that

$$E(2n,0) - E(2n,1) + \cdots + (-1)^{n-1}E(2n,n-1) + (-1)^n E(2n,n)$$
$$+ \cdots + E(2n,2n-2) - E(2n,2n-1) + E(2n,2n)$$
$$= E(2n,0) - E(2n,1) + \cdots + (-1)^{n-1}E(2n,n-1)$$
$$- (-1)^{n-1}E(2n,n-1) + \cdots + E(2n,1) - E(2n,0) = 0. \qquad \square$$

For odd n, after the summation of the numbers, we get $1, -2, 16, -272, \ldots$. At first glance, we find no remarkable information. However, it can be proved that all of the obtained numbers can be calculated using the formula $2^{n+1}(2^{n+1}-1)\frac{B_{n+1}}{n+1}$, where n is the number of the considered row of Euler's triangle and B_{n+1} is the corresponding *Bernoulli number*. Since the Bernoulli numbers with odd indices are equal to zero, the above formula is true for all rows of the triangle.

- *The alternating sum of the elements of the nth row of Euler's triangle, $n = 0, 1, 2, \ldots$, is equal to $2^{n+1}(2^{n+1}-1)\frac{B_{n+1}}{n+1}$, where B_{n+1} is the $(n+1)$th Bernoulli number:*

$$E(n,0) - E(n,1) + \cdots + (-1)^k E(n,k) + (-1)^{k+1}E(n,k+1)$$
$$+ \cdots + (-1)^n E(n,n) = 2^{n+1}(2^{n+1}-1)\frac{B_{n+1}}{n+1}.$$

□ In fact, the sequence of Bernoulli numbers starts with the elements $1, \pm\frac{1}{2}, \frac{1}{6}, 0, -\frac{1}{30}, 0, \frac{1}{42}, 0, -\frac{1}{30}, \ldots, n \geq 0$ (sequences A027641, A164555 / A027642 in the OEIS).

For $n = 0$, the value $2^{n+1}(2^{n+1}-1)\frac{B_{n+1}}{n+1}$ is equal to $2 \cdot 1 \cdot \frac{1}{2} = 1$ (here, we use the *second* (or *original*) Bernoulli numbers with $B_1^+ = \frac{1}{2}$).

For $n = 1$, the value $2^{n+1}(2^{n+1}-1)\frac{B_{n+1}}{n+1}$ is equal to $2^2 \cdot 3 \cdot \frac{1}{12} = 1$.

For $n = 2$, the value $2^{n+1}(2^{n+1}-1)\frac{B_{n+1}}{n+1}$ is equal to $2^3 \cdot 7 \cdot \frac{0}{3} = 0$.

For $n = 3$, the value $2^{n+1}(2^{n+1}-1)\frac{B_{n+1}}{n+1}$ is equal to $-2^4 \cdot 15 \cdot \frac{1}{120} = -2$.

For $n = 4$, the value $2^{n+1}(2^{n+1}-1)\frac{B_{n+1}}{n+1}$ is equal to $2^5 \cdot 31 \cdot \frac{0}{5} = 0$.

For $n = 5$, the value $2^{n+1}(2^{n+1} - 1)\frac{B_{n+1}}{n+1}$ is equal to $2^6 \cdot 63 \cdot \frac{1}{42 \cdot 6} = 16$.

For $n = 6$, the value $2^{n+1}(2^{n+1} - 1)\frac{B_{n+1}}{n+1}$ is equal to $2^7 \cdot 127 \cdot \frac{0}{7} = 0$.

For $n = 7$, the value $2^{n+1}(2^{n+1} - 1)\frac{B_{n+1}}{n+1}$ is equal to $-2^8 \cdot 255 \cdot \frac{1}{30 \cdot 8} = -272$.

In other words, the identity holds for all $n \leq 7$ if, for $n = 0$, we use the second Bernoulli number $B_1^+ = \frac{1}{2}$. Moreover, we proved it for all even n. The general proof will be presented later. □

3.1.18. Constructing the triangle of Eulerian numbers of the first order modulo two (using black points for odd numbers and white points for even numbers), we can obtain several simple properties:

- *For Euler's triangle modulo two, we have the following statements:*
- *the left-hand side of the triangle consists of black points:* $E(n, 0) = 1$, $n \geq 0$;
- *the right-hand side of the triangle consists of white points, except the vertex of the triangle:* $E(0, 0) = 1$ *and* $E(n, n) = 0$, $n \in \mathbb{N}$;
- *the first-right descending diagonal of the triangle consists of black points:* $E(n, n-1) = 1$, $n \geq 1$.

It is also easy to prove the following fact:

- *In the first vertical of Euler's triangle, as well as in its second-right descending diagonal, even and odd numbers alternate.*

□ The first vertical of the triangle is formed from the numbers $E(n, 1), n = 1, 2, 3, \ldots$: $E(1, 1) = 0$, $E(2, 1) = 1$, $E(3, 1) = 4, \ldots$. As $E(n, 1) = 2^n - n - 1$, $n \in \mathbb{N}$, and 2^n is even for all positive integers n, then the parity of the number $E(n, 1)$ coincides with the parity of the number $n - 1$. So, $E(n, 1)$ is even for odd n, and $E(n, 1)$ is odd for even n.

The second-right descending diagonal of Euler's triangle is formed from the numbers $E(n, n-2)$, $n = 2, 3, 4, \ldots$: $E(2, 0) = 1$, $E(3, 1) = 4$, $E(4, 2) = 11, \ldots$. As $E(n, n-2) = E(n, 1)$, we have the same result: $E(n, n-2)$ is even for odd n, and $E(n, n-2)$ is odd for even n. □

Considering the isosceles triangle of Eulerian numbers of the first order, we can now get similar results:

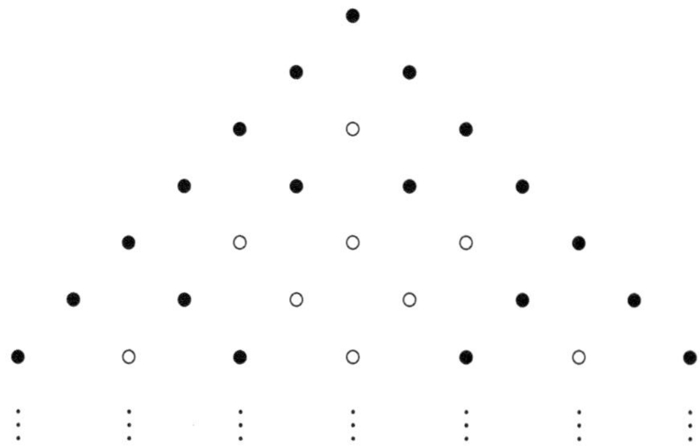

- *For the isosceles triangle of Eulerian numbers of the first order modulo two, we have the following statements:*
  - *the triangle is symmetric with respect to the main diagonal:* $E(n, k) = E(n, n-1-k)$, $n \geq 1$, $1 \leq k \leq n-1$;
  - *the sides of the triangle consist of black points:* $E(n, 0) = 1$, $n \geq 1$;
  - *in the first left and right descending diagonals of the triangle the black and the white points alternate:* $E(n, 1) = E(n, n-2)$ *are even for odd* n *and are odd for even* n;
  - *the main diagonal of this triangle consists of white points: the central Eulerian numbers of the first order* $E(2n+1, n)$ *are even.*

Exercises

1. Using the definition of the Eulerian numbers of the first order, find $E(n, k)$ for $n = 1, 2, 3, 4, 5, 6$ and $k = 0, 1, \ldots, n$.
2. Using the recurrent relation for the Eulerian numbers of the first order, find $E(n, k)$ for $n = 2, 3, 4, 5, 6$ and $k = 1, \ldots, n$. Continue the construction of the triangle of Eulerian numbers of the first order up to $n = 12$. Continue the construction of the isosceles triangle of Eulerian numbers of the first order up to $n = 12$.
3. Check that there are exactly 26 permutations of the fifth order with three descents: 15432, 21543, 25431, 31542, 32154, 32541,

35421, 41532, 42153, 42531, 43152, 43215, 43251, 43521, 45321, 51432, 52143, 52431, 53142, 53214, 53241, 53421, 54132, 54213, 54231, and 54312. How many ascents (excedances, maximal increasing runs, or readings) does each such permutation have? Find all permutations of the fifth order with one, two, and four descents.

4. Let $A_{n,k}$ be the number of permutations of the set $\{1, 2, \ldots, n\}$ with k maximal increasing runs. Find $A_{n,k}$ for $n = 1, 2, 3, 4$ and $k = 0, 1, \ldots, n$. Check that $A_{n,k} = E(n, k-1)$. Prove this fact.
5. Find the alternating sum of the elements of the nth row of Euler's triangle, $n = 0, 1, 2, \ldots, 12$. Check that the summation gives the numbers $2^{n+1}(2^{n+1}-1)\frac{B_{n+1}}{n+1}$, where B_{n+1} is the $(n+1)$th Bernoulli number. Prove this fact.
6. Check that the first column of Euler's triangle contains the numbers $2^n - n - 1$ if $n = 1, 2, \ldots, 12$. Prove this fact for any positive integer n using some combinatorial considerations for counting the number $E(n, 1)$ of permutations of the nth order with exactly one ascent.
7. Find the recurrent relation for the numbers $E(n, 2)$ of permutations of the nth order with exactly two ascents; prove that $E(n, 2) = 3^n - (n+1)2^n + \frac{n(n+1)}{2}$ using this recurrent relation. Find the recurrent relation for the numbers $E(n, 3)$ of permutations of the nth order with exactly three ascents; find the closed formula for $E(n, 3)$ using this recurrent relation.
8. Construct the triangle of Eulerian numbers of the first order modulo 3, modulo 4, and modulo 5. What can you say about the distribution of its elements?

3.2. Problems Related to Eulerian Numbers of the First Order

3.2.1. The Eulerian numbers of the first order arise in many combinatorial problems. The most important is the connection of Eulerian numbers of the first order $E(n, k)$, $k = 0, 1, \ldots, n-1$, with representations of ordinary powers x^n.

It is well known that the *Stirling numbers of the second kind* $S(n,k)$, as well as the *Stirling numbers of the first kind* $s(n,k)$, help to construct formulas expressing ordinary powers x^n in terms of *falling powers* (or *falling factorials*), $x^{\underline{n}} = x(x-1)\cdots(x-n+1)$, and *rising powers* (or *rising factorials*), $x^{\overline{n}} = x(x+1)\cdots(x+n-1)$, as well as vice versa (for proofs, see [Deza24]).

- *Using the Stirling numbers of the second kind, we can express the ordinary powers x^n of x through the falling powers $x^{\underline{k}}$ of x:*

$$x^n = S(n,0)x^{\underline{0}} + S(n,1)x^{\underline{1}} + S(n,2)x^{\underline{2}} \\ + \cdots S(n,3)x^{\underline{3}} + \cdots + S(n,n)x^{\underline{n}}.$$

In particular, it holds that

$$x^1 = S(1,0)x^{\underline{0}} + S(1,1)x^{\underline{1}};$$
$$x^2 = S(2,0)x^{\underline{0}} + S(2,1)x^{\underline{1}} + S(2,2)x^{\underline{2}};$$
$$x^3 = S(3,0)x^{\underline{0}} + S(3,1)x^{\underline{1}} + S(3,2)x^{\underline{2}} + S(3,3)x^{\underline{3}};$$
$$x^4 = S(4,0)x^{\underline{0}} + S(4,1)x^{\underline{1}} + S(4,2)x^{\underline{2}} + S(4,3)x^{\underline{3}} + S(4,4)x^{\underline{4}}.$$

As the coefficients of the right-hand side are the elements of the nth row of the triangle of Stirling numbers of the second kind, this triangle can be used to express the ordinary powers x^n through the falling powers. Exactly, for $n = 1, 2, 3, 4$, we get

$$x^1 = 0 \cdot x^{\underline{0}} + 1 \cdot x^{\underline{1}};$$
$$x^2 = 0 \cdot x^{\underline{0}} + 1 \cdot x^{\underline{1}} + 1 \cdot x^{\underline{2}};$$
$$x^3 = 0 \cdot x^{\underline{0}} + 1 \cdot x^{\underline{1}} + 3 \cdot x^{\underline{2}} + 1 \cdot x^{\underline{3}};$$
$$x^4 = 0 \cdot x^{\underline{0}} + 1 \cdot x^{\underline{1}} + 7 \cdot x^{\underline{2}} + 6 \cdot x^{\underline{3}} + 1 \cdot x^{\underline{4}}.$$

3.2.2. On the other hand, the ordinary powers can be represented through the rising powers.

- *Using the Stirling numbers of the second kind, we can express the ordinary powers x^n of x through the rising powers $x^{\overline{k}}$ of x:*

$$x^n = (-1)^n(S(n,0)x^{\overline{0}} - S(n,1)x^{\overline{1}} + S(n,2)x^{\overline{2}} - S(n,3)x^{\overline{3}} \\ + \cdots + (-1)^n S(n,n)x^{\overline{n}}).$$

In particular, it holds that

$$x^1 = -S(1,0)x^{\overline{0}} + S(1,1)x^{\overline{1}};$$

$$x^2 = S(2,0)x^{\overline{0}} - S(2,1)x^{\overline{1}} + S(2,2)x^{\overline{2}};$$

$$x^3 = -S(3,0)x^{\overline{0}} + S(3,1)x^{\overline{1}} - S(3,2)x^{\overline{2}} + S(3,3)x^{\overline{3}};$$

$$x^4 = S(4,0)x^{\overline{0}} - S(4,1)x^{\overline{1}} + S(4,2)x^{\overline{2}} - S(4,3)x^{\overline{3}} + S(4,4)x^{\overline{4}}.$$

Using the elements of the nth row of the triangle of Stirling numbers of the second kind for $n = 1, 2, 3, 4$, we get

$$x^1 = -0 \cdot x^{\underline{0}} + 1 \cdot x^{\underline{1}};$$

$$x^2 = 0 \cdot x^{\underline{0}} - 1 \cdot x^{\underline{1}} + 1 \cdot x^{\underline{2}};$$

$$x^3 = -0 \cdot x^{\underline{0}} + 1 \cdot x^{\underline{1}} - 3 \cdot x^{\underline{2}} + 1 \cdot x^{\underline{3}};$$

$$x^4 = 0 \cdot x^{\underline{0}} - 1 \cdot x^{\underline{1}} + 7 \cdot x^{\underline{2}} - 6 \cdot x^{\underline{3}} + 1 \cdot x^{\underline{4}}.$$

3.2.3. The Stirling numbers of the first kind can be used for representing the falling and rising powers through the ordinary powers.

- *Using the Stirling numbers of the first kind, we can express the falling powers $x^{\underline{n}}$ of x through the ordinary powers $x^0, x^1, \ldots, x^n$ of x:*

$$x^{\underline{n}} = s(n,0) + s(n,1)x + s(n,2)x^2 + \cdots + s(n,n-1)x^{n-1} + s(n,n)x^n.$$

In particular, it holds that

$$x^{\underline{1}} = s(1,0)x^0 + s(1,1)x^1;$$

$$x^{\underline{2}} = s(2,0)x^0 + s(2,1)x^1 + s(2,2)x^2;$$

$$x^{\underline{3}} = s(3,0)x^0 + s(3,1)x^1 + s(3,2)x^2 + s(3,3)x^3;$$

$$x^{\underline{4}} = s(4,0)x^0 + s(4,1)x^1 + s(4,2)x^2 + s(4,3)x^3 + s(4,4)x^4.$$

Using the triangle of Stirling numbers of the first kind, we obtain

$$x^{\underline{1}} = 0 \cdot x^0 + 1 \cdot x^1;$$

$$x^{\underline{2}} = 0 \cdot x^0 - 1 \cdot x^1 + 1 \cdot x^2;$$

$$x^{\underline{3}} = 0 \cdot x^0 2 \cdot x^1 - 3 \cdot x^2 + 1 \cdot x^3;$$
$$x^{\underline{4}} = 0 \cdot x^0 - 6 \cdot x^1 + 11 \cdot x^2 - 6 \cdot x^3 + 1 \cdot x^4.$$

- *Using the Stirling numbers of the first kind, we can express the rising powers $x^{\overline{n}}$ of x through the ordinary powers $x^0, x^1, \ldots, x^n$ of x:*

$$x^{\overline{n}} = (-1)^n(s(n,0) - s(n,1)x + s(n,2)x^2 - \cdots + (-1)^{n-1}s(n,n-1)x^{n-1} + (-1)^n s(n,n)x^n.$$

As, by definition, $s(n,k) = (-1)^{n+k}|s(n,k)|$, where $|s(n,k)|$ are the unsigned Stirling numbers of the first kind, we get that

$$x^{\overline{n}} = |s(n,0)| + |s(n,1)|x + |s(n,2)|x^2 - \cdots + |s(n,n-1)|x^{n-1} + |s(n,n)|x^n.$$

In particular, it holds that

$$x^{\overline{1}} = |s(1,0)|x^0 + |s(1,1)|x^1;$$
$$x^{\overline{2}} = |s(2,0)|x^0 + |s(2,1)|x^1 + |s(2,2)|x^2;$$
$$x^{\overline{3}} = |s(3,0)|x^0 + |s(3,1)|x^1 + |s(3,2)|x^2 + |s(3,3)|x^3;$$
$$x^{\overline{4}} = |s(4,0)|x^0 + |s(4,1)|x^1 + |s(4,2)|x^2 + |s(4,3)|x^3 + |s(4,4)|x^4.$$

Using the triangle of unsigned Stirling numbers of the first kind, we obtain

$$x^{\overline{1}} = 0 \cdot x^0 + 1 \cdot x^1,$$
$$x^{\overline{2}} = 0 \cdot x^0 + 1 \cdot x^1 + 1 \cdot x^2;$$
$$x^{\overline{3}} = 0 \cdot x^0 + 2 \cdot x^1 + 3 \cdot x^2 + 1 \cdot x^3;$$
$$x^{\overline{4}} = 0 \cdot x^0 + 6 \cdot x^1 + 11 \cdot x^2 + 6 \cdot x^3 + 1 \cdot x^4.$$

3.2.4. So, we can state that the Stirling numbers of the first and second kinds are coefficients of the changes between the standard basis $x^0, x^1, x^2, \ldots$, and the basis $x^{\underline{0}}, x^{\underline{1}}, x^{\underline{2}}, \ldots$, as well as the basis

$x^{\overline{0}}, x^{\overline{1}}, x^{\overline{2}}, \ldots$:

$$x^n = \sum_{k=0}^{n} S(n,k) x^{\underline{k}}, \quad \text{and} \quad x^n = \sum_{k=0}^{n} (-1)^{n+k} S(n,k) x^{\overline{k}};$$

$$x^{\underline{n}} = \sum_{k=0}^{n} s(n,k) x^k, \quad \text{and} \quad x^{\overline{n}} = \sum_{k=0}^{n} (-1)^{n+k} s(n,k) x^k.$$

If we add to the consideration the *Lah numbers* (or the *Stirling numbers of the third kind*) $L(n,k)$ (see, for example, [Deza24]), we get the possibility to represent the falling powers through the rising powers, and vice versa:

$$x^{\overline{n}} = \sum_{k=0}^{n} L(n,k) x^{\underline{k}}, \quad \text{and} \quad x^{\overline{n}} = \sum_{k=0}^{n} (-1)^{n+k} L(n,k) x^{\underline{k}}.$$

The main changes among the three considered systems x^n, $x^{\overline{n}}$, and $x^{\overline{n}}$ of powers are presented in the following table:

Change of basis	Coefficients
$x^n \to x^{\underline{n}}$	$S(n,k)$
$x^n \to x^{\overline{n}}$	$(-1)^{n+k} S(n,k)$
$x^{\underline{n}} \to x^n$	$s(n,k)$
$x^{\overline{n}} \to x^n$	$(-1)^{n+k} s(n,k)$
$x^{\overline{n}} \to x^{\underline{n}}$	$L(n,k)$
$x^{\underline{n}} \to x^{\overline{n}}$	$(-1)^{n+k} L(n,k)$

3.2.5. It is easy to see that the falling powers $x^{\underline{k}}$, $x \in \mathbb{R}$, define, up to scaling, the same polynomials as the *generalised binomial coefficients* $\binom{x}{k}$, $x \in \mathbb{R}$:

$$\binom{x}{k} = \frac{x(x-1)\cdots(x-k+1)}{1 \cdot 2 \cdot \cdots \cdot k} = \frac{x^{\underline{k}}}{k!}.$$

So, we can consider one more basis: $\binom{x}{0}, \binom{x}{1}, \binom{x}{2}, \ldots$. The changes between the standard basis $x^0, x^1, x^2, \ldots$ and the basis $\binom{x}{0}, \binom{x}{1}, \binom{x}{2}, \ldots$ are described by similar formulas, using the Stirling numbers of the first and second kinds:

$$x^n = \sum_{k=0}^{n} S(n,k) k! \binom{x}{k}, \quad \text{and} \quad \binom{x}{n} = \sum_{k=0}^{n} \frac{s(n,k)}{n!} x^k.$$

Moreover, the *binomial theorem* states that the binomial coefficients $\binom{n}{k}$, $n \in \mathbb{N}$, allow us to represent the *shifted power* $(x+1)^n$ through the ordinary powers x^k, $k = 0, 1, 2, \ldots, n$:

$$(x+1)^n = \sum_{k=0}^{n} \binom{n}{k} x^k.$$

3.2.6. It turns out that the Eulerian numbers of the first order allow us to represent the ordinary powers x^n of x as linear combinations of the generalised binomial coefficients

$$\binom{x+k}{n} = \frac{(x+k)(x+k-1)\cdots(x+k-n+1)}{1 \cdot 2 \cdot \cdots \cdot n} = \frac{(x+k)^{\underline{n}}}{n!},$$

where $x \in \mathbb{R}$, and $k = 0, 1, 2, \ldots, n-1$. This connection is given by the well-known *Worpitzky's identity.*

- *For any $x \in \mathbb{R}$ and any positive integer $n \geq 0$, Worpitzky's identity holds*:

$$x^n = E(n,0)\binom{x}{n} + E(n,1)\binom{x+1}{n} + E(n,2)\binom{x+2}{n} + \cdots + E(n,n-1)\binom{x+(n-1)}{n}.$$

□ The identity

$$x^n = \sum_{k=0}^{n-1} E(n,k)\binom{x+k}{n}, n \geq 0,$$

can be proved by induction using the recurrent relation for the Eulerian numbers of the first order.

For $n = 0$, we have that

$$E(0,0)\binom{x}{0} = 1 = x^0,$$

i.e., the statement holds. For $n = 1$, we have that

$$E(1,0)\binom{x}{1} = 1 \cdot x = x,$$

i.e., the statement holds. For $n = 2$, we have that

$$E(2,0)\binom{x}{2} + E(2,1)\binom{x+1}{2} = 1 \cdot \frac{x(x-1)}{2} + 1 \cdot \frac{(x+1)x}{2} = x^2,$$

i.e., the statement also holds.

Assume that the equality holds for a positive integer n:

$$x^n = E(n,0)\binom{x}{n} + E(n,1)\binom{x+1}{n} + E(n,2)\binom{x+2}{n} + \cdots + E(n,n-1)\binom{x+(n-1)}{n}.$$

Now, try to prove this equality for $n+1$. As

$$\binom{x+k}{n} = \frac{(x+k)(x+k-1)\cdots(x+k-n+1)}{1 \cdot 2 \cdot \cdots \cdot n} = \frac{(x+k)^{\underline{n}}}{n!},$$

we can rewrite the equality

$$x^n = E(n,0)\binom{x}{n} + E(n,1)\binom{x+1}{n} + E(n,2)\binom{x+2}{n} + \cdots + E(n,n-1)\binom{x+(n-1)}{n}$$

in the form of

$$n!x^n = E(n,0)x^{\underline{n}} + E(n,1)(x+1)^{\underline{n}} + E(n,2)(x+2)^{\underline{n}} + \cdots + E(n,n-1)(x+n)^{\underline{n}}.$$

Multiplying both parts of the last equality by $(n+1)x$, we get the identity

$$(n+1)!x^{n+1} = \sum_{k=0}^{n-1} E(n,k)(n+1)x \cdot (x+k)^{\underline{n}}$$
$$= \sum_{k=0}^{n-1} E(n,k)(n+1)((x+k-n)+(n-k)) \cdot (x+k)^{\underline{n}}$$

$$= \sum_{k=0}^{n-1} E(n,k)(n+1) \cdot (x+k)^{\underline{n+1}}$$

$$+ \sum_{k=0}^{n-1} E(n,k)(n+1)(n-k) \cdot (x+k)^{\underline{n}}.$$

It is easy to check that

$$(n+1)(x+k)^{\underline{n}} = (x+k+1)^{\underline{n+1}} - (x+k)^{\underline{n+1}}.$$

Therefore, we have that

$$\begin{aligned}(n+1)!x^{n+1} &= \sum_{k=0}^{n-1} E(n,k)(n+1) \cdot (x+k)^{\underline{n+1}} \\ &\quad + \sum_{k=0}^{n-1} E(n,k)(n-k) \cdot (x+k+1)^{\underline{n+1}} \\ &\quad - \sum_{k=0}^{n-1} E(n,k)(n-k) \cdot (x+k)^{\underline{n+1}} \\ &= \sum_{k=0}^{n-1} E(n,k)(k+1) \cdot (x+k)^{\underline{n+1}} \\ &\quad + \sum_{k=0}^{n-1} E(n,k)(n-k) \cdot (x+k+1)^{\underline{n+1}}.\end{aligned}$$

As $E(n,0) = E(n+1,0) = 1$, $E(n,n) = 0$ for $n \geq 0$, and

$$\begin{aligned}&\sum_{k=0}^{n-1} E(n,k)(n-k) \cdot (x+k+1)^{\underline{n+1}} \\ &\quad = \sum_{m=1}^{n} E(n,m-1)(n-m+1) \cdot (x+m)^{\underline{n+1}},\end{aligned}$$

we can state, using for $n \geq 1$ and $0 \leq k \leq n-1$ the recurrent relation

$$E(n+1,k) = (k+1)E(n,k) + (n+1-k)E(n,k-1),$$

that

$$
\begin{aligned}
(n+1)!x^{n+1} &= \sum_{k=0}^{n-1} E(n,k)(k+1)\cdot(x+k)^{\underline{n+1}} \\
&\quad + \sum_{k=1}^{n} E(n,k-1)(n-k+1)\cdot(x+k)^{\underline{n+1}} \\
&= E(n,0)x^{\underline{n+1}} + \sum_{k=1}^{n} E(n,k)(k+1)\cdot(x+k)^{\underline{n+1}} \\
&\quad + \sum_{k=1}^{n} E(n,k-1)(n-k+1)\cdot(x+k)^{\underline{n+1}} \\
&= E(n+1,0)x^{\underline{n+1}} + \sum_{k=1}^{n} ((k+1)E(n,k) \\
&\quad + (n-k+1)E(n,k-1))\cdot(x+k)^{\underline{n+1}} \\
&= \sum_{k=0}^{n} E(n+1,k)\cdot(x+k)^{\underline{n+1}}.
\end{aligned}
$$

So, we have proven that

$$
\begin{aligned}
x^{n+1} &= \frac{1}{(n+1)!}\sum_{k=0}^{n} E(n+1,k)\cdot(x+k)^{\underline{n+1}} \\
&= \sum_{k=0}^{n} E(n+1,k)\binom{x+k}{n+1}.
\end{aligned}
$$

□

In particular, for $n = 1, 2, 3, 4$, we have the following decompositions:

$$x^1 = E(1,0)\binom{x}{1};$$

$$x^2 = E(2,0)\binom{x}{2} + E(2,1)\binom{x+1}{2};$$

$$x^3 = E(3,0)\binom{x}{3} + E(3,1)\binom{x+1}{3} + E(3,2)\binom{x+2}{3};$$

$$x^4 = E(4,0)\binom{x}{4} + E(4,1)\binom{x+1}{4} + E(4,2)\binom{x+2}{4}$$
$$+E(4,3)\binom{x+3}{4}.$$

Using the elements of the nth row of Euler's triangle for $n = 1, 2, 3, 4$, we get that

$$x^1 = 1 \cdot \binom{x}{1};$$

$$x^2 = 1 \cdot \binom{x}{2} + 1 \cdot \binom{x+1}{2};$$

$$x^3 = 1 \cdot \binom{x}{3} + 4 \cdot \binom{x+1}{3} + 1 \cdot \binom{x+2}{3};$$

$$x^4 = 1 \cdot \binom{x}{4} + 11 \cdot \binom{x+1}{4} + 11 \cdot \binom{x+2}{4} + 1 \cdot \binom{x+3}{4}.$$

3.2.7. As $\binom{x+k}{n} = \frac{(x+k)(x+k-1)\cdots(x+k-n+1)}{n!} = \frac{(x+k)^{\underline{n}}}{n!}$, we get from Worpitzky's identity an expression that allows us to represent an ordinary power x^n of x using the Eulerian numbers of the first order $E(n,k)$ and the *shifted* falling powers $(x+k)^{\underline{n}}$, $k = 0, 1, 2, \ldots, n-1$.

- *The following identity holds*:

$$x^n = \frac{1}{n!}\left(E(n,0)x^{\underline{n}} + E(n,1)(x+1)^{\underline{n}} + E(n,2)(x+2)^{\underline{n}}\right.$$
$$\left. + \cdots + E(n,n-1)(x+n-1)^{\underline{n}}\right).$$

□ The decomposition

$$x^n = \frac{1}{n!}\sum_{k=0}^{n-1} E(n,k)(x+k)^{\underline{n}}$$

is an alternative form of Worpitzky's identity; we used this approach in the previous proof. □

In particular, for $n = 1, 2, 3, 4$, it holds that

$$x^1 = \frac{1}{1!} E(1,0) x^{\underline{1}};$$

$$x^2 = \frac{1}{2!} \left(E(2,0) x^{\underline{2}} + E(n,1)(x+1)^{\underline{2}} \right);$$

$$x^3 = \frac{1}{3!} \left(E(3,0) x^{\underline{3}} + E(3,1)(x+1)^{\underline{3}} + E(3,2)(x+2)^{\underline{3}} \right);$$

$$x^4 = \frac{1}{4!} \left(E(4,0) x^{\underline{4}} + E(4,1)(x+1)^{\underline{4}} + E(4,2)(x+2)^{\underline{4}} + E(4,3)(x+3)^{\underline{n}} \right).$$

In other words, we have

$$x^1 = 1 \cdot \frac{x}{1} x;$$

$$x^2 = 1 \cdot \frac{x(x-1)}{2!} + 1 \cdot \frac{(x+1)x}{2!};$$

$$x^3 = 1 \cdot \frac{x(x-1)(x-2)}{3!} + 4 \cdot \frac{(x+1)x(x-1)}{3!} + 1 \cdot \frac{(x+2)(x+1)x}{3!};$$

$$x^4 = 1 \cdot \frac{x(x-1)(x-2)(x-3)}{4!} + 11 \cdot \frac{(x+1)x(x-1)(x-2)}{4!} + 11 \cdot \frac{(x+2)(x+1)x(x-1)}{4!} + 1 \cdot \frac{(x+3)(x+2)(x+1)x}{4!}.$$

3.2.8. Going from the falling powers to the rising powers, we can obtain a modification of the last identity:

- *The following identity holds*:

$$x^n = \frac{1}{n!} \left(E(n,0) x^{\overline{n}} + E(n,1)(x-1)^{\overline{n}} + E(n,2)(x-2)^{\overline{n}} + \cdots + E(n,n-1)(x+n-1)^{\overline{n}} \right).$$

□ As $(x+k)^{\underline{n}} = (x+k)(x+k-1)\cdots(x+k-n+1) = (x+k-n+1)^{\overline{n}}$ and $E(n,k) = E(n, n-1-k)$, we have that

$$n! x^n = \sum_{k=0}^{n-1} E(n,k)(x+k)^{\underline{n}} = \sum_{k=0}^{n-1} E(n, n-1-k)(x+k-n+1)^{\overline{n}}$$

$$= \sum_{j=0}^{n-1} E(n,j)(x-j)^{\overline{n}}.$$

However, the same result can be obtained using other reasons. In fact,

$$\begin{aligned}(-x+k)^{\underline{n}} &= (-x+k)(-x+k-1)\cdots(-x+k-n+1)\\ &= (-1)^n(x-k)(x-k+1)\cdots(x-k+n-1)\\ &= (-1)^n(x-k)^{\overline{n}},\end{aligned}$$

and we get that

$$\begin{aligned}n!x^n = n!(-1)^n(-x)^n &= (-1)^n\sum_{k=0}^{n-1}E(n,k)(-x+k)^{\underline{n}}\\ = (-1)^n\sum_{k=0}^{n-1}E(n,k)(-1)^n(x-k)^{\overline{n}} &= \sum_{k=0}^{n-1}E(n,k)(x-k)^{\overline{n}}.\end{aligned}$$

So, we have proven that

$$x^n = \frac{1}{n!}\sum_{k=0}^{n-1}E(n,k)(x-k)^{\overline{n}}. \qquad \square$$

3.2.9. In terms of the generalised binomial coefficients, we obtain the following decomposition:

- *The following identity holds:*

$$x^n = E(n,0)\binom{x+n-1}{n} + E(n,1)\binom{x+n-2}{n} + \cdots + E(n,n-1)\binom{x}{n}.$$

$\square$ As $\binom{x+n-1-k}{n} = \frac{(x+n-1-k)(x+n-1-k-1)\cdots(x-k)}{n!} = \frac{(x-k)^{\overline{n}}}{n!}$, the equality

$$x^n = \sum_{k=0}^{n-1}E(n,k)\binom{x+n-1-k}{n}$$

is simply an alternative version of the equality

$$x^n = \frac{1}{n!}\sum_{k=0}^{n-1}E(n,k)(x-k)^{\overline{n}}. \qquad \square$$

3.2.10. Using Worpitzky's identity for a positive integer x, we get the expansion of the power $x^k, k \in \mathbb{N}$, of a positive integer x as a linear combination of binomial coefficients. This gives an interesting way to study the sums $1^k + 2^k + \cdots + n^k$ of positive integer powers of the first n positive integer numbers.

In fact, as

$$x^2 = \binom{x}{2} + \binom{x+1}{2},$$

then to find the sum $1^2+2^2+\cdots+n^2$ of the first n squares of positive integers, we can use the formula

$$\sum_{k=1}^{n} k^2 = \sum_{k=1}^{n} \left(\binom{k}{2} + \binom{k+1}{2}\right).$$

Since $\binom{n}{k} = \binom{n-1}{k-1} + \binom{n-1}{k}$ and $\binom{k}{k} = \binom{k-1}{k-1} = 1$, we have

$$\begin{aligned}\binom{n}{k} &= \binom{n-1}{k-1} + \binom{n-2}{k-1} + \binom{n-3}{k-1} + \cdots + \binom{k}{k-1} + \binom{k}{k} \\ &= \binom{n-1}{k-1} + \binom{n-2}{k-1} + \binom{n-3}{k-1} + \cdots + \binom{k}{k-1} + \binom{k-1}{k-1}.\end{aligned}$$

In particular,

$$\binom{n+1}{3} = \binom{n}{2} + \binom{n-1}{2} + \binom{n-3}{2} + \cdots + \binom{3}{2} + \binom{2}{2},$$

as well as

$$\binom{n+2}{3} = \binom{n+1}{2} + \binom{n}{2} + \binom{n-2}{2} + \cdots + \binom{3}{2} + \binom{2}{2}.$$

Therefore, it holds that

$$\begin{aligned}\sum_{k=1}^{n} \left(\binom{k}{2} + \binom{k+1}{2}\right) &= \left(\binom{2}{2} + \binom{3}{2} + \cdots + \binom{n}{2}\right) \\ &\quad + \left(\binom{2}{2} + \binom{3}{2} + \cdots + \binom{n+1}{2}\right) = \binom{n+1}{3} + \binom{n+2}{3} \\ &= \frac{(n+1)n(n-1)}{3!} + \frac{(n+2)(n+1)n}{3!} = \frac{n(n+1)(2n+1)}{6},\end{aligned}$$

and we get the classic formula

$$1^2 + 2^2 + \cdots + n^2 = \frac{n(n+1)(2n+1)}{6}.$$

Similarly, as

$$x^3 = \binom{x}{3} + 4 \cdot \binom{x+1}{3} + \binom{x+2}{3},$$

then to find the sum $1^3 + 2^3 + \cdots + n^3$ of cubes of the first n positive integers $1, 2, \ldots, n$ we can use the equality

$$\sum_{k=1}^{n} k^3 = \sum_{k=1}^{n} \left(\binom{k}{3} + 4 \cdot \binom{k+1}{3} + \binom{k+2}{2} \right).$$

Since it holds that

$$\binom{n+1}{4} = \binom{n}{3} + \binom{n-1}{3} + \binom{n-2}{3} + \cdots + \binom{3}{3},$$

$$\binom{n+2}{4} = \binom{n+1}{3} + \binom{n}{3} + \binom{n-1}{3} + \cdots + \binom{3}{3},$$

$$\binom{n+3}{4} = \binom{n+2}{3} + \binom{n}{3} + \binom{n-1}{3} + \cdots + \binom{3}{3},$$

we obtain that

$$\begin{aligned}
&\sum_{k=1}^{n} \left(\binom{k}{3} + 4 \cdot \binom{k+1}{3} + \binom{k+2}{2} \right) \\
&= \left(\binom{3}{3} + \binom{4}{3} + \cdots + \binom{n}{3} \right) + 4 \left(\binom{3}{3} + \binom{4}{3} \right. \\
&\quad \left. + \cdots + \binom{n+1}{3} \right) + \left(\binom{3}{3} + \binom{4}{3} + \cdots + \binom{n+2}{3} \right) \\
&= \binom{n+1}{4} + 4\binom{n+2}{4} + \binom{n+3}{4} \\
&= \frac{(n+1)n(n-1)(n-2)}{4!} + 4\frac{(n+2)(n+1)n(n-1)}{4!} \\
&\quad + \frac{(n+3)(n+2)(n+1)n}{4!} = \frac{n^2(n+1)^2}{4},
\end{aligned}$$

and we get the well-known formula

$$1^3+2^3+\cdots+n^3=\frac{n^2(n+1)^2}{4}=(1+2+\cdots+n)^2.$$

In general, we obtain the following identity:

- *For the sum of the* k*th powers of the first* n *positive integers* $1, 2, \ldots, n$, *the following identity holds:*

$$\begin{aligned}&1^k+2^k+\cdots+n^k\\&\quad=E(k,0)\binom{n+1}{k+1}+E(k,1)\binom{n+2}{k+1}+E(k,2)\binom{n+3}{k+1}\\&\qquad+\cdots+E(k,k-1)\binom{n+k}{k+1}.\end{aligned}$$

□ The proof is similar. As

$$\begin{aligned}\sum_{t=1}^{n}\binom{t+m}{k}&=\sum_{t=k-m}^{n}\binom{t+m}{k}\\&=\binom{k}{k}+\binom{k+1}{k}+\cdots+\binom{n+m}{k}=\binom{n+m+1}{k+1},\end{aligned}$$

we get that

$$\begin{aligned}\sum_{t=1}^{n}t^k&=\sum_{t=1}^{n}\sum_{m=0}^{k-1}E(k,m)\binom{t+m}{k}\\&=\sum_{m=0}^{k-1}E(k,m)\sum_{t=1}^{n}\binom{t+m}{k}=\sum_{m=0}^{k-1}E(k,m)\sum_{t=k}^{n}\binom{t+m}{k}\\&=\sum_{m=0}^{k-1}E(k,m)\binom{n+m+1}{k+1}=E(k,0)\binom{n+1}{k+1}\\&\quad+E(k,1)\binom{n+2}{k+1}+E(k,2)\binom{n+3}{k+1}\\&\quad+\cdots+E(k,k-1)\binom{n+k}{k+1}.\end{aligned}$$

□

Recall that the famous formula for the sum $1^k + 2^k + \cdots + n^k$ is the *Bernoulli formula* (or *Faulhaber's formula*):

$$1^k + 2^k + \cdots + n^k = B_k n + \frac{B_{k-1}}{2}\binom{k}{k-1}n^2 + \frac{B_{k-2}}{3}\binom{k}{k-2}n^3 + \cdots + \frac{B_1}{k}\binom{k}{1}n^k + \frac{n^{k+1}}{k+1},$$

where $B_1, \ldots, B_k$ are the *Bernoulli numbers* (in fact, *second*, or *original*, or B^+-numbers): $B_0 = 1$, $B_1 = \frac{1}{2}$, $B_2 = \frac{1}{6}$, $B_3 = 0$, $B_4 = -\frac{1}{30}$, $B_5 = 0$, $B_6 = \frac{1}{42}, \ldots$ (sequences A164555 / A027642 in the OEIS).

3.2.11. We can find another connection between the powers of positive integers and the Eulerian numbers of the first order. Now, let us consider the sequences

$$1, 1, 1, \ldots, 1, \ldots, \quad 1, 2, 3, \ldots, k, \ldots, \quad 1^2, 2^3, 3^2, \ldots, k^2, \ldots, \quad \ldots, \quad 1^n, 2^n, 3^n, \ldots, k^n, \ldots,$$

i.e., the sequences of non-negative powers of positive integers. It is easy to prove that the generating functions of these sequences have a very specific form.

Precisely, using the formula of the sum of infinite geometric progression, we get that

$$\sum_{k=1}^{\infty} 1 \cdot x^k = \frac{x}{1-x}, \quad |x| < 1.$$

The differentiation of the last equality gives $\sum_{k=1}^{\infty} k \cdot x^{k-1} = \frac{1}{(1-x)^2}$, and hence, we have that

$$\sum_{k=1}^{\infty} k^n \cdot x^k = \frac{x}{(1-x)^2}, \quad |x| < 1.$$

By repeating the differentiation and multiplication operations, we can check that

$$\sum_{k=1}^{\infty} k^2 \cdot x^k = \frac{x}{(1-x)^3}(1+x), \quad |x| < 1,$$

$$\sum_{k=1}^{\infty} k^3 \cdot x^k = \frac{x}{(1-x)^4}(1+4x+x^2), \quad |x|<1,$$

$$\sum_{k=1}^{\infty} k^4 \cdot x^k = \frac{x}{(1-x)^5}(1+11x+11x^2+x^3), \quad |x|<1.$$

The calculations can be continued, but from the existing expansions, it is already clear that the coefficients of the polynomials found in the numerators on the right-hand side are the Eulerian numbers of the first order. The corresponding expansions have the form

$$\sum_{k=1}^{\infty} k^n \cdot x^k = \frac{x}{(1-x)^{n+1}}(E(n,0)+E(n,1)x+E(n,2)x^2 + \cdots + E(n,n-1)x^{n-1}).$$

The polynomials

$$A_n(x) = E(n,0)+E(n,1)x+E(n,2)x^2+\cdots+E(n,n-1)x^{n-1}$$

are called the *Eulerian polynomials* and have many useful properties. We will talk about them later.

So, we have the following property:

- *For any positive integer n, there exists the identity*

$$\sum_{k=0}^{\infty} k^n x^k = \frac{E(n,0)x+E(n,1)x^2+E(n,2)x^3+\cdots+E(n,n-1)x^n}{(1-x)^{n+1}},$$

$$|x|<1.$$

□ Let us prove the identity

$$\sum_{k=0}^{\infty} k^n x^k = \frac{\sum_{j=0}^{n-1} E(n,j)x^j}{(1-x)^{n+1}} = \frac{xA_n(x)}{(1-x)^{n+1}}, \quad n \geq 1,$$

by induction. In fact, we have checked this identity earlier for small values of n:

$$\sum_{k=1}^{\infty} k \cdot x^k = \sum_{k=0}^{\infty} k \cdot x^k = \frac{x}{(1-x)^2} = \frac{xE(1,0)}{(1-x)^2}, \quad |x|<1;$$

$$\sum_{k=0}^{\infty} k^2 \cdot x^k = \frac{x(1+x)}{(1-x)^2} = \frac{x(E(2,0)+E(2,1)x)}{(1-x)^2}, \quad |x|<1,$$

$$\sum_{k=0}^{\infty} k^3 \cdot x^k = \frac{x}{(1-x)^4}(1+4x+x^2)$$

$$= \frac{x}{(1-x)^4}(E(3,0)+E(3,1)x+E(3,2)x^2), \quad |x|<1.$$

So, we have proven for $n = 1, 2, 3$ that

$$\sum_{k=0}^{\infty} k^n \cdot x^k = \frac{xA_n(x)}{(1-x)^{n+1}},$$

where $A_n(x) = E(n,0) + E(n,1)x + \cdots + E(n,n-1)x^{n-1} = \sum_{k=0}^{n-1} E(n,k)x^k$ is the nth Eulerian polynomial.

Note that for $n = 0$, it holds that $\sum_{k=1}^{\infty} 1 \cdot x^k = \sum_{k=1}^{\infty} k^0 \cdot x^k = x + x^2 + x^3 + \cdots + x^k + \cdots = \frac{x}{1-x}$, $|x| < 1$. As $E(0,0) = 1$, we get the equality $\sum_{k=1}^{\infty} k^0 \cdot x^k = \frac{xE(0,0)}{(1-x)^1} = \frac{xA_0(x)}{(1-x)^1}$. If we suppose that $0^0 = 0$, then we will have the identity $\sum_{k=0}^{\infty} k^n \cdot x^k = \frac{xA_n(x)}{(1-x)^{n+1}}$ also for $n = 0$; otherwise, the result should be corrected.

Now, let us suppose that the identity is true for a positive integer n:

$$\sum_{k=0}^{\infty} k^n \cdot x^k = \sum_{k=1}^{\infty} k^n \cdot x^k = \frac{xA_n(x)}{(1-x)^{n+1}}.$$

The differentiation of the last equality gives

$$\sum_{k=1}^{\infty} k^{n+1} \cdot x^{k-1} = \frac{(A_n(x) + xA_n'(x))(1-x) + xA_n(x)(n+1)}{(1-x)^{n+2}}$$

$$= \frac{x(1-x)A_n'(x) + (1+nx)xA_n(x)}{(1-x)^{n+2}}.$$

As $A_n(x) = \sum_{k=0}^{n-1} E(n,k)x^k$, we have $A_n'(x) = \sum_{k=1}^{n-1} kE(n,k)x^{k-1}$, and the polynomial

$$x\,(1-x)\,A_n'(x) + A_n(x)\,(1+n\,x)$$

takes the form

$$
\begin{aligned}
x(1-x)\sum_{k=1}^{n-1} kE(n,k)x^{k-1} &+ (1+nx)\sum_{k=0}^{n-1} E(n,k)x^k \\
&= \sum_{k=0}^{n-1} kE(n,k)x^k - \sum_{k=0}^{n-1} kE(n,k)x^{k+1} + \sum_{k=0}^{n-1} E(n,k)x^k \\
&\quad + n\sum_{k=0}^{n-1} E(n,k)x^{k+1} = \sum_{k=0}^{n-1}(k+1)E(n,k)x^k \\
&\quad + \sum_{k=0}^{n-1}((n-k)E(n,k)x^{k+1} \\
&= E(n,0) + \sum_{k=1}^{n}(k+1)E(n,k)x^k + \sum_{k=1}^{n}(n+1-k)E(n,k-1)x^k \\
&= E(n,0) + \sum_{k=1}^{n}((k+1)E(n,k) + (n+1-k)E(n,k-1))x^k \\
&= E(n+1,0) + \sum_{k=1}^{n} E(n+1,k)x^k = \sum_{k=0}^{n} E(n+1,k)x^k = A_{n+1}(x).
\end{aligned}
$$

So, we have that

$$\sum_{k=1}^{\infty} k^{n+1}\cdot x^{k-1} = \frac{A_{n+1}(x)}{(1-x)^{n+2}}.$$

Then, multiplying both sides of this equality by x, we obtain that

$$
\begin{aligned}
\sum_{k=0}^{\infty} k^{n+1}\cdot x^k = \sum_{k=1}^{\infty} k^{n+1}\cdot x^k &= \frac{xA_{n+1}(x)}{(1-x)^{n+2}} \\
&= \frac{E(n+1,0)x + E(n+1,1)x^2 + \cdots + E(n+1,n)x^n}{(1-x)^{n+2}}.
\end{aligned}
$$

□

3.2.12. Now, we can prove Worpitzky's identity using the last equality. Since

$$(1+x)^\alpha = \sum_{k=0}^{\infty}\binom{\alpha}{k}x^k = \sum_{k=0}^{\infty}\frac{\alpha\cdot(\alpha-1)\cdot\cdots\cdot(\alpha-k+1)}{k!}x^k$$

and

$$\binom{-n-1}{k} = \frac{(-n-1)\cdot(-n-2)\cdot\dots\cdot(-n-k)}{k!}$$
$$= (-1)^k \frac{(n+1)\cdot\dots\cdot(n+k)}{k!} = (-1)^k \binom{n+k}{k},$$

we get that

$$(1-x)^{-n-1} = \sum_{k=0}^{\infty} \binom{n+k}{k} x^k = \sum_{k=0}^{\infty} \binom{n+k}{n} x^k,$$

and hence, it holds that

$$\sum_{k=0}^{\infty} k^n \cdot x^k = \frac{xA_{n+1}(x)}{(1-x)^{n+1}} = (1-x)^{-n-1} \sum_{j=0}^{n-1} E(n,j) x^{j+1}$$
$$= \left(\sum_{k=0}^{\infty} \binom{n+k}{n} x^k\right) \left(\sum_{j=0}^{n-1} E(n,j) x^{j+1}\right)$$
$$= \left(\binom{0+n}{n} + \binom{1+n}{n} x + \dots + \binom{k+n}{n} x^k + \dots\right)$$
$$\cdot \left(E(n,0)x + E(n,1)x^2 + \dots + E(n,n-1)x^n\right)$$
$$= \left(\binom{n+0}{n} E(n,0)\right) x + \left(\binom{n+0}{n} E(n,1)\right.$$
$$\left. + \binom{n+1}{n} E(n,0)\right) x^2 + \dots + \left(\binom{n+0}{n} E(n,k-1)\right.$$
$$\left. + \binom{n+1}{n} E(n,k-2) + \dots + \binom{n+k-1}{n} E(n,0)\right) x^k$$
$$+ \dots = \sum_{k=0}^{\infty} \left(\sum_{j=0}^{k-1} \binom{n+k-j-1}{n} E(n,j)\right) x^k.$$

Comparing the coefficients at x^k and using the symmetry property $E(n,j) = E(n-j-1)$, as well as the property $\binom{k}{n} = 0$ for $0 \le k < n$, we obtain that

$$\begin{aligned}
k^n &= \sum_{j=0}^{k-1} \binom{n+k-j-1}{n} E(n,j) \\
&= \sum_{j=0}^{n-1} \binom{n+k-j-1}{n} E(n,n-j-1) = \sum_{t=0}^{n-1} \binom{k-t}{n} E(n,t) \\
&= \binom{k}{n} E(n,0) + \binom{k+1}{n} E(n,1) + \cdots + \binom{k+n-1}{n} E(n,n-1).
\end{aligned}$$

□

3.2.13. We can rewrite the identity

$$\sum_{k=0}^{\infty} k^n x^k = \frac{xA_n(x)}{(1-x)^{n+1}}$$

in the form

$$\sum_{k=1}^{\infty} k^n x^{k-1} = \frac{A_n(x)}{(1-x)^{n+1}}.$$

In other words, we can easily get the *Carlitz identity*

$$A_n(x) = (1-x)^{n+1} \sum_{k=0}^{\infty} (k+1)^n x^k,$$

which is closely connected with the theory of *Eulerian fractions*.

The classical *Eulerian fraction* $\alpha_n(z)$ is defined as

$$\alpha_n(x) = \sum_{k=0}^{\infty} (k+1)^n x^k = 1 + 2^n x + 3^n x^2 + \cdots + (k+1)^n x^k + \cdots,$$

$$n \in \mathbb{Z}, n \geq 0, x \in \mathbb{R}, |x| < 1.$$

It can be obtained by inverting the Carlitz identity:

$$\alpha_n(x) = \sum_{k=0}^{\infty} (k+1)^n x^k = \frac{A_n(x)}{(1-x)^{n+1}},$$

where $A_n(x) = E(n,0) + E(n,1)x + \cdots + E(n,n-1)x^{n-1} = (1 - x)^{n+1}\alpha_n(x)$.

The functions $\alpha_n(x)$ serve as a powerful tool in many applications, including number theory, combinatorics, and computational mathematics.

Exercises

1. Express Worpitzky's identity for $n = 4, 5, 6$ and check it by direct computations.
2. Write out the formulas expressing the ordinary power x^n, $n = 5, 6$ using the shifted falling powers and the Eulerian numbers of the first order.
3. Find the sum of the fourth powers of the first n positive integers $1, 2, \ldots, n$ using Worpitzky's identity.
4. Find the sum of the fifth powers of the first n positive integers $1, 2, \ldots, n$ using Worpitzky's identity.
5. Compare the complexity of calculating the sum of the natural powers of the first n positive integers using Worpitzky's identity and using the Bernoulli formula.
6. For $n = 4, 5, 6, 7$, check that $\sum_{k=1}^{\infty} k^n \cdot x^k = \frac{x}{(1-x)^{n+1}}(E(n, 0) + E(n, 1)x + \cdots + E(n, n-1)x^{n-1})$.
7. Find the values of the sums $\sum_{k=1}^{\infty} \frac{k}{2^k}$, $\sum_{k=1}^{\infty} \frac{k^3}{3^k}$, and $\sum_{k=1}^{\infty} \frac{k^5}{5^k}$.

3.3. Closed Formula for Eulerian Numbers of the First Order

3.3.1. For the Eulerian numbers of the first order, there exists the following closed formula:

- *For any $n \geq 0$ and any $0 \leq k \leq n-1$, it holds that*

$$E(n, k) = \binom{n+1}{0}(k+1)^n - \binom{n+1}{1}k^n + \cdots + (-1)^m \binom{n+1}{m}(k+1-m)^n + \cdots + (-1)^k \binom{n+1}{k}1^n.$$

□ The formula

$$E(n,k) = \sum_{j=0}^{k} (-1)^j \binom{n+1}{j} (k+1-j)^n, \quad n \geq 1, 0 \leq k \leq n-1,$$

can be easily proved by induction. Indeed, the formula holds for $E(n,0)$:

$$E(n,0) = \binom{n+1}{0} \cdot 1^n = 1.$$

It is easy to check that it also holds for $E(n,1)$. It was proven earlier that $E(n,1) = 2^n - n - 1$ for any positive integer n. So, it holds that

$$\begin{aligned}\binom{n+1}{0}2^n - \binom{n+1}{1}1^n &= 1 \cdot 2^n - (n+1) \cdot 1^n \\ &= 2^n - n - 1 = E(n,1).\end{aligned}$$

Now we can use the induction step for the general case. Assuming that we have the relations

$$\begin{aligned}E(n,k) &= \binom{n+1}{0}(k+1)^n - \binom{n+1}{1}k^n \\ &\quad + \cdots + (-1)^m \binom{n+1}{m}(k+1-m)^n \\ &\quad + \cdots + (-1)^k \binom{n+1}{k}, \\ E(n,k-1) &= \binom{n+1}{0}k^n - \binom{n+1}{1}(k-1)^n \\ &\quad + \cdots + (-1)^{m-1}\binom{n+1}{m-1}(k+1-m)^n \\ &\quad + \cdots + (-1)^{k-1}\binom{n+1}{k-1},\end{aligned}$$

and using the recurrent relation $E(n+1,k) = (k+1)E(n,k) + (n+1-k)E(n,k-1)$, let us prove that

$$
\begin{aligned}
E(n+1,k) = \binom{n+2}{0}(k+1)^{n+1} - \binom{n+2}{1}k^{n+1} \\
+ \cdots + (-1)^m \binom{n+2}{m}(k+1-m)^{n+1} \\
+ \cdots + (-1)^k \binom{n+2}{k}.
\end{aligned}
$$

In fact, it holds that

$$
\begin{aligned}
&(k+1)E(n,k) + (n+1-k)E(n,k-1) \\
&= \binom{n+1}{0}(k+1)^n(k+1) - k^n\left(\binom{n+1}{1}(k+1)\right. \\
&\quad \left. - \binom{n+1}{0}(n+1-k)\right) + (k-1)^n\left(\binom{n+1}{2}(k+1)\right. \\
&\quad \left. - \binom{n+1}{1}(n+1-k)\right) + \cdots + (-1)^j(k+1-j)^n \\
&\quad \times \left(\binom{n+1}{j}(k+1) - \binom{n+1}{j-1}(n+1-k)\right) \\
&\quad + \cdots + (-1)^k\left(\binom{n+1}{k}(k+1) - \binom{n+1}{k-1}(n+1-k)\right).
\end{aligned}
$$

Since

$$
\begin{aligned}
&\binom{n+1}{j}(k+1) - \binom{n+1}{j-1}(n+1-k) \\
&\quad = \frac{(n+1)!(k+1)}{j!(n+1-j)!} - \frac{(n+1)!(n+1-k)}{(j-1)!(n+2-j)!} \\
&\quad = \frac{(n+1)!}{j!(n+2-j)!}((k+1)(n+2-j) - (n+1-k)j),
\end{aligned}
$$

and

$$\begin{aligned}
&(k+1)(n+2-j)-(n+1-k)j\\
&\quad=(k+1)(n+2-j)-(n+2-k-1)j\\
&\quad=(n+2)(k+1)-j(k+1)-(n+2)j+(k+1)j\\
&\quad=(n+2)(k+1-j),
\end{aligned}$$

we get that

$$\begin{aligned}
&(-1)^j(k+1-j)^n\left(\binom{n+1}{j}(k+1)-\binom{n+1}{j-1}(n+1-k)\right)\\
&\quad=(-1)^j(k+1-j)^{n+1}\binom{n+2}{j}.
\end{aligned}$$

Using also that

$$\binom{n+1}{0}(k+1)^n(k+1)=\binom{n+2}{0}(k+1)^{n+1},$$

we get the equation

$$\begin{aligned}
&(k+1)E(n,k)+(n+1-k)E(n,k-1)\\
&\quad=(k+1)^{n+1}\binom{n+2}{0}-k^{n+1}\binom{n+2}{1}+(k-1)^{n+1}\binom{n+2}{2}\\
&\qquad+\cdots+(-1)^j(k+1-j)^{n+1}\binom{n+2}{j}+\cdots+(-1)^k\binom{n+2}{k}.
\end{aligned}$$

In other words, it holds that

$$\begin{aligned}
&E(n+1,k)\\
&\quad=(k+1)^{n+1}\binom{n+2}{0}-k^{n+1}\binom{n+2}{1}+(k-1)^{n+1}\binom{n+2}{2}\\
&\qquad+\cdots+(-1)^j(k+1-j)^{n+1}\binom{n+2}{j}+\cdots+(-1)^k\binom{n+2}{k}.
\end{aligned}$$

So, we have proved that the formula is true for $n+1$: $E(n+1,k)=\sum_{j=0}^{k}(-1)^j\binom{n+2}{j}(k+1-j)^n$. □

3.3.2. However, we can consider an alternative proof for this formula, which uses the decomposition

$$\sum_{k=0}^{\infty} k^n x^k = \frac{xA_n(x)}{(1-x)^{n+1}}.$$

In fact, as

$$(1-x)^{n+1} = \sum_{k=0}^{n+1} \binom{n+1}{k} (-1)^k x^k = \sum_{k=0}^{\infty} \binom{n+1}{k} (-1)^k x^k.$$

Then, it holds that

$$\begin{aligned} xA_n(x) &= (1-x)^{n+1} \sum_{k=0}^{\infty} k^n x^k \\ &= \left(\sum_{k=0}^{\infty} \binom{n+1}{k} (-1)^k x^k \right) \left(\sum_{k=0}^{\infty} k^n x^k \right), \end{aligned}$$

or

$$\begin{aligned} &E(n,0)x + E(n,1)x^2 + \cdots + E(n,n-1)x^n \\ &= \left(\binom{n+1}{0} \cdot 1^n \right) x + \left(\binom{n+1}{0} \cdot 2^n - \binom{n+1}{2} \cdot 1^n \right) x^2 \\ &\quad + \cdots + \left(\binom{n+1}{0} (k+1)^n - \binom{n+1}{1} k^n \right. \\ &\quad \left. + \cdots + (-1)^k \binom{n+1}{k} \cdot 1^n \right) x^{k+1} + \cdots . \end{aligned}$$

Comparing the coefficients at x^{k+1}, $k = 0, 1, 2, \ldots, n-1$, we get that

$$\begin{aligned} E(n,k) &= \binom{n+1}{0} (k+1)^n - \binom{n+1}{1} k^n + \cdots + (-1)^k \binom{n+1}{k} \cdot 1^n \\ &= \sum_{j=0}^{k} (-1)^j \binom{n+1}{j} (k+1-j)^n. \end{aligned}$$

3.3.3. For small values of k, the closed formula for the Eulerian numbers of the first order allows us to obtain relatively simple expressions for $E(n,k)$. In fact, we have already checked that the numbers $E(n,1)$ can be calculated using the formula

$$E(n,1) = \binom{n+1}{0}2^n - \binom{n+1}{1} = 2^n - n - 1.$$

For $k = 2$, we get the formula

$$E(n,2) = \binom{n+1}{0}3^n - \binom{n+1}{1}2^n + \binom{n+1}{2}.$$

As

$$\binom{n+1}{0}3^n - \binom{n+1}{1}2^n + \binom{n+1}{2} = 1\cdot 3^n - (n+1)2^n + \frac{(n+1)n}{2},$$

the numbers $E(n,2)$ can be calculated using the formula

$$E(n,2) = 3^n - (n+1)2^n + \frac{n^2+n}{2}.$$

For $k = 3$, we have the formula

$$E(n,3) = \binom{n+1}{0}4^n - \binom{n+1}{1}3^n + \binom{n+1}{2}2^n - \binom{n+1}{3}.$$

As

$$\binom{n+1}{0}4^n - \binom{n+1}{1}3^n + \binom{n+1}{2}2^n - \binom{n+1}{3}$$
$$= 4^n - (n+1)3^n + \frac{(n+1)n}{2}2^n - \frac{(n+1)n(n-1)}{6},$$

the numbers $E(n,3)$ can be calculated using the formula

$$E(n,3) = 4^n - (n+1)3^n + (n^2+n)2^{n-1} - \frac{(n+1)n(n-1)}{6}.$$

3.3.4. Alternatively, we can obtain the considered closed formulas using Worpitzky's identity:

$$x^n = E(n,n-1)\binom{x}{n} + E(n,n-2)\binom{x+1}{n}$$
$$+ \cdots + E(n,0)\binom{x+n-1}{n}$$

for small values of x. For $x = 1$, we get $1^n = E(n,0)\binom{n}{n}$, i.e., $E(n,0) = 1^n = 1$. For $x = 2$, we get

$$2^n = E(n,0)\binom{n+1}{n} + E(n,1)\binom{n}{n},$$

and as $E(n,0) = 1$, it holds that

$$E(n,1) = 2^n - E(n,0)\binom{n+1}{n} = 2^n - (n+1).$$

Similarly, for $x = 3$, we get that

$$3^n = E(n,0)\binom{n+2}{n} + E(n,1)\binom{n+1}{n} + E(n,2)\binom{n}{n},$$

and as $E(n,0) = 1$, $E(n,1) = 2^n - (n+1)$, we have

$$\begin{aligned} E(n,2) &= 3^n - E(n,1)\binom{n+1}{1} - E(n,0)\binom{n+2}{2} \\ &= 3^n - (2^n - (n+1))(n+1) - 1 \cdot \frac{(n+2)(n+1)}{2} \\ &= 3^n - (n+1)2^n + \frac{n^2+n}{2}. \end{aligned}$$

This approach gives another proof for the closed formula of the Eulerian numbers of the first order:

$$\begin{aligned} E(n,k) = (k+1)^n - (n+1)k^n + \frac{(n+1)n}{2}(k-1)^n \\ + \cdots + (-1)^k \frac{(n+1)n\cdots(n-k+2)}{k!}1^n. \end{aligned}$$

3.3.5. A list of simple closed formulas for the numbers $E(n,k)$ can be compiled using the symmetry property of the Eulerian numbers of the first order. Recall that $E(n,k) = E(n, n-1-k)$, i.e., $E(n,1) = E(n-2)$, $E(n,2) = E(n,n-3)$, $E(n,3) = E(n,n-4)$, and using the list of the simplest values of Eulerian numbers, we can obtain the following useful information:

- $E(0,0) = 1$; $E(n,0) = 1$ for $n \geq 1$;
- $E(n,n) = 0$ for $n \geq 1$, and $E(n,n+k) = 0$ for $n \geq 0$, $k \geq 1$;
- $E(n,n-1) = 1$ for $n \geq 1$;

- $E(n,1) = 2^n - n - 1$ for $n \geq 1$;
- $E(n,n-2) = 2^n - n - 1$ for $n \geq 2$;
- $E(n,2) = 3^n - (n+1)2^n + \frac{n^2+n}{2}$ for $n \geq 2$;
- $E(n,n-3) = 3^n - (n+1)2^n + \frac{n^2+n}{2}$ for $n \geq 3$;
- $E(n,3) = 4^n - (n+1)3^n + (n^2+n)2^{n-1} - \frac{(n+1)n(n-1)}{6}$ for $n \geq 3$;
- $E(n,n-4) = 4^n - (n+1)3^n + (n^2+n)2^{n-1} - \frac{(n+1)n(n-1)}{6}$ for $n \geq 4$.

Exercises

1. Check the closed formulas for $E(n,1)$, $E(n,2)$, and $E(n,3)$, for $n = 1,2,3,4,5$.
2. Find the closed formula for the numbers $E(n,4)$.
3. Find the recurrent relation for the numbers $E(n,4)$, and prove the closed formula for $E(n,4)$ using this recurrent relation.
4. Calculate $E(n,k)$, $n = 5,6,7,8$, $k = 1,2,3,4$, using the obtained closed formulas. Compare the complexity of these calculations with that of the calculations when considering $E(n,k)$ using the recurrent approach.
5. Using the closed formula for the Eulerian numbers of the first order $E(n,k)$, prove that $E(n,k) = E(n,n-1-k)$.
6. Formulate and check the rule for calculating the numbers $E(n,k)$ using the $(n+1)$th row of Pascal's triangle and the sequence of nth powers of positive integers.
7. Prove that $E(n,3) = \frac{6\cdot 4^n - 6(n+1)3^n + 3n(n+1)2^n - (n-1)n(n+1)}{6}$.

3.4. Generating Function of the Sequence of Eulerian Numbers of the First Order

3.4.1. Now, we try to find the generating function of the array of elements of Euler's triangle. It turns out that this problem is relatively simple if we consider the generating functions of finite sequences of the Eulerian numbers of the first order $E(n,k)$, $k = 0,1,2,\ldots,n$

with a fixed n, that is, polynomials of the form

$$A_n(x) = \sum_{k=0}^{n} E(n,k)x^k = E(n,0) + E(n,1)x + \cdots + E(n,n-1)x^{n-1} + E(n,n)x^n.$$

These polynomials are called the *Eulerian polynomials* and have a long and interesting history. We will talk about them in detail in a later section.

Using Euler's triangle, we can easily get these polynomials for small n:

$$A_0(x) = E(0,0) \equiv 1;$$
$$A_1(x) = E(1,0) + E(1,1)x \equiv 1;$$
$$A_2(x) = E(2,0) + E(2,1)x + E(2,2)x^2 = 1 + x;$$
$$A_3(x) = E(3,0) + E(3,1)x + E(3,2)x^2 + E(3,3)x^3 = 1 + 4x + x^2;$$
$$A_4(x) = E(4,0) + E(4,1)x + E(4,2)x^2 + E(4,3)x^3 + E(4,4)x^4 = 1 + 11x + 11x^2 + x^3.$$

For $n > 0$, the values $E(n,n)$ are equal to zero. It means that the sum over k in the definition of $A_n(x)$, $n > 0$, can be written with an upper index only up to $n-1$. It also means that the polynomials $A_n(x)$ are really of degree $n-1$ for $n > 0$. Moreover, as $E(n,0) = 1$ for any $n \geq 0$, the free term of any Eulerian polynomial is equal to 1.

3.4.2. Unfortunately, a simple compact formula for the Eulerian polynomials does not exist, but there are convenient recurrent relations for the construction of these polynomials.

- *The Eulerian polynomials $A_n(x)$, $n \geq 1$, can be computed using the recurrent relation*

$$A_n(x) = x\,(1-x)\,A'_{n-1}(x) + A_{n-1}(x)\,(1 + (n-1)\,x),$$
$$A_0(x) \equiv 1.$$

☐ This linear polynomial recurrence can be obtained by induction. In fact, as $A_0(x) \equiv 1$, $A_0'(x) \equiv 0$, and

$$\begin{aligned} A_1(x) &\equiv 1 = x(1-x)\cdot 0 + 1(1+0\cdot x) \\ &= x\,(1-x)\,A_0'(x) + A_0(x)\,(1+(1-1)\,x). \end{aligned}$$

Similarly, as $A_1(x) \equiv 1$, $A_1'(x) \equiv 0$, and

$$\begin{aligned} A_2(x) &= 1+x = x(1-x)\cdot 0 + 1(1+x) \\ &= x\,(1-x)\,A_1'(x) + A_1(x)\,(1+(2-1)\,x). \end{aligned}$$

As $A_2(x) = 1+x$, $A_2'(x) \equiv 1$, and

$$\begin{aligned} A_3(x) &= 1+4x+x^2 = x(1-x)\cdot 1 + (1+x)(1+2x) \\ &= x\,(1-x)\,A_2'(x) + A_2(x)\,(1+(3-1)\,x). \end{aligned}$$

As $A_3(x) = 1+4x+x^2$, $A_3'(x) = 4+2x$, and

$$\begin{aligned} A_4(x) &= 1+11x+11x^2+x^3 = x(1-x)\cdot(4+2x) \\ &\quad + (1+4x+x^2)(1+3x) \\ &= (1-x)\,A_3'(x) + A_3(x)\,(1+(4-1)\,x). \end{aligned}$$

In general, as $A_n(x) = \sum_{k=0}^{n} E(n,k)x^k$, $A_n'(x) = \sum_{k=0}^{n} kE(n, k)x^{k-1}$, and the polynomial $x\,(1-x)\,A_n'(x) + A_n(x)\,(1+n\,x)$ takes the form

$$\begin{aligned} &x(1-x)\sum_{k=0}^{n} kE(n,k)x^{k-1} + (1+nx)\sum_{k=0}^{n} E(n,k)x^k \\ &= \sum_{k=0}^{n} kE(n,k)x^k - \sum_{k=0}^{n} kE(n,k)x^{k+1} + \sum_{k=0}^{n} E(n,k)x^k \\ &\quad + n\sum_{k=0}^{n} E(n,k)x^{k+1} \\ &= \sum_{k=0}^{n}(k+1)E(n,k + \sum_{k=0}^{n}((n-k)E(n,h)x^{k+1} \end{aligned}$$

$$= E(n,0) + \sum_{k=1}^{n}(k+1)E(n,k)x^k + \sum_{k=1}^{n+1}(n+1-k)E(n,k-1)x^k$$

$$= E(n,0) + \sum_{k=1}^{n}(k+1)E(n,k)x^k + \sum_{k=1}^{n}(n+1-k)E(n,k-1)x^k$$

$$= E(n,0) + \sum_{k=1}^{n}((k+1)E(n,k) + (n+1-k)E(n,k-1))x^k$$

$$= E(n+1,0) + \sum_{k=1}^{n} E(n+1,k)x^k = \sum_{k=0}^{n} E(n+1,k)x^k$$

$$= \sum_{k=0}^{n+1} E(n+1,k)x^k = A_{n+1}(x).$$

So, we have proven that the recurrence holds for $n+1$:

$$A_{n+1}(x) = x(1-x)A_n^{'}(x) + A_n(x)(1+nx). \qquad \square$$

3.4.3. The last relation can be written in a more compact form:

$$\frac{A_{n+1}(x)}{(1-x)^{n+2}} = \left(\frac{xA_n(x)}{(1-x)^{n+1}}\right)^{'}.$$

In other words, we can easily remember this recurrent relation by considering a rational function $\alpha_n(x) = \frac{A_n(x)}{(x-1)^{n+1}}$. This function satisfies the simple recurrent relation

$$\alpha_{n+1}(x) = (x\alpha_n(x))', \quad \alpha_0(x) = \frac{1}{1-x}.$$

In fact, the function $\alpha_n(x)$ is simply the Eulerian fraction defined earlier:

$$\alpha_n(x) = \frac{A_n(x)}{(1-x)^{n+1}} = \sum_{k=0}^{\infty}(k+1)^n x^k, \quad n \geq 0.$$

In particular, we have that

$$\alpha_0(x) = \frac{A_0(x)}{1-x} = \frac{1}{1-x}$$
$$= 1 + 1\cdot x + 1\cdot x^2 + \cdots 1\cdot x^k + \cdots, \quad |x| < 1,$$

$$\alpha_1(x) = \frac{A_1(x)}{(1-x)^2} = \frac{1}{(1-x)^2}$$
$$= 1 + 2x + 3x^2 + \cdots (k+1)x^k + \cdots, \quad |x| < 1,$$
$$\alpha_2(x) = \frac{A_2(x)}{(1-x)^2} = \frac{1+x}{(1-x)^2}$$
$$= 1^2 + 2^2x + 3^2x^2 + \cdots (k+1)^2x^k + \cdots \; |x| < 1.$$

3.4.4. However, the numbers $E(n, k)$ depend on two parameters n and k, and it is interesting to get a two-dimensional function $A(x, y)$ which generates all triangle array of Eulerian numbers of the first order.

- *The two-dimensional exponential generating function $A(x, y)$ of the sequence of Eulerian numbers of the first order has the form $\frac{1-x}{e^{(x-1)y}-x}$, i.e., it holds that*

$$\frac{1-x}{e^{(x-1)y} - x} = \sum_{n,k=0}^{\infty} E(n,k)x^k \frac{y^n}{n!}.$$

It is easy to see that

$$\sum_{n,k=0}^{\infty} E(n,k)x^k \frac{y^n}{n!} = \sum_{n=0}^{\infty} \left(\sum_{k=0}^{n} E(n,k)x^k \right) \frac{y^n}{n!} = \sum_{n=0}^{\infty} A_n(x) \frac{y^n}{n!},$$

where $A_n(x) = \sum_{k=0}^{n} E(n,k)x^k$ is the nth Eulerian polynomial. So, in order to obtain the two-dimensional exponential generating function of the array of Eulerian numbers of the first order, we only need to prove the following statement.

- *The exponential generating function $A_x(y) = A(x, y)$ of the sequence $A_n(x)$, $n = 0, 1, 2, \ldots$ of Eulerian polynomials $A_n(x) = \sum_{k=0}^{n} E(n,k)x^k$ has the form $\frac{1-x}{e^{(x-1)y}-x}$, i.e., it holds that*

$$\frac{1-x}{e^{(x-1)y} - x} = \sum_{n=0}^{\infty} A_n(x) \frac{y^n}{n!}.$$

□ We have proven that

$$\frac{xA_n(x)}{(1-x)^{n+1}} = \sum_{k=0}^{\infty} k^n x^k, \quad n \geq 1, \quad \text{and} \quad \frac{xA_0(x)}{1-x} = \sum_{k=1}^{\infty} x^k.$$

Then, as $e^t = \sum_{j=0}^{\infty} \frac{t^j}{j!}$, we get that

$$\begin{aligned}
&\sum_{n=0}^{\infty} \frac{xA_n(x)}{(1-x)^{n+1}} \frac{y^n}{n!} \\
&\quad = (x + x^2 + \cdots + x^k + \cdots)\frac{y^0}{0!} + (1 \cdot x + 2x^2 \\
&\quad + \cdots + kx^k + \cdots)\frac{y^1}{1!} + (1^2 \cdot x + 2^2x^2 + \cdots + k^2x^k + \cdots)\frac{y^2}{2!} \\
&\quad + \cdots + (1^n \cdot x + 2^n x^2 + \cdots + k^n x^k + \cdots)\frac{y^n}{n!} \\
&\quad + \cdots = \left(1^0\frac{y^0}{0!} + 1^1\frac{y^1}{1!} + 1^2\frac{y^2}{2!} + \cdots + 1^n\frac{y^n}{n!} + \cdots\right)x \\
&\quad + \left(2^0\frac{y^0}{0!} + 2^1\frac{y^1}{1!} + 2^2\frac{y^2}{2!} + \cdots + 2^n\frac{y^n}{n!} + \cdots\right)x^2 \\
&\quad + \cdots + \left(k^0\frac{y^0}{0!} + k^1\frac{y^1}{1!} + k^2\frac{y^2}{2!} + \cdots + k^n\frac{y^n}{n!} + \cdots\right)x^k \\
&\quad + \cdots = e^y x + e^{2y}x^2 + \cdots + e^{ky}x^k + \cdots = 1 + e^y x + (e^y x)^2 \\
&\quad + \cdots + (e^y x)^k + \cdots = \frac{e^y x}{1 - e^y x}.
\end{aligned}$$

So, for $t = \frac{y}{1-x}$, we have that

$$\begin{aligned}
\sum_{n=0}^{\infty} A_n(x)\frac{t^n}{n!} &= \sum_{n\geq 0} \frac{A_n(x)}{(1-x)^n}\frac{y^n}{n!} = \frac{1-x}{x}\sum_{n\geq 0}\frac{xA_n(x)}{(1-x)^{n+1}}\frac{y^n}{n!} \\
&= \frac{1-x}{x} \cdot \frac{e^y x}{1-e^y x} = \frac{(1-x)e^{t(1-x)}}{1-e^{t(1-x)}x} = \frac{1-x}{e^{(x-1)t} - x}. \quad \square
\end{aligned}$$

3.4.5. The above-considered generating function $A(x, y)$ for the array of Eulerian numbers of the first order have the disadvantage of being asymmetric. Everything becomes easier if we introduce the *symmetric Eulerian numbers of the first order* $E^s(n,k)$, defined by

$$E^s(n,k) = E(n+k+1,k).$$

By definition, $E^s(n,0) = E(n+1,0)$, $E^s(n,1) = E(n+2,1)$, $E^s(n,2) = E(n+3,2), \ldots$, i.e., the triangle of symmetric Eulerian numbers $E^s(n,k)$, $n,k \geq 0$, is obtained from Euler's triangle by sliding all columns (including the zeroth column) upwards:

$n \backslash k$	0	1	2	3	4	5
0	1	1	1	1	1	1
1	1	4	11	26	57	120
2	1	11	66	302	1191	4293
3	1	26	302	2416	15619	88234

- *For the array* $E^s(n,k)$, $n,k \geq 0$, *of symmetric Eulerian numbers of the first order, it holds that*

$$\sum_{n,k\geq 0} E^s(n,k)\frac{x^k y^n}{(n+k+1)!} = \sum_{n,k\geq 0} E(n+k+1,k)\frac{x^k y^n}{(n+k+1)!}$$
$$= \frac{e^x - e^y}{xe^y - ye^x}.$$

□ In fact,

$$\sum_{n,k=0}^{\infty} E^s(n,k)\frac{x^k y^n}{(n+k+1)!} = \sum_{n,k\geq 0} E(n+k+1,k)\frac{x^k y^n}{(n+k+1)!}.$$

Let $t = n+k+1$. As $n \geq 0$, $t \geq k+1 \geq 1$. So, in order to obtain the summation over all $t \geq 0$, we should add the summand ± 1, corresponding to $E(0,0) = 1$. Then, it holds that

$$\sum_{n,k=0}^{\infty} E^s(n,k)\frac{x^k y^n}{(n+k+1)!} = \frac{1}{y}\sum_{k\geq 0, t\geq k+1} E(t,k)\cdot\left(\frac{x}{y}\right)^k \frac{y^t}{t!}$$
$$= \frac{1}{y}\left(\sum_{k\geq 0,t\geq 0} E(t,k)\cdot\left(\frac{x}{y}\right)^k \frac{y^t}{t!} - 1\right) = \frac{1}{y}\left(A\left(\frac{x}{y}, y\right) - 1\right)$$

$$= \frac{1}{y}\left(\frac{1-\frac{x}{y}}{e^{(\frac{x}{y}-1)y}-\frac{x}{y}}-1\right) = \frac{1}{y}\left(\frac{y-x}{ye^{x-y}-x}-1\right)$$

$$= \frac{1}{y}\cdot\frac{y-ye^{x-y}}{ye^{x-y}-x} = \frac{e^y-e^x}{ye^x-xe^y} = \frac{e^x-e^y}{xe^y-ye^x}. \qquad \square$$

3.4.6. We have considered well-known generating functions connected with Euler's triangle. However, there are other generating functions representing some special sequences of Eulerian numbers of the first order.

For the zeroth column of the triangle, i.e., for the sequence $E(n,0)$, $n \geq 0$, consisting of only unities (sequence A000012 in the OEIS), the generating function has the form $\frac{1}{1-x}$, i.e., it holds that

$$\frac{1}{1-x} = 1 + 1\cdot x + 1\cdot x^2 + \cdots + 1\cdot x^n + \cdots, \quad |x| < 1,$$

while the exponential generation function of this sequence is e^x:

$$e^x = 1 + 1\cdot\frac{x}{1!} + 1\cdot\frac{x^2}{2} + \cdots + 1\cdot\frac{x^n}{n!} + \cdots, \quad x \in \mathbb{R}.$$

For the first column of the triangle, i.e., for the sequence $E(n,1)$, $n \geq 2$, starting with the elements $1, 4, 11, 26, 57, 120, 247, 502, 1013, 2036, \ldots$ (sequence A000295 in the OEIS), we have the following result:

- *The generating function of the sequence $E(n,1)$, $n \geq 2$, has the form $\frac{1}{(1-2x)(x-1)^2}$, i.e., it holds that*

$$\frac{1}{(1-2x)(x-1)^2} = 1 + 4x + 11x^2 + 26x^3 + \cdots, \quad |x| < \frac{1}{2}.$$

$\square$ We have proven that $E(n+1,1) = 2E(n,1) + n$. Therefore, $E(n+2,1) = 2E(n+1,1) + (n+1)$, and $E(n+2,1) = 3E(n+1,1) - 2E(n,1) + 1$. Similarly, $E(n+3,1) = 3E(n+2,1) - 2E(n+1,1) + 1$, and $E(n+3,1) = 4E(n+2,1) - 5E(n,1) + 2E(n,1)$. So, we have obtained for the sequence $E(n,1)$, $n \geq 2$, the linear recurrent relation of the third order

$$E(n+3,1) - 4E(n+2,1) + 5E(n,1) - 2E(n,1) = 0$$

with constant coefficients $b_0 = 1$, $b_1 = -4$, $b_2 = 5$, $b_3 = -2$, and with initial conditions $E(2,1) = 1$, $E(3,1) = 4$, $E(4,1) = 11$. Therefore (see Chapter 1), the generating function of the sequence $E(n,1)$, $n \geq 2$, has the form $\frac{f(x)}{g(x)}$, where $g(x) = b_0 + b_1x + b_2x^2 + b_3x^3$, and $f(x) = a_0 + a_1x + a_2x^2$ with $a_0 = b_0E(2,1) = 1$, $a_1 = b_0E(3,1) + b_1E(2,1) = 0$, $a_2 = b_0E(4,1) + b_1E(3,1) + b_2E(2,1) = 0$. As $g(x) = 1 - 4x + 5x^2 - 2x^3 = (x-1)^2(1-2x)$, we obtain that $\frac{f(x)}{g(x)} = \frac{1}{(x-1)^2(1-2x)}$. The decomposition $\frac{1}{(x-1)^2(1-2x)} = \sum_{n=0}^{\infty} E(n+2,1)x^n$ holds for $|x| < \frac{1}{2}$ since $\frac{1}{2}$ is the minimal root of the polynomial $g(x)$. □

Similarly, for the second column of the triangle, i.e., for the sequence $E(n,2)$, $n \geq 3$, starting with the elements 1, 11, 66, 302, 1191, 4293, 14608, 47840, 152637, 478271, ... (sequence A000460 in the OEIS), the generating function has the form $\frac{1+x-4x^2}{(1-x)^3(1-2x)^2(1-3x)}$, i.e, it holds that

$$\frac{1+x-4x^2}{(1-x)^3(1-2x)^2(1-3x)}$$
$$= 1 + 11x + 66x^2 + 302x^3 + \cdots, |x| < \frac{1}{3},$$

while for the third column of the triangle, i.e., for the sequence $E(n,3)$, $n \geq 4$, starting with the elements $1, 26, 302, 2416, 15619, 88234, 455192, 2203488, 10187685, 45533450, \ldots$ (sequence A000498 in the OEIS, see also A066912), we obtain the generating function $\frac{1+6x-43x^2+44x^3+52x^4-72x^5}{(x-1)^4(1-2x)^3(1-3x)^2(1-4x)}$:

$$\frac{1+6x-43x^2+44x^3+52x^4-72x^5}{(x-1)^4(1-2x)^3(1-3x)^2(1-4x)}$$
$$= 1 + 26x + 302x^2 + 2416x^3 + \cdots, \quad |x| < \frac{1}{4}.$$

Exercises

1. Find the value of the Euler polynomial $A_n(x)$, $n = 0, 1, 2, 3, 4$,5, at $x = 0$, x =1, $x = -1$, x =0.5, and $x = -0,5$. Prove that $A_n(1) = n!$. Prove that $A_n(-1) = \frac{2^{n+1}(2^{n+1}-1)B_{n+1}}{n+1}$, where B_{n+1} is the $(n+1)$th Bernoulli number.

2. Find several values of the variables x and y for which the series $\sum_{n=0}^{\infty} A_n(x)\frac{y^n}{n!}$ is convergent. Construct infinite sums corresponding to several values of the variables x and y, for which the series $\sum_{n=0}^{\infty} A_n(x)\frac{y^n}{n!}$ is convergent, and find their values.
3. Find the generating function for the following sequences: $E(n,2)$, $n = 3,4,5,\ldots$; $E(n,n-3)$, $n = 3,4,5,\ldots$; $E(n,n-2)$, $n = 2,3,4,\ldots$; and $E(n,n-1)$, $n = 1,2,3,\ldots$.
4. Prove that the generating function of the sequence $E(n,1)$, $n \geq 0$, starting with the elements $0,0,1,4,11,26,57,120,247,502,\ldots$, has the form $\frac{x^2}{(1-2x)(1-x)^2}$, while the exponential generating function of this sequence is $e^x(e^x-1-x)$.
5. Prove that the generating function of the sequence $E(n,2)$, $n \geq 0$, starting with the elements 0, 0, 0, 1, 11, 66, 302, 1191, 4293, 14608, ..., has the form $\frac{x^3(1+x-4x^2)}{(1-x)^3(1-2x)^2(1-3x)}$, while the exponential generating function of this sequence is $e^x(e^{2x}-(1+2x)e^x+x+\frac{x^2}{2})$.
6. Prove that the generating function of the sequence $E(n,3)$, $n \geq 0$, starting with the elements 0, 0, 0, 0, 1, 26, 302, 2416, 15619, 88234, ..., has the form $\frac{x^4(1+6x-43x^2+44x^3+52x^4-72x^5)}{(x-1)^4(1-2x)^3(1-3x)^2(1-4x)}$, while the exponential generating function of this sequence is $e^x(e^{3x}-(1+3x)e^{2x}+2(x+\frac{2x^2}{2!})e^x-\frac{x^2}{2!}-\frac{x^3}{3!})$.

3.5. Properties of Eulerian Numbers of the First Order

Sums with Eulerian numbers of the first order

3.5.1. In this section, we summarise the already known and some new properties of the Eulerian numbers of the first order. Let's start with the simplest summation formulas.

- *The sum of the numbers $E(n,0)$, $E(n,1),\ldots,E(n,n)$, located in the nth row of Euler's triangle, is equal to $n!$:*

$$E(n,0)+E(n,1)+E(n,2)+\cdots+E(n,n-1)=n!.$$

□ We have proven this property earlier; this statement for a positive integer n follows from the fact that the considered sum lists all permutations of the nth order. Moreover, we put $E(0,0) = 1$ to avoid conflict with the empty sum convention, so $E(0,0) = 0!$. □

3.5.2. For the alternating sum of numbers $E(n,0)$, $E(n,1), \ldots,$ $E(n,n)$, the situation is much more complicated.

- *The alternating sum of numbers* $E(n,0)$, $E(n,1), \ldots, E(n,n)$, *located in the* n*th row of Euler's triangle, is related to the* $(n+1)$*th Bernoulli number* B_{n+1}:

$$E(n,0) - E(n,1) + E(n,2) - \cdots + (-1)^{n-1}E(n,n-1)$$
$$= \frac{2^{n+1}(2^{n+1}-1)B_{n+1}}{n+1}, \quad n \geq 1.$$

□ In fact, for $n = 1$, we get that

$$\frac{2^2(2^2-1)B_2}{2} = 2 \cdot 3 \cdot \frac{1}{6} = 1 = E(1,0);$$

for $n = 2$, the situation is trivial:

$$\frac{2^3(2^3-1)B_3}{3} = 0 = E(2,0) - E(2,1);$$

for $n = 3$, we get the equation

$$\frac{2^4(2^4-1)B_4}{4} = -4 \cdot 15 \cdot \frac{1}{30} = -2 = E(3,0) - E(3,1) + E(3,2);$$

for $n = 4$, we get again zeros on both sides of the equation, etc.

Note that, for $n = 0$, we have the equality $\frac{2^1(2^1-1)B_1^+}{1} = 2 \cdot 1 \cdot \frac{1}{2} = 1 = E(0,0)$; therefore, this formula can be generalised to the zero value of the index, but in this case, we should use the sequence of second Bernoulli numbers.

In general, it is enough to take a look at the exponential generating function of the sequence of Eulerian numbers of the first order.

Since

$$E(n,0) - E(n,1) + E(n,2) - \cdots + (-1)^{n-1}E(n,n-1) = A_n(-1),$$

then using the value $x = -1$ in the formula $\frac{1-x}{e^{(x-1)y}-x} = \sum_{n=0}^{\infty} A_n(x)\frac{y^n}{n!}$, we get that

$$\frac{2}{e^{-2y}+1} = \sum_{n=0}^{\infty} A_n(-1)\frac{y^n}{n!}.$$

On the other hand, the exponential generating function of the sequence of Bernoulli numbers has the form $\frac{x}{e^x-1}$, i.e.,

$$\frac{x}{e^x-1} = B_0 + \frac{B_1}{1!}x + \frac{B_2}{2!}x^2 + \cdots + \frac{B_k}{k!}x^k + \cdots, \quad |x| < 2\pi.$$

As

$$\frac{x}{e^x+1} = \frac{x(e^x-1)}{e^{2x}-1} = \frac{xe^x+x-2x}{e^{2x}-1} = \frac{x}{e^x-1} - \frac{2x}{e^{2x}-1},$$

$$\begin{aligned}
\frac{x}{e^x+1} &= \left(1 + B_1x + \frac{B_2}{2!}x^2 + \frac{B_3}{3!}x^3 + \frac{B_4}{4!}x^4 + \cdots\right) \\
&\quad - \left(1 + B_1(2x) + \frac{B_2}{2!}(2x)^2 + \frac{B_3}{3!}(2x)^3 + \frac{B_4}{4!}(2x)^4 + \cdots\right) \\
&\quad - B_1x - \frac{B_2(2^2-1)}{2!}x^2 - \frac{B_3(2^3-1)}{3!}x^3 - \frac{B_4(2^4-1)}{4!}x^4 + \cdots \\
&= \frac{x}{2} - \frac{B_2(2^2-1)}{2!}x^2 - \frac{B_4(2^4-1)}{4!}x^4 - \frac{B_6(2^6-1)}{6!}x^6 \\
&\quad + \cdots + (-1)^k\frac{B_{2k}(2^{2k}-1)}{(2k)!}x^{2k} + \cdots.
\end{aligned}$$

So, it holds that

$$\begin{aligned}
\frac{y}{e^y+1} &= \frac{y}{2} - \frac{B_2(2^2-1)}{2!}y^2 - \frac{B_4(2^4-1)}{4!}y^4 - \frac{B_6(2^6-1)}{6!}y^6 \\
&\quad + \cdots + (-1)^k\frac{B_{2k}(2^{2k}-1)}{(2k)!}y^{2k} + \cdots.
\end{aligned}$$

Therefore, we have that

$$\frac{1}{e^y+1} = \frac{1}{2} - \frac{B_2(2^2-1)}{2!}y - \frac{B_4(2^4-1)}{4!}y^3 - \frac{B_6(2^6-1)}{6!}y^5 + \cdots + (-1)^k\frac{B_{2k}(2^{2k}-1)}{(2k)!}y^{2k-1} + \cdots.$$

Replacing y with $-2y$, multiplying both sides of the resulting identity by 2, and conducting several simple transformations, we get the following result:

$$\frac{2}{e^{-2y}+1} = 1 + \frac{B_2 2^2(2^2-1)}{1!2}y + \frac{B_4 2^4(2^4-1)}{3!4}y^3 + \frac{B_6 2^6(2^6-1)}{5!6}y^5 + \cdots + \frac{B_{2k}2^{2k}(2^{2k}-1)}{(2k-1)!2k}y^{2k-1} + \cdots.$$

As all Bernoulli numbers with odd indices greater than one are equal to zero, we can state that

$$\frac{2}{e^{-2y}+1} = 1 + \sum_{n=1}^{\infty}\frac{B_{n+1}2^{n+1}(2^{n+1}-1)}{n+1}\cdot\frac{y^n}{n!}.$$

Comparing the coefficients at the same powers y^n, $n = 1, 2, 3, \ldots$, of the two available decompositions,

$$\frac{2}{e^{-2y}+1} = \sum_{n=0}^{\infty} A_n(-1)\frac{y^n}{n!} = 1 + \sum_{n=1}^{\infty}\frac{B_{n+1}2^{n+1}(2^{n+1}-1)}{n+1}\cdot\frac{y^n}{n!},$$

of the function $\frac{2}{e^{-2y}+1}$, we get the following result:

$$\frac{B_{n+1}2^{n+1}(2^{n+1}-1)}{n+1} = A_n(-1) = \sum_{k=0}^{n}(-1)^k E(n,k), \quad n \geq 1.$$

As was noted above, this statement is also true for $n = 0$ if we use the second Bernoulli numbers B^+_{n+1}, with $B^+_1 = \frac{1}{2}$: $E(0,0) = 1 = 2\cdot\frac{1}{2} = \frac{2^{0+1}(2^{0+1}-1)B^+_{0+1}}{0+1}$. □

3.5.3. There is another summation formula unexpectedly connecting the Eulerian numbers of the first order and the Bernoulli numbers; it contains binomial coefficients.

- *The alternating sum of the values* $\frac{E(n,k)}{\binom{n}{k}}$ *obtained by sequential division of the elements of the nth row of Euler's triangle by the corresponding elements of the nth row of Pascal's triangle, can be expressed using the nth Bernoulli number:*

$$\frac{E(n,0)}{\binom{n}{0}} - \frac{E(n,1)}{\binom{n}{1}} + \frac{E(n,2)}{\binom{n}{2}} - \cdots + (-1)^k \frac{E(n,k)}{\binom{n}{k}} + \cdots + (-1)^n \frac{E(n,n)}{\binom{n}{n}} = (n+1)B_n, \quad n \geq 2.$$

□ In fact, for $n = 2$, we have that

$$\frac{E(2,0)}{\binom{2}{0}} - \frac{E(2,1)}{\binom{2}{1}} + \frac{E(2,2)}{\binom{2}{2}} = 1 - \frac{1}{2} + 0 = \frac{1}{2} = 3 \cdot \frac{1}{6} = 3 \cdot B_2.$$

For $n = 3$, we get that

$$\frac{E(3,0)}{\binom{3}{0}} - \frac{E(3,1)}{\binom{3}{1}} + \frac{E(3,2)}{\binom{3}{2}} - \frac{E(3,3)}{\binom{3}{3}} = 1 - \frac{4}{3} + \frac{1}{3} - 0 = 0 = 4B_3.$$

However, the equation holds also for $n = 0$: in this case, we have that $\frac{E(0,0)}{\binom{0}{0}} = 1 = 1 \cdot B_0$. For $n = 1$, it holds that $\frac{E(1,0)}{\binom{1}{0}} - \frac{E(1,1)}{\binom{1}{1}} = 1 - 0 = 1 = 2 \cdot B_1^+$. So, the equality

$$\sum_{k=0}^{n} (-1)^k \frac{E(n,k)}{\binom{n}{k}} = (n+1)B_n$$

holds for any non-negative integer n if we use the sequence of second Bernoulli numbers. □

3.5.4. As $B_n = 0$ for odd $n \geq 3$, we can say that all the sums $\sum_{k=0}^{n} (-1)^k \frac{E(n,k)}{\binom{n}{k}}$, $n \geq 2$, obtained using the odd rows of Euler's triangle and Pascal's triangle, are equal to zero, i.e., zeros and "non-zeros" alternate, when we calculate such sums (compare with the alternating sums of the elements of the rows of Euler's triangle).

3.5.5. If we perform a similar operation with the nth row of Euler's triangle and with the $(n-1)$th row of Pascal's triangle, we will always have zero as a result; this is the statement of the following property.

- *The alternating sum of the values* $\frac{E(n,k)}{\binom{n-1}{k}}$, *obtained by sequential division of the elements of the nth row of the triangle of Eulerian numbers of the first order by the corresponding elements of the* $(n-1)$*th row of Pascal's triangle, is equal to zero*:

$$\frac{E(n,0)}{\binom{n-1}{0}} - \frac{E(n,1)}{\binom{n-1}{1}} + \frac{E(n,2)}{\binom{n-1}{2}} - \cdots + (-1)^k \frac{E(n,k)}{\binom{n-1}{k}} + \cdots + (-1)^{n-1} \frac{E(n,n-1)}{\binom{n-1}{n-1}} = 0, \ n \geq 2.$$

□ In fact, for $n = 2$, we have that

$$\frac{E(2,0)}{\binom{1}{0}} - \frac{E(2,1)}{\binom{1}{1}} = 1 - 1 = 0.$$

For $n = 3$, the equality takes the form

$$\frac{E(3,0)}{\binom{2}{0}} - \frac{E(3,1)}{\binom{2}{1}} + \frac{E(3,2)}{\binom{2}{2}} = 1 - \frac{4}{2} + 1 = 0.$$

For $n = 4$, we get that

$$\frac{E(4,0)}{\binom{3}{0}} - \frac{E(4,1)}{\binom{3}{1}} + \frac{E(4,2)}{\binom{3}{2}} - \frac{E(4,3)}{\binom{3}{3}} = 1 - \frac{11}{3} + \frac{11}{3} - 1 = 0.$$

Note that, for even n, the symmetry of the two arithmetic triangles used makes the property trivial. In fact, as $E(n,k) = E(n-1-k)$ and $\binom{n-1}{k} = \binom{n-1}{n-1-k}$, then for an even $n \geq 2$, we have that

$$\frac{E(n,0)}{\binom{n-1}{0}} - \frac{E(n,1)}{\binom{n-1}{1}} + \frac{E(n,2)}{\binom{n-1}{2}} - \cdots + (-1)^k \frac{E(n,k)}{\binom{n-1}{k}} + \cdots (-1)^{n-1-k} \frac{E(n,n-1-k)}{\binom{n-1}{n-1-k}} + \frac{E(n-1,n-2)}{\binom{n-1}{n-2}}$$

$$
\begin{aligned}
&- \frac{E(n-1,n-1)}{\binom{n-1}{n-1}} \\
&= \left(\frac{E(n,0)}{\binom{n-1}{0}} - \frac{E(n-1,n-1)}{\binom{n-1}{n-1}} \right) \\
&- \left(\frac{E(n,1)}{\binom{n-1}{1}} - \frac{E(n-1,n-2)}{\binom{n-1}{n-2}}) + \frac{E(n,2)}{\binom{n-1}{2}} \right) \\
&+ \cdots + (-1)^k \left(\frac{E(n,k)}{\binom{n-1}{k}} - \frac{E(n,n-1-k)}{\binom{n-1}{n-1-k}} \right) + \cdots = 0, \ n \geq 2.
\end{aligned}
$$

Unlike the two previous properties, for $n = 0$ and $n = 1$, the identity does not hold. □

3.5.6. If we multiply the Eulerian numbers of the first order by binomial coefficients, we can find the connections of our objects with the *Stirling numbers of the second kind.*

- *It holds the following identity, connecting the Eulerian numbers of the first order with the Stirling numbers of the second kind:*

$$
\begin{aligned}
&E(n,n-m)\binom{n-m}{n-m} + E(n,n-m+1)\binom{n-m+1}{n-m} \\
&+ E(n,n-m+2)\binom{n-m+2}{n-m} + \cdots + E(n,n)\binom{n}{n-m} \\
&= m!S(n,m), \quad n \geq 0.
\end{aligned}
$$

□ Consider the identity

$$
\sum_{k=n-m}^{n} E(n,k)\binom{k}{n-m} = \sum_{k=0}^{n} E(n,k)\binom{k}{n-m} = m!S(n,m)
$$

for small values of parameters. For $n = 0$ and $k = 0$, we get a degenerate sum,

$$
E(0,0)\binom{0}{0} = 1 = 0!S(0,0).
$$

For a positive integer n and $k = 0$, we get that

$$E(n,n)\binom{n}{n} = 0 = 0!S(n,0).$$

For a positive integer n and $k = n$, the relationship takes the form

$$E(n,0)\binom{0}{0} + E(n,1)\binom{1}{0} + E(n,2)\binom{2}{0} + \cdots + E(n,n)\binom{n}{0}$$
$$= E(n,0) + E(n,1) + E(n,2) + \cdots + E(n,n) = n! = n!S(n,n).$$

If $n = 2$ and $k = 1$, we get the relationship

$$E(2,1)\binom{1}{1} + E(2,2)\binom{2}{1} = 1 = 1!S(2,1).$$

For $n = 3$ and $k = 1$, the equation takes the form

$$E(3,2)\binom{2}{2} + E(3,3)\binom{3}{2} = 1 = 1!S(3,1).$$

If $n = 3$ and $k = 2$, the situation is thus:

$$E(3,1)\binom{1}{1} + E(3,2)\binom{2}{1} + E(3,3)\binom{3}{1} = 4 + 1 \cdot 2 = 6 = 2!S(3,2).$$

In order to prove this property in the general case, let us to use Newton's formula for falling powers:

$$(x+y)^{\underline{n}} = \sum_{m=0}^{n} \binom{n}{m} x^{\underline{m}} y^{\underline{n-m}}.$$

Applying it to the falling powers on the right-hand side of the formula

$$n!x^n = \sum_{k=0}^{n} E(n,k)(x+k)^{\underline{n}},$$

we get that

$$n!x^n = \sum_{k=0}^{n}\sum_{m=0}^{n} E(n,k)\binom{n}{m} x^{\underline{m}} k^{\underline{n-m}} = \sum_{k=0}^{n}\sum_{m=0}^{n} E(n,k)\frac{n!x^{\underline{m}}k^{\underline{n-m}}}{m!(n-m)!}.$$

It is easy to see that $k^{\underline{n-m}} = 0$ for $k < n - m$: in this case, the product $k^{\underline{n-m}} = k \cdot (k-1) \cdot \cdots \cdot (k - (n-m) + 1)$ contains a zero term. So,

$$n!x^n = \sum_{m=0}^{n} \left(\frac{1}{m!} \sum_{k=n-m}^{n} E(n,k) \frac{n!k^{\underline{n-m}}}{(n-m)!} \right) x^{\underline{m}}.$$

Dividing both parts of the equality by $n!$ and noting that $\frac{k^{\underline{n-m}}}{(n-m)!} = \frac{k!}{(n-m)!(n-m-k)!} = \binom{k}{n-m}$, we get that

$$x^n = \sum_{m=0}^{n} \left(\frac{1}{m!} \sum_{k=n-m}^{n} E(n,k) \binom{k}{n-m} \right) x^{\underline{m}}.$$

But on the other hand, it is well known that

$$x^n = \sum_{m=0}^{n} S(n,m) x^{\underline{m}};$$

therefore, it holds that

$$S(n,m) = \frac{1}{m!} \sum_{k=n-m}^{n} E(n,k) \binom{k}{n-m},$$

or, in another form,

$$\begin{aligned} &\sum_{k=n-m}^{n} E(n,k) \binom{k}{n-m} \\ &\quad = E(n,n-m)\binom{n-m}{n-m} + E(n,n-m+1)\binom{n-m+1}{n-m} \\ &\qquad + E(n,n-m+2)\binom{n-m+2}{n-m} + \cdots + E(n,n)\binom{n}{n-m} \\ &\quad = m!S(n,m). \end{aligned}$$

□

3.5.7. Using the last property, we can represent the Stirling numbers of the second kind using the Eulerian numbers of the first order and the Stirling numbers of the first kind.

- *The following identity holds:*

$$S(n,m) = \frac{1}{n!}\binom{n}{m}\sum_{j=0}^{n-m} s(n-m,j)\sum_{k=n-m}^{n} E(n,k)k^j.$$

□ In fact, we have proven that

$$m!S(n,m) = \sum_{k=n-m}^{n} E(n,k)\binom{k}{n-m}.$$

It is well known (see, for example, Section 3.2) that $x^{\underline{n}} = \sum_{j=0}^{n} s(n,j)x^j$, where $s(n,j)$ are the Stirling numbers of the first kind. As $\binom{x}{n} = \frac{x^{\underline{n}}}{n!}$, we can state that

$$\binom{x}{n} = \frac{1}{n!}\sum_{j=0}^{n} s(n,j)x^j.$$

Therefore, we have that

$$\begin{aligned} m!S(n,m) &= \sum_{k=n-m}^{n} E(n,k)\binom{k}{n-m} \\ &= \sum_{k=n-m}^{n} E(n,k)\left(\frac{1}{(n-m)!}\sum_{j=0}^{n-m} s(n-m,j)k^j\right) \\ &= \frac{1}{(n-m)!}\sum_{j=0}^{n-m} s(n-m,j)\left(\sum_{k=n-m}^{n} E(n,k)k^j\right). \end{aligned}$$

As $\frac{1}{m!(n-m)!} = \frac{1}{n!}\binom{n}{m}$, we get for $S(n,m)$ the required result:

$$\begin{aligned} S(n,m) &= \frac{(1}{m!(n-m)!}\sum_{j=0}^{n-m} s(n-m,j)\sum_{k=n-m}^{n} E(n,k)k^j \\ &= \frac{(1}{n!}\binom{n}{m}\sum_{j=0}^{n-m} s(n-m,j)\sum_{k=n-m}^{n} E(n,k)k^j. \end{aligned}$$ □

3.5.8. Consider another property connecting the Eulerian numbers of the first order and the Stirling numbers of the second kind.

It includes, in addition to these two classes of special numbers, factorials and the powers of two.

- *The sum of the elements $E(n,k)$ of the nth row of Euler's triangle with weight coefficients 2^k is equal to the sum of the elements $S(n,k)$ of the nth row of the triangle of Stirling numbers of the second kind with weight coefficients $k!$:*

$$E(n,0) + 2E(n,1) + 2^2E(n,2) + \cdots + 2^nE(n,n)$$
$$= S(n,0) + 1!S(n,1) + 2!S(n,2) + \cdots + n!S(n,n).$$

☐ In fact, summing the zero rows of the two triangles, we get unities in each case:

$$2^0E(0,0) = 1 = 0!S(0,0).$$

The same situation occurs for the first row:

$$E(1,0) + 2E(1,1) = 1 + 2 \cdot 0 = 0 + 1 \cdot 1 = S(1,0) + 1!S(1,1).$$

For $n = 2$, the situation is thus:

$$E(2,0) + 2E(2,1) + 2^2E(2,2) = 1 + 2 \cdot 1 + 4 \cdot 0$$
$$= 3 = 0 + 1! \cdot 1 + 2! \cdot 1 = S(2,0) + 1!S(2,1) + 2!S(2,2).$$

Now, we can note that, on the one hand, the right-hand side of the considered identity is the sums of the right-hand side $m!S(n,m)$ of the previous identity, considered for $m = 0, 1, , \ldots, n$. On the other hand, the numbers 2^k, $k = 0, 1, 2, \ldots, n$, are the sums of the elements of the kth row of Pascal's triangle. These considerations can help us derive the proof of the last identity for any positive integer n. Really, it holds that

$$2^0E(n,0) + 2^1E(n,1) + 2^2E(n,2) + \cdots + 2^nE(n,n)$$
$$= \binom{0}{0}E(n,0) + \left(\binom{1}{0} + \binom{1}{1}\right)E(n,1) + \left(\binom{2}{0} + \binom{2}{1}\right.$$
$$\left. + \binom{2}{2}\right)E(n,2) + \cdots + \left(\binom{n-1}{0} + \binom{n-1}{1}\right.$$

$$+\cdots+\binom{n-1}{n-1}\Bigg)E(n-1,n-1)+\left(\binom{n}{0}+\binom{n}{1}\right.$$
$$\left.+\cdots+\binom{n}{n}\right)E(n,n)$$
$$=E(n,n)\binom{n}{n}+\left(E(n,n-1)\binom{n-1}{n-1}+E(n,n)\binom{n}{n-1}\right)$$
$$+\left(E(n,n-2)\binom{n-2}{n-2}+E(n,n-1)\binom{n-1}{n-2}\right.$$
$$\left.+E(n,n)\binom{n}{n-2}\right)+\cdots+\left(E(n,0)\binom{0}{0}+E(n,1)\binom{1}{0}\right.$$
$$\left.+\cdots+E(n,n)\binom{n}{0}\right)$$
$$=0!S(n,0)+1!S(n,1)+2!S(n,2)+\cdots+n!S(n,n). \qquad \square$$

3.5.9. The last identity can be rewritten differently: looking closely, we will find that its left-hand side is the value of the Eulerian polynomial $A_n(x)$ at $x=2$:

$$E(n,0)+E(n,1)2+E(n,2)2^2+\cdots+E(n,n)2^n=A_n(2).$$

So, we can state that

$$A_n(2)=\sum_{k=0}^{n}k!S(n,k).$$

In our arsenal, there are already several statements concerning the behaviour of the Eulerian polynomials. One of them we obtained by looking at the generating functions of the sequences of non-negative integer powers of positive integers:

$$\sum_{k=1}^{\infty}k^n\cdot x^k=\frac{xA_n(x)}{(1-x)^{n+1}}.$$

In terms of the Eulerian numbers of the first order, it can be written in the following form:

- *For any positive integer n, there exists the identity*

$$\frac{E(n,0)x + E(n,1)x^2 + E(n,2)x^3 + \cdots + E(n,n-1)x^n}{(1-x)^{n+1}} = \sum_{k=0}^{\infty} k^n x^k, \quad |x| < 1.$$

3.5.10. Taking $x = 2$ in the identity $\frac{xA_n(x)}{(1-x)^{n+1}} = \sum_{k=1}^{\infty} k^n x^k$, we get in the numerator on the left-hand side the expression $2E(n,0) + 2^2E(n,1) + 2^3E(n,2) + \cdots + 2^nE(n,n-1)$, which is equal to $2\sum_{k=0}^{n} k!S(n,k)$. However, in this case, the right-hand side will obviously diverge. On the other hand, the value $x = \frac{1}{2}$ can be used; this choice of x gives a convergent series $\sum_{k=0}^{\infty} \frac{k^n}{2^k}$. On the left-hand side, for $x = \frac{1}{2}$, we get the expression

$$\begin{aligned}
&\frac{\frac{1}{2}A_n(\frac{1}{2})}{(1-\frac{1}{2})^{n+1}} \\
&\quad = \frac{E(n,0)2^{-1} + E(n,1)2^{-2} + E(n,2)2^{-3} + \cdots + E(n,n-1)2^{-n}}{(1-\frac{1}{2})^{n+1}} \\
&\quad = 2^{n+1}\left(E(n,0)2^{-1} + E(n,1)2^{-2} + E(n,2)2^{-3}\right. \\
&\qquad \left. + \cdots + E(n,n-1)2^{-n}\right) \\
&\quad = E(n,0)2^n + E(n,1)2^{n-1} + E(n,2)2^{n-2} + \cdots + E(n,n-1)2^1.
\end{aligned}$$

As $E(n,k) = E(n,n-1-k)$, changing $E(n,0)$ to $E(n,n-1)$, $E(n,1)$ to $E(n,n-2),\ldots,E(n,n-1)$ to $E(n,0)$ gives

$$\begin{aligned}
\sum_{k=0}^{\infty} \frac{k^n}{2^k} &= 2\left(E(n,n-1)2^{n-1} + E(n,n-2)2^{n-2}\right. \\
&\qquad \left. + E(n,n-3)2^{n-3} + \cdots + E(n,0)2^0\right).
\end{aligned}$$

So, we have proved a new property.

- *For any positive integer n, we have the identity*

$$E(n,0) + 2E(n,1) + 2^2E(n,2) + \cdots + 2^nE(n,n) = \frac{1^n}{2^2} + \frac{2^n}{2^3} + \frac{3^n}{2^4} + \cdots + \frac{k^n}{2^{k+1}} + \cdots.$$

3.5.11. Now, when we know what to do, we can derive many such identities.

- *For any positive integer n, it holds that*

$$E(n,0) + 3E(n,1) + 3^2E(n,2) + \cdots + 3^nE(n,n) = 2^{n+1}\left(\frac{1^n}{3^2} + \frac{2^n}{3^3} + \frac{3^n}{3^4} + \cdots + \frac{k^n}{3^{k+1}} + \cdots\right).$$

□ To prove this identity, it is enough to take $x = \frac{1}{3}$ in the equality $\sum_{k=0}^{\infty} k^n \cdot x^k = \frac{xA_n(x)}{(1-x)^{n+1}}$:

$$\sum_{k=0}^{\infty} \frac{k^n}{3^k} = \frac{3^{n+1}\sum_{m=0}^{n-1} 3^{-m-1}E(n,m)}{2^{n+1}} = \frac{1}{2^{n+1}}\sum_{m=0}^{n-1} 3^{n-m}E(n,m)$$

$$= \frac{1}{2^{n+1}}\sum_{m=0}^{n-1} 3^{n-m}E(n,n-m-1) = \frac{3}{2^{n+1}}\sum_{t=0}^{n-1} 3^tE(n,t).$$

So,

$$2^{n+1}\left(\frac{1^n}{3^2} + \frac{2^n}{3^3} + \frac{3^n}{3^4} + \cdots + \frac{k^n}{3^{k+1}} + \cdots\right) = 2^{n+1}\sum_{k=0}^{\infty} \frac{k^n}{3^{k+1}}$$

$$= \sum_{t=0}^{n-1} 3^tE(n,t) = E(n,0) + 3E(n,1) + 3^2E(n,2) + \cdots + 3^nE(n,n).$$ □

3.5.12. These formulas have a simple generalisation.

- *For any positive integer $a \geq 2$, it holds that*

$$E(n,0) + aE(n,1) + a^2E(n,2) + \cdots + a^nE(n,n) = (a-1)^{n+1}\left(\frac{1^n}{a^2} + \frac{2^n}{a^3} + \frac{3^n}{a^4} + \cdots + \frac{k^n}{a^{k+1}} + \cdots\right).$$

□ The proof of this identity is quite similar; we should take $x = \frac{1}{a}$:

$$\sum_{k=0}^{\infty} \frac{k^n}{a^k} = \frac{a^{n+1}\sum_{m=0}^{n-1} a^{-m-1}E(n,m)}{(a-1)^{n+1}}$$

$$= \frac{1}{(a-1)^{n+1}} \sum_{m=0}^{n-1} a^{n-m}E(n,m)$$

$$= \frac{1}{(a-1)^{n+1}} \sum_{m=0}^{n-1} a^{n-m}E(n,n-m-1)$$

$$= \frac{a}{(a-1)^{n+1}} \sum_{t=0}^{n-1} 3^t E(n,t).$$

So, we have that

$$(a-1)^{n+1}\left(\frac{1^n}{a^2} + \frac{2^n}{a^3} + \frac{3^n}{a^4} + \cdots + \frac{k^n}{a^{k+1}} + \cdots\right)$$

$$= (a-1)^{n+1}\sum_{k=0}^{\infty} \frac{k^n}{a^{k+1}} = \sum_{t=0}^{n-1} a^t E(n,t)$$

$$= E(n,0) + aE(n,1) + a^2E(n,2) + \cdots + a^nE(n,n). \qquad \square$$

Closed formulas for Eulerian numbers of the first order

3.5.13. In Section 3.3, we proved the most widely known closed formula for $E(n,k)$:

- *For any non-negative integer* n *and any non-negative integer* k, $0 \le k \le n-1$, *it holds that*

$$E(n,k) = \binom{n+1}{0}(k+1)^n - \binom{n+1}{1}k^n$$
$$+ \cdots + (-1)^m \binom{n+1}{m}(k+1-m)^n$$
$$+ \cdots + (-1)^k \binom{n+1}{k}1^n.$$

However, there exist other formulas that allow us to obtain the values of $E(n,k)$ as functions of the parameters n and k.

3.5.14. For example, we can represent the Eulerian numbers of the first order using the Stirling numbers of the second kind:

- *For any non-negative integer n and any non-negative integer k, $0 \le k \le n-1$, it holds that*

$$E(n,k) = (-1)^{n-k}\left(0!S(n,0)\binom{n}{k} - 1!S(n,1)\binom{n-1}{k}\right.$$
$$\left. + \cdots + (-1)^k S(n,k)\binom{k}{k}\right).$$

□ Consider the equality

$$E(n,k) = \sum_{m=0}^{n-k} (-1)^{n-m-k} m! S(n,m)\binom{n-m}{k}$$

for small values of the parameters.

For $n = 0, k = 0$, the result holds: $E(0,0) = 1 = 0!S(0,0)$. For $n = 1, k = 0$, we have $E(1,0) = 1 = (-1)^0 1!S(1,1)\binom{0}{0}$. For $n = 2, k = 0$, we have $E(2,0) = 1 = (-1)^1 1!S(2,1)\binom{1}{0} + (-1)^0 2!S(2,2)\binom{0}{0}$. For $n = 2, k = 1$, we have $E(2,1) = 1 = (-1)^0 1!S(2,1)\binom{1}{1}$. For $n = 3, k = 0$, we have

$$E(3,0) = 1 = 1 - 2\cdot 3 + 6\cdot 1 = (-1)^2 1!S(3,1)\binom{2}{0}$$
$$+ (-1)^1 2!S(3,2)\binom{1}{0} + (-1)^0 3!S(3,3)\binom{0}{0}.$$

For $n = 3, k = 1$, we have

$$E(3,1) = 4 = -1\cdot 2 + 2\cdot 6 = (-1)^1 1!S(3,1)\binom{2}{1} + (-1)^0 2!S(3,2)\binom{1}{1}.$$

For $n = 3, k = 2$, we have $E(3,2) = 1 = (-1)^0 1!S(3,1)\binom{2}{1}$.

Note that, for $k = 0$, it holds that $E(n,0) = 1 = \sum_{m=1}^{n}(-1)^{n-m} m!S(n,m)$; therefore, we obtain an interesting identity for the Stirling numbers of the second kind:

$$n!S(n,n) - (n-1)!S(n,n-1) + \cdots (-1)^{n-1} 1!S(n,1) = 1,$$

i.e., *the alternating sum of the elements $S(n,k)$ of the nth row of the triangle of Stirling numbers of the second kind with the coefficients $k!$ of their weights is equal to unity.*

For $n \geq 1$, it holds that $S(n,0) = 0$, and we can prove that

$$E(n,k) = (-1)^{n-k}\left(-1!S(n,1)\binom{n-1}{k} + 2!S(n,2)\binom{n-2}{k} - \cdots + (-1)^k S(n,n-k)\binom{k}{k}\right).$$

The proof is very simple. As

$$m!S(n,m) = E(n,n-m)\binom{n-m}{n-m} + E(n,n-m+1)\binom{n-m+1}{n-m} + \cdots + E(n,n-1)\binom{n-1}{n-m},$$

then through multiplication by x^{n-m} and summation over m, $1 \leq m \leq n$, we have that

$$\sum_{m=0}^{n} m!S(n,m)x^{n-m}$$

$$= \sum_{m=1}^{n} m!S(n,m)x^{n-m} = 1!S(n,1)x^n + 2!S(n,2)x^{n-1}$$

$$+ \cdots + n!S(n,n) = \left(E(n,n-1)\binom{b-1}{n-1}\right)x^{n-1}$$

$$+ \left(E(n,n-2)\binom{n-2}{n-2} + E(n,n-2)\binom{n-1}{n-2}\right)x^{n-2}$$

$$+ \cdots + \left(E(n,0)\binom{0}{0} + E(n,1)\binom{1}{0}\right.$$

$$\left. + \cdots + E(n,n-1)\binom{n-1}{0}\right)x^0 = E(n,0)\binom{0}{0}x^0$$

$$+ E(n,1)\left(\binom{1}{0}x^0 + \binom{1}{1}x^1\right) + E(n,2)\left(\binom{2}{0}x^0 + \binom{2}{1}x^1\right.$$

$$+\binom{2}{2}x^2\bigg) + E(n,n-1)\left(\binom{n-1}{0}x^0 + \binom{n-1}{1}x^1\right.$$

$$\left. + \cdots + \binom{n-1}{n-1}x^{n-1}\right) = E(n,0)(x+1)^0 + E(n,1)(1+x)^1$$

$$+ E(n,2)(1+x)^2 + \cdots + E(n,n-1)(1+x)^{n-1}.$$

So, we have proven that

$$\sum_{m=0}^{n} m!S(n,m)x^{n-m} = \sum_{k=0}^{n-1} E(n,k)(1+x)^k.$$

Going from x to $x-1$, we get the equality

$$\sum_{m=0}^{n} m!S(n,m)(x-1)^{n-m} = \sum_{k=0}^{n-1} E(n,k)x^k.$$

Furthermore, it holds that

$$\sum_{m=0}^{n} m!S(n,m)(x-1)^{n-m}$$

$$= \sum_{m=0}^{n} m!S(n,m)(-1)^{n-m}(1-x)^{n-m}$$

$$= \sum_{m=0}^{n} m!S(n,m)\sum_{k=0}^{n-m}\binom{n-m}{k}(-1)^k x^k$$

$$= (-1)^{n-1}1!S(n,1)\left(\binom{n-1}{0}x^0 - \binom{n-1}{1}x^1\right.$$

$$\left. + \cdots + (-1)^{n-1}\binom{n-1}{n-1}x^{n-1}\right)$$

$$+ (-1)^{n-2}2!S(n,2)\left(\binom{n-2}{0}x^0 - \binom{n-2}{1}x^1 + \cdots + (-1)^{n-2}\right.$$

$$\left. \times \binom{n-2}{n-2}x^{n-2}\right) + \cdots + (-1)^1(n-1)!S(n,n-1)$$

$$\times \left(\binom{1}{0}x^0 - \binom{1}{1}x^1\right) + (-1)^0 n!S(n,n)\left(\binom{0}{0}x^0\right)$$

$$= x^0 \left((-1)^{n-1} 1! S(n,1) \binom{n-1}{0} + (-1)^{n-2} 2! S(n,n-2) \binom{n-2}{0} \right.$$
$$\left. + \cdots + (-1)^0 n! S(n,n) \binom{0}{0} \right) - x^1 \left((-1)^{n-1} 1! S(n,1) \binom{n-1}{1} \right.$$
$$\left. + \cdots + (-1)^1 (n-1)! S(n,n-1) \binom{1}{1} \right) + x^2 \left((-1)^{n-1} 1! S(n,1) \right.$$
$$\left. \times \binom{n-1}{2} + \cdots + (-1)^2 (n-2)! S(n,n-2) \binom{2}{2} \right)$$
$$+ \cdots + (-1)^{n-2} x^{n-2} \left((-1)^{n-1} 1! S(n,1) \binom{n-1}{n-2} \right.$$
$$\left. + (-1)^{n-2} 2! S(n,2) \binom{n-2}{n-2} \right) + (-1)^{n-1} x^{n-1} (-1)^{n-1} 1! S(n,1)$$
$$\times \binom{n-1}{n-1}.$$

So, we get that

$$\sum_{k=0}^{n-1} E(n,k) x^k = \sum_{m=0}^{n} m! S(n,m) (x-1)^{n-m}$$
$$= \sum_{k=0}^{n-1} \left(\sum_{m=1}^{n-k} (-1)^{n-k-m} m! S(n,m) \binom{n-m}{k} \right) x^k.$$

Comparing the coefficients at x^k, we have that

$$E(n,k) = \sum_{m=1}^{n-k} (-1)^{n-k-m} m! S(n,m) \binom{n-m}{k}. \qquad \square$$

3.5.15. In [He23], a new interesting formula for calculating the Eulerian numbers is given:

- *For any $n \geq 1$ and any $0 \leq k \leq n-1$, it holds that*

$$E(n,k) = \sum_{t_1+t_2+\cdots+t_{k+1}=n-k-1, t_i \geq 0} 1^{t_1} 2^{t_2} \cdots (k+1)^{t_{k+1}} (t_1+1)$$
$$\times (t_1+t_2+1) \cdots (t_1+\cdots+t_k+1).$$

For example, we have that

$$E(5,2) = \sum_{t_1+t_2+t_3=2, t_i \geq 0} 1^{t_1}2^{t_2}3^{t_3}(t_1+1)(t_1+t_2+1)$$

$$= 1^0 2^0 3^2 (0+1)(0+0+1) + 1^0 2^2 3^0 (0+1)(0+2+1)$$

$$+ 1^2 2^0 3^0 (2+1)(2+0+1) + 1^0 2^1 3^1 (0+1)(0+1+1)$$

$$+ 1^1 2^0 3^1 (1+1)(1+0+1) + 1^1 2^1 3^0 (1+1)(1+1+1)$$

$$= 9 + 12 + 9 + 12 + 12 + 12 = 66.$$

□ Let us prove this identity by induction. It is true for small values of n and k:

$$E(1,0) = 1 = 1^0 = \sum_{t_1=0} 1^{t_1}; \quad E(2,0) = 1 = 1^1 = \sum_{t_1=1} 1^{t_1};$$

$$E(3,0) = 1 = 1^2 = \sum_{t_1=2} 1^{t_1};$$

$$E(2,1) = 1^0 \cdot 2^0 (0+1) = \sum_{t_1+t_2=0} 1^{t_1}2^{t_2}(t_1+1);$$

$$E(3,1) = 1^0 \cdot 2^1 (0+1) + 1^1 \cdot 2^0 \cdot (1+1) = \sum_{t_1+t_2=1} 1^{t_1}2^{t_2}(t_1+1);$$

$$E(3,2) = 1^0 \cdot 2^0 \cdot 3^0 (0+1)(0+0+1)$$

$$= \sum_{t_1+t_2+t_3=0} 1^{t_1}2^{t_2}3^{t_3}(t_1+1)(t_1+t_2+1).$$

Moreover, the identity holds for two extreme values of k. In fact, for $k = 0$, we have that

$$E(n,0) = 1 = 1^{n-1} = \sum_{t_1=n-1} 1^{t_1}.$$

For $k = n-1$, we get that

$$E(n,n-1) = 1 = 1^0 2^0 \cdots n^0 (0+1)(0+0+1)\cdots(0+0\cdots+0+1)$$

$$= \sum_{t_1+\cdots+t_n=0} 1^{t_1}2^{t_2}\cdots n^{t_n}(t_1+1)(t_1+t_2+1)$$

$$\cdots(t_1+t_2+\cdots+t_{n-1}+1).$$

Suppose that the identity is true for a positive integer n and all $0 \le k \le n-1$, and we try to prove it for $n+1$ and all $0 \le k \le n$. As $E(n+1,k) = (k+1)E(n,k) + (n-k+1)E(n,k-1)$, we have that

$$\begin{aligned}
E(n+1,k) &= (k+1)E(n,k) + (n-k+1)E(n,k-1) \\
&= (k+1) \sum_{t_1+\cdots+t_{k+1}=n-k-1,t_i\ge 0} 1^{t_1}2^{t_2}\cdots(k+1)^{t_{k+1}} \\
&\quad \times (t_1+1)(t_1+t_2+1)\cdots(t_1+t_2+\cdots+t_k+1) \\
&\quad + (n-k+1) \sum_{t_1+\cdots+t_k=n-k,t_i\ge 0} 1^{t_1}2^{t_2}\cdots k^{t_k}(t_1+1) \\
&\quad \times (t_1+t_2+1)\cdots(t_1+t_2+\cdots+t_{k-1}+1) \\
&= \sum_{t_1+\cdots+t_{k+1}=n-k-1,t_i\ge 0} 1^{t_1}2^{t_2}\cdots(k+1)^{t_{k+1}+1}(t_1+1) \\
&\quad \times (t_1+t_2+1)\cdots(t_1+t_2+\cdots+t_k+1) \\
&\quad + \sum_{t_1+\cdots+t_k=n-k,t_i\ge 0} 1^{t_1}2^{t_2}\cdots k^{t_k}(t_1+1)(t_1+t_2+1) \\
&\quad \times \cdots(t_1+t_2+\cdots+t_{k-1}+1)(n-k+1) \\
&= \sum_{t_1+\cdots+t_k+s_{k+1}=n-k,t_i\ge 0,s_{k+1}\ge 1} 1^{t_1}2^{t_2}\cdots(k+1)^{s_{k+1}} \\
&\quad \times (t_1+1)(t_1+t_2+1)\cdots(t_1+t_2+\cdots+t_k+1) \\
&\quad + \sum_{t_1+\cdots+t_k+s_{k+1}=n-k,t_i\ge 0,s_{k+1}=0} 1^{t_1}2^{t_2}\cdots k^{t_k}(k+1)^{s_{k+1}} \\
&\quad \times (t_1+1)(t_1+t_2+1)\cdots(t_1+t_2+\cdots+t_{k-1}+1) \\
&\quad \times (t_1+t_2+\cdots+t_k+1) \\
&= \sum_{t_1+\cdots+t_k+t_{k+1}=n-k,t_i\ge 0} 1^{t_1}2^{t_2}\cdots k^{t_k}(k+1)^{t_{k+1}}(t_1+1) \\
&\quad \times (t_1+t_2+1)\cdots(t_1+t_2+\cdots+t_{k-1}+1) \\
&\quad \times (t_1+t_2+\cdots+t_k+1).
\end{aligned}$$

The statement is proven. □

Divisibility properties of the Eulerian numbers of the first order

3.5.16. In the beginning of this chapter, we considered some simple divisibility properties of the elements $E(n,k)$, $n \geq 1, 0 \leq k \leq n-1$, of Euler's triangle.

As $E(n,0) = E(n,n-1) = 1$ for any positive integer n, we can state that the sides of the triangle are formed by odd numbers.

As $E(n,1) = E(n,n-2)$ are even for odd n and are odd for even n, it holds that the parities of the elements of the first vertical and the first descending diagonal of the triangle alternate.

Moreover, all central Eulerian numbers of the first order $E(2n+1,n)$ are even. They form the main diagonal of the isosceles Euler's triangle.

3.5.17. However, there are deeper properties of Euler's triangle related to the divisibility of the Eulerian numbers by primes.

It's easy to see that two non-zero numbers 1 and 1 in the second row of the triangle have the form $2m+1$, i.e., the numbers $E(2,k)-1$, $k = 0,1$, are divisible by 2; the three non-zero numbers $1, 4$, and 1 in the third row have the form $3m+1$, i.e., the numbers $E(3,k)-1$, $k = 0,1,2$, are divisible by 3; the five non-zero numbers $1, 26, 66, 26$, and 1 in the fifth row have the form $5m+1$, i.e., the numbers $E(5,k)-1$, $k = 0,1,2,3,4$, are divisible by 5; the seven non-zero numbers $1, 120, 1191, 2416, 1191, 120$, and 1 in the seventh row have the form $7m+1$, i.e., the numbers $E(7,k)-1$, $k = 0,1,\ldots,6$, are divisible by 7.

All other rows up to and including the ninth row have no such property. Which condition combines the numbers 2, 3, 5, and 7, and distinguishes them from other numbers up to 10? Yes, they are prime numbers.

For the convenience of further presentation, we state that *the nth row of Euler's triangle is congruent to r modulo n* if all non-zero elements of this row are congruent to r modulo n, i.e., if $E(n,k) \equiv r(\bmod n)$, $n \geq 2$, $0 \leq k \leq n-1$.

- *For a prime number p, the pth row of Euler's triangle is congruent to* 1 *modulo p.*

Recall that a similar property holds for Pascal's triangle. It's easy to see that only the internal element 2 of the second row $1, 2, 1$ of Pascal's triangle is divisible by 2; the two internal numbers of the third row $1, 3, 3$, and 1 of the triangle are divisible by 3; the four internal numbers $5, 10, 10$, and 5 of the fifth row $1, 5, 10, 10, 5, 1$ are divisible by 5, etc. Let us say that *the nth row of Pascal's triangle is divisible by n if all the internal elements of this row are divisible by n.* Then, we can prove the following statement:

- *For a prime number p, the pth row of Pascal's triangle is divisible by p.*

In fact (see [Uspe76] and [Deza21]), there exist stronger properties: *the nth row of Pascal's triangle is divisible by n if and only if n is a prime number*, and *the mth row of Pascal's triangle is divisible by a prime p if and only if $m = p^k$, $k \in \mathbb{N}$.*

The proof is based on the following simple fact: for a prime number p, the binomial coefficient $\binom{p}{k}$ is divisible by p for all $1 \le k \le p - 1$. In fact, it is obvious that for such k, it holds that $\binom{p}{k} = \frac{p!}{k!(n-k)!} = p \cdot K$, where $K \in \mathbb{N}$. We can prove the above-mentioned property of Euler's triangle (see [Sloa25]) using similar arguments.

□ For a prime number p, consider the sequence $E(p, k)$, $k = 0, 1, 2, \ldots, p - 1$. As $E(p, 0) = E(p, p - 1) = 1$ and $1 \equiv 1(mod \quad p)$, the statement holds for $k = 0, p - 1$. On the other hand, using Fermat's little theorem, stating that $a^p \equiv a(mod \quad p)$ for any prime p and any integer a, we have

$$E(p, 1) = E(p, p - 2) = 2^p - (p + 1) \equiv 2 - p - 1 \equiv 1(mod\, p).$$

For any $1 < k < p - 2$, we have that

$$E(p, k) = \binom{p+1}{0}(k + 1)^p - \binom{p+1}{1}k^p + \binom{p+1}{2}(k - 1)^p$$
$$- \cdots + (-1)^k \binom{p+1}{k} 1^p = (k + 1)^p - \left(\binom{p}{1} + \binom{p}{0}\right) k^p$$
$$+ \left(\binom{p}{2} + \binom{p}{2}\right)(k - 1)^p - \cdots + (-1)^k \left(\binom{p}{l}\right.$$

$$+\left(\binom{p}{k-1}\right)1^p = \left(k^p + \binom{p}{1}k^{p-1} + \cdots + \binom{p}{p-1}k + 1\right)$$
$$-\left(\binom{p}{1}+1\right)k^p + \left(\binom{p}{2}+\binom{p}{1}\right)(k-1)^p$$
$$-\cdots+(-1)^k\left(\binom{p}{k}+\binom{p}{k-1}\right)1^p = 1 + \binom{p}{1}k^{p-1}$$
$$+\cdots+\binom{p}{p-1}k - \binom{p}{1}k^p + \left(\binom{p}{2}+\binom{p}{1}\right)(k-1)^p$$
$$-\cdots+(-1)^k\left(\binom{p}{l}+\binom{p}{k-1}\right)1^p.$$

As all binomial coefficients, represented in the last sum, are divisible by p, we get that

$$E(p,k) \equiv 1 (mod\, p). \qquad \square$$

Eulerian numbers of the first order and enumeration of trees

3.5.18. In Chapter 2, we considered several ways to represent a permutation π of the set $\{1,2,\ldots,n\}$ as a rooted tree.

The first method gives a bijection between the set S_n of all permutations of $\{1,2,\ldots,n\}$ and the set of all *increasing binary trees* on n vertices, i.e., rooted binary trees with n vertices labelled $1,2,\ldots,n$ such that the labels along any path from the root are increasing.

The increasing binary tree $T(\pi)$, corresponding to a permutation π, can be constructed inductively: first, represent π as $\pi = \rho 1 \delta$, where ρ and δ are permutations of the sets A_1 and A_2, forming a partition (A_1, A_2) of the set $\{2,3,\ldots,n\}$; choose 1 as the root of $T(\pi)$; use the trees $T(\rho)$ and $T(\delta)$ as the left and right subtrees of $T(\pi)$; the tree $T(\rho)$ (as well as $T(\delta)$) is constructed using the same algorithm if we choose as the root of $T(\rho)$ the least element of ρ: $T(\emptyset) = \emptyset$. To obtain π from $T(\pi)$, we can simply read the numbers of vertices in the symmetric order, i.e., first the labels of the left subtree (in symmetric order, recursively), then the label of the root, and then the labels of the right subtree.

By its construction, the tree $T(\pi)$, corresponding to a permutation $\pi = (\pi_1, \pi_2, \ldots, \pi_n)$, has the following properties (for convenience, consider the permutation π as the string $(0, \pi_1, \pi_2, \ldots, \pi_n, 0)$, i.e., set $\pi_0 = \pi_{n+1} = 0$):

- a vertex i of $T(\pi)$ has precisely the successor below right if $\pi_{i-1} < \pi_i < \pi_{i+1}$ (the element i is a *double ascent*);
- a vertex i of $T(\pi)$ has precisely the successor below left if $\pi_{i-1} > \pi_i > \pi_{i+1}$ (the element i is a *double descent*);
- a vertex i of $T(\pi)$ has precisely the successors below left and below right if $\pi_{i-1} > \pi_i < \pi_{i+1}$ (the element i is a *valley*);
- a vertex i of $T(\pi)$ has no successors if $\pi_{i-1} < \pi_i > \pi_{i+1}$ (the element i is a *peak*).

So, we can state that there are exactly $n!$ rooted binary trees on n vertices, and the Eulerian numbers $E(n,k)$ count the numbers of such trees with exactly k vertices having left successors.

- *The number of increasing binary trees on n vertices with exactly k vertices having left successors is the Eulerian number $E(n,k)$.*

3.5.19. A second way to represent a permutation of the set $\{1, 2, \ldots, n\}$ by a tree uses the bijection between the set S_n of all n-permutations and the set of all *increasing trees* on $n+1$ vertices, labelled as $0, 1, 2, \ldots, n$.

In fact, given $\pi = (\pi_1, \pi_2, \ldots, \pi_n) \in S_n$, we construct an (unordered) tree $T(\pi)$ with vertices $0, 1, \ldots, n$ by defining the vertex i to be the successor of the rightmost element j of π, which precedes i, and which is less than i. If there is no such element j, then let i be the successor of the root 0.

It is easy to see that the successors of the root 0 are simply the *left-to-right minima* (*retreating elements*) of π, i.e., the elements π_i, such that $\pi_i < \pi_j$ for all $j < i$. Moreover, the leaves of the tree $T(\pi)$ are simply the elements π_i for which $\pi_i > \pi_{i+1}$ (including $\pi_n > \pi_{n+1} = 0$).

So, we can state that the number of unordered increasing trees on $n+1$ vertices is $n!$, and the number of such trees with k leaves is the Eulerian number $E(n, k-1)$.

- *The number of unordered increasing trees on $n+1$ vertices with exactly k leaves is the Eulerian number $E(n, k-1)$.*

3.5.20. The third way to connect the Eulerian numbers of the first order with some special trees is given in [Pete15]. Define a *full increasing binary tree of size n* as a rooted planar full binary tree with n internal vertices, labelled as $1, 2, \ldots, n$, such that the labels along any path from the root are increasing.

There is an obvious bijection between the set of all full increasing binary trees of size n and the set of all increasing binary trees on n vertices: we should simply delete in a given full increasing binary tree of size n all (unlabelled) leaves.

So, we can state that there are exactly $n!$ full increasing binary trees of size n, and the Eulerian numbers $E(n, k)$ count the numbers of such trees with exactly k internal vertices having left successors.

- *The number of full increasing binary trees of size n with exactly k internal vertices having left successors is the Eulerian number $E(n, k)$.*

Eulerian numbers of the first order and probability theory

3.5.21. It is proven that

$$E(n,0) + E(n,1) + \cdots + E(n,n-1) + E(n,n) = n!, \quad n \geq 0.$$

We can rewrite this equality in the form

$$\frac{E(n,0)}{n!} + \frac{E(n,1)}{n!} + \frac{E(n,2)}{n!} + \cdots + \frac{E(n,n)}{n!} = 1.$$

For such approach, the values $\frac{E(n,k)}{n!}$ can be considered as a system of probabilities of some random variable. Moreover, as $\int_0^1 \cdots \int_0^1 dx_1 \ldots dx_n = 1$, the last equality can be represented as

$$\frac{E(n,0)}{n!} + \frac{E(n,1)}{n!} + \frac{E(n,2)}{n!} + \cdots + \frac{E(n,n)}{n!} = \int_0^1 \cdots \int_0^1 dx_1 \ldots dx_n.$$

In fact, there exists a much more general result (see, for example, [GKP94], [JaWa24], and [Pete15]):

- *For a fixed function* $f\colon \mathbb{R} \to \mathbb{C}$, *integrable on the interval* $(0,1)$, *it holds that*

$$\frac{E(n,0)}{n!}f(0) + \frac{E(n,1)}{n!}f(1) + \frac{E(n,2)}{n!}f(2) + \cdots + \frac{E(n,n)}{n!}f(n)$$
$$= \int_0^1 \cdots \int_0^1 f(\lfloor x_1 + \cdots + x_n \rfloor) dx_1 \cdots dx_n.$$

□ In order to prove the equality

$$\sum_{k=0}^{n} E(n,k)\frac{f(k)}{n!} = \int_0^1 \cdots \int_0^1 f(\lfloor x_1 + \cdots + x_n \rfloor) dx_1 \cdots dx_n,$$

or, in other words, the equality

$$\sum_{k=0}^{n} E(n,k)f(k) = n! \int_0^1 \cdots \int_0^1 f\left(\lfloor x_1 + \cdots + x_n \rfloor\right) dx_1 \cdots dx_n,$$

we should note that it equivalent to the statement that $\frac{E(n,k)}{n!}$ is the probability of $\lfloor x_1 + \cdots + x_n \rfloor$ being equal to k if $x_1, \ldots, x_n$ are independent random variables, uniformly distributed on $[0,1]$. In fact, for given random variables $x_1, \ldots, x_n$, define the random variables $y_1 = x_1$, $y_2 = x_1 + x_2 (mod\ 1)$, $y_3 = x_1 + x_2 + x_3 (mod\ 1), \ldots, y_n = x_1 + x_2 + \cdots + x_n (mod\ 1)$. Then, $y_1, y_2, \ldots, y_n$ are independent random variables, uniformly distributed on $[0,1]$, and $\lfloor x_1 + \cdots + x_n \rfloor$ is the number of descents in the sequence $y_1, \ldots, y_n$. So, the probability that $\lfloor x_1 + \cdots + x_n \rfloor$ is equal to k can be calculated as the number $E(n,k)$ of sequences $y_1, \ldots, y_n$ with k descents, divided by the number $n!$ of all sequences $y_1, \ldots, y_n$. □

3.5.22. Upon proving the previous statement, we get the following probabilistic interpretation of the value $\frac{E(n,k)}{n!}$:

- *If* $x_1, \ldots, x_n$ *are independent random variables, uniformly distributed on* $[0,1]$, *then* $\frac{E(n,k)}{n!}$ *is the probability that* $\lfloor x_1 + \cdots + x_n \rfloor$ *is equal to* k.

In other words, the number $\frac{E(n,k)}{n!}$ gives the probability that the sum of n independent uniformly distributed on the segment $[0, 1]$ random variables lies between $k - 1$ and k.

This property leads to a natural geometric interpretation of the Eulerian numbers of the first order (see [Pete15] and [Stan97]).

- *The value of* $\frac{E(n,k)}{n!}$ *expresses the volume of the part of the* n*-dimensional hypercube* $[0,1]^n$*, bounded by the hyperplanes* $x_1+x_2 + \cdots + x_n = k$ *and* $x_1 + x_2 + \cdots + x_n = k + 1$.

□ This statement immediately follows from the previous properties: $\lfloor x_1 + \cdots + x_n \rfloor = k$ if and only if $k \leq x_1 + \cdots + x_n < k + 1$. □

3.5.23. If we consider a random variable η_n, counting the number of ascents in a random permutation of the set $\{1, 2, \ldots, n\}$ (see, for example, [Sach97]), we obviously get that η_n can take values $k = 0, 1, \ldots, n - 1$, with probabilities $\frac{E(n,0)}{n!}, \frac{E(n,1)}{n!}, \ldots, \frac{E(n,n-1)}{n!}$, respectively.

As $n \to \infty$, the distribution of η_n approaches the normal distribution. More exactly,

$$\lim_{n\to\infty} \sum_{k=0}^{\lfloor \beta_n \rfloor} \frac{E(n,k)}{n!} = \frac{1}{\sqrt{2\pi}} \int_{-\infty}^{x} e^{-\frac{t^2}{2}} dt, \quad \text{where}$$

$$\beta_n = \sqrt{\frac{n+1}{12}} x + \frac{n-1}{2}$$

and $\Phi(x) = \frac{1}{\sqrt{2\pi}} \int_{-\infty}^{x} e^{-\frac{t^2}{2}} dt$ is the cumulative distribution function of the standard normal distribution $N(0,1)$.

For such β_n, we also have

$$\frac{E(n, \lfloor \beta_n \rfloor)}{n!} = \sqrt{\frac{12}{n+1}} \cdot \frac{e^{-\frac{x^2}{2}}}{\sqrt{2\pi}} + O(n^{-\frac{3}{4}})$$

uniformly for all x (see [CKSS72]).

In other words, for $\beta_n = \sqrt{\frac{n+1}{12}}x + \frac{n-1}{2}$, it holds that

$$\lim_{n\to\infty} \sqrt{\frac{n+1}{12}} \cdot \frac{E(n, \lfloor \beta_n \rfloor)}{n!} = \frac{1}{\sqrt{2\pi}} e^{-\frac{x^2}{2}},$$

where $\varphi(x) = \frac{1}{\sqrt{2\pi}} e^{-\frac{x^2}{2}}$ is the probability density function of the standard normal distribution $N(0, 1)$.

Estimations of Eulerian numbers of the first order

3.5.24. Continuing the study of Euler's triangle, we see that the same pattern occurs in each row: when we move from the left element to the right element, the numbers at first increase and then start to decrease. It turns out that this pattern has a general character.

One can prove (see [LeNi92] and [CKSS72]) that:

- *for a fixed positive integer n, the Eulerian numbers of the first order $E(n, k)$, $k = 0, 1, 2, \ldots, n-1$, have a single maximum, which is achieved for either one or two consecutive values $k \le \frac{n-1}{2}$.*

More exactly, for odd $n = 2m+1$, the sequence $E(n, k) = E(2m+1, k), k = 0, 1, 2, \ldots, n-1$, has a unique maximum at $k = m = \frac{n-1}{2}$; for even $n = 2m$, the maximum occurs at $k = m - 1 = \frac{n-2}{2}$ and at $k = m = \frac{n}{2}$. In other words, the following property holds:

- *For any $n \ge 1$, there exists a positive integer $K_n \le \frac{n-1}{2}$ such that*

  $$1 = E(n, 0) < E(n, 1) < \cdots < E(n, K_n),$$

  $$E(n, K_n) \ge E(n, K_n + 1) > E(n, K_n + 2) > \cdots > E(n, n-1) = 1.$$

  Moreover, it holds that $K_{n+1} = K_n$ for odd n and $K_{n+1} = K_n + 1$ for even n.

□ We can prove this property using the strong logarithmic concavity property. However, the proof can also be obtained using the recurrent relation for the Eulerian numbers of the first order and can be derived by induction.

A direct consideration shows that, for $n = 1, 2, 3, 4, 5$, the statement is true: $K_1 = 0, K_2 = 0, K_3 = 1$, $K_4 = 1$, and $K_5 = 2$. In fact, $K_n = \frac{n-1}{2}$ for $n = 1, 3, 5$, and $K_n = \frac{n-2}{2}$ for $n = 2, 4$.

Suppose that the statement is true for any positive integer less than or equal to n, and try to prove it for $n+1$. Consider two possible cases.

For $n = 2m$, the induction's assertion means that

$$\begin{aligned} 1 &= E(n,0) < E(n,1) < \cdots < E(n,m-1) \\ &= E(n,m) > E(n,m+1) > \cdots > E(n,n-1) = 1. \end{aligned}$$

So, using the recurrent relation $E(n+1,k) = (k+1)E(n,k) + (n-k+1)E(n,k-1)$ and the induction's assertion, we get that, for any $2 \le k \le m$, it holds that

$$\begin{aligned} &E(n+1,k) - E(n+1,k-1) \\ &\quad = ((k+1)E(n,k) + (n-k+1)E(n,k-1)) - (kE(n,k-1) \\ &\qquad + (n-k+2)E(n,k-2)) \\ &\quad = k(E(n,k) - E(n,k-1)) \\ &\qquad + (n-k+2)(E(n,k-1) - E(n,k-2)) + (E(n,k) \\ &\qquad - E(n,k-1) > 0. \end{aligned}$$

For $k = 1$, the statement is obvious: $E(n+1,1) = 2^{n+1} - n > 1 = E(n+1,0)$. So, we have proven that $1 = E(n+1,0) < E(n,1) < \cdots < E(n+1,m)$. As $E(n+1,k) = E(n-k-1)$, we get that, for $n = 2m$, $E(n+1,m+1) = E(n+1,m-1)$, $E(n+1,m+2) = E(n+1,m-2), \ldots, E(n+1,n) = E(n+1,0)$ and, therefore, $E(n+1,m) > E(n+1,m+1) > \cdots > E(n+1,n) = 1$. In other words, we have proven that, for $n+1 = 2m+1$, the sequence $E(n+1,k)$ has the only maximum at $E(n+1,m)$, and

$$\begin{aligned} 1 &= E(n+1,0) < E(n+1,1) < \cdots < E(n+1,m) > E(n+1,m+1) \\ &> \cdots > E(n+1,n) = 1. \end{aligned}$$

For $n = 2m+1$, the induction's assertion means that

$$\begin{aligned} 1 &= E(n,0) < E(n,1) < \cdots < E(n,m) > E(n,m+1) \\ &> \cdots > E(n,n-1) = 1. \end{aligned}$$

So, using similar reasons, we get that for any $1 \le k \le m$, it holds that $E(n+1,k) - E(n+1,k-1) > 0$, i.e., it holds that $1 = E(n+1,0) < E(n,1) < \cdots < E(n+1,m)$. As $E(n+1,k) = E(n-k-1)$, we get that, for $n = 2m+1$, $E(n+1,m+1) = E(n+1,m)$, $E(n+1,m+2) = E(n+1,m-1), \ldots, E(n+1,n) = E(n+1,0)$ and, therefore, $E(n+1,m) = E(n+1,m+1) > \cdots > E(n+1,n) = 1$. In other words, we have proven that, for $n+1 = 2m+2$, the sequence $E(n+1,k)$ has the only maximum at $E(n+1,m)$ and $E(n+1,m+1)$, and

$$\begin{aligned} 1 = E(n+1,0) &< E(n+1,1) < \cdots < E(n+1,m) \\ &= E(n+1,m+1) > \cdots > E(n+1,n) = 1. \end{aligned}$$ □

3.5.25. A similar result can be obtained if we consider the sequence $\frac{E(n,k)}{n!}$, $n = k+1, k+2, \ldots$, for a fixed k; that is, the sequence of elements $E(n,k)$ of the kth column of Euler's triangle with weights $\frac{1}{n!}$. In turns out (see [LeNi92]) that, for a fixed $k \ge 1$, the sequence $\frac{E(n,k)}{n!}$, $n = k+1, k+2, \ldots$ increases from $\frac{1}{(k+1)!}$ to $\frac{E(2k+1,k)}{(2k+1)!}$ for $k+1 \le n \le 2n+1$ and decreases from $\frac{E(2k+1,k)}{(2k+1)!}$ to zero for $n \ge 2n+1$.

3.5.26. Returning to the considered property of the rows of Euler's triangle, we can see that such "increasing-decreasing" behaviour is shown by the rows of other number triangles, including Pascal's triangle, the triangles of Stirling numbers of the first and second kinds, and the triangle of Narayana numbers. This situation is closely connected with the so-called *strong logarithmic concavity* (*SLC*) *property* (see [Kurt72] and [CKSS72]) for triangular arrays.

- *Euler' triangle has the strong logarithmic concavity property, i.e., it holds that*

$$(E(n,k))^2 > E(n,k-1)E(n,k+1) \quad \text{for } n \ge 2, 0 < k < n.$$

□ Let us prove this property by induction. It is easy to see that the property holds for $k = n-1$. In this case, $E(n,k+1) = E(n,n) = 0$, while $E(n,k) \ne 0$; therefore, $(E(n,k))^2 > E(n,k-1)E(n,k)$. Moreover, the property holds for all $E(n,k)$ with small values of n.

In fact,

$$\begin{aligned}(E(3,1))^2 &= 4^2 > 1\cdot 1 = E(3,0)\cdot E(3,2),\\ (E(4,1))^2 &= 11^2 > 1\cdot 11 = E(4,0)\cdot E(4,2),\\ (E(4,2))^2 &= 11^2 > 11\cdot 1 = E(4,1)\cdot E(4,3),\\ (E(5,1))^2 &= 26^2 > 1\cdot 66 = E(5,0)\cdot E(5,2),\\ (E(5,2))^2 &= 66^2 > 26\cdot 26 = E(5,1)\cdot E(5,3),\\ (E(5,3))^2 &= 26^2 > 66\cdot 1 = E(5,2)\cdot E(5,4).\end{aligned}$$

Suppose, that the SLC property holds for all rows of Euler's triangle up to the nth row, i.e., $(E(m,k))^2 > E(m,k-1)E(m,k+1)$ for all $2 \le m \le n$ and for all $0 < k < m$. Consider the $(n+1)$th row of the triangle and prove that

$$(E(n+1,k))^2 > E(n+1,k-1)E(n+1,k+1) \quad \text{for} \quad 0 < k < n+1.$$

For $k = 1$ and $k = n$, the situation is obvious.

Let $1 < k < n$. As $E(n+1,k) = (k+1)E(n,k) + (n-k+1)E(n,k-1)$, then using the induction's assertion, we get that

$$\begin{aligned}&(E(n+1,k))^2 - E(n+1,k-1)E(n+1,k+1)\\ &\quad = ((k+1)E(n,k) + (n-k+1)E(n,k-1))^2 - (kE(n,k-1)\\ &\qquad + (n-k+2)E(n,k-1))\,((k+1)E(n,k-1)\\ &\qquad + (n-k)E(n,k-1)) \ge (E(n,k-1))^2((n-k+1)^2\\ &\qquad - (n-k+2)(n-k)) + (E(n,k))^2((k+1)^2 - (k+2)k)\\ &\qquad + E(n,k)E(n,k-1)(2(k+1)(n+1-k) - k(n-k))\\ &\qquad - E(n,k-2)E(n,k+1)(n-k+2)(k+2)\\ &\quad = (E(n,k-1))^2 + (E(n,k))^2 + (E(n,k))^2((k+1)^2 - (k+2)k)\\ &\qquad + E(n,k)E(n,k-1)(2(k+1)(n+1-k) - k(n-k))\\ &\qquad - E(n,k-2)E(n,k+1)(n-k+2)(k+2)\end{aligned}$$

$$= (E(n,k-1) - E(n,k))^2 + E(n,k)E(n,k-1)(2(k+1)$$
$$\times (n+1-k) - k(n-k) + 2) - E(n,k-2)E(n,k+1)$$
$$\times (n-k+2)(k+2)$$
$$\geq (E(n,k)E(n,k-1)$$
$$- E(n,k-2)E(n,k+1))(n-k+2)(k+2).$$

Here, we used the equality $2(k+1)(n+1-k) - k(n-k)) + 2 = (n-k+2)(k+2)$. As

$$(E(n,k))^2 > E(n,k-1)E(n,k+1) \quad \text{and}$$
$$(E(n,k-1))^2 > E(n,k-2)E(n,k),$$

we get that

$$(E(n,k)E(n,k-1))^2 > E(n,k-1)E(n,k+1)E(n,k-2)E(n,k);$$

therefore,

$$E(n,k)E(n,k-1) > E(n,k-2)E(n,k+1).$$

So, the value $(E(n,k)E(n,k-1) - E(n,k-2)E(n,k+1))(n-k+2)(k+2)$ is positive, and hence, it holds that $(E(n+1,k))^2 > E(n+1,k-1)E(n+1,k+1)$. □

The same approach allows us to obtain a more general result (see [CKSS72]):

- *For $n \geq 2, 0 < r < k < n$, it holds that*

$$(E(n,k))^2 > E(n,k-r)E(n,k+r).$$

3.5.27. Consider now some other estimations of the Eulerian numbers of the first order. The simplest upper bound for the Eulerian numbers of the first order has the following form:

- *For any $n \geq 0$ and any $0 \leq k \leq n-1$, it holds that*

$$E(n,k) \leq (k+1)^n \cdot (n+2)^k.$$

□ It can be obtained by induction, using the recurrent relation

$$E(n,k) = (k+1)E(n-1,k) + (n-k)E(n-1,k-1). \qquad \square$$

In [LeNi92], we can find a more accurate estimation:

- *For $n \geq 2k-3$ and $k \geq 1$, it holds that*

$$E(n, k-1) \leq \left(\frac{n-k-1}{n-k+1}\right)^{n-2k} E(n, k).$$

However, there exists also (see [LeNi92]) a weaker form of the previous estimation:

- *For $n \geq 2k-3$ and $k \geq 1$, it holds that*

$$E(n, k-1) \leq \frac{n-k-1}{3n-5k-1} \cdot E(n, k).$$

Exercises

1. Prove that the sum of elements of the nth row of Euler's triangle, $n \geq 2$, is an even number.
2. Prove that the sum of elements of the nth row of Euler's triangle, $n \geq 2$, is divisible by all prime numbers, not exceeding n; by the product of all prime numbers, not exceeding n.
3. Find the alternating sum of the elements of the nth row of Euler's triangle for $n = 5, 6, 7, 8, 9, 10$ directly and using the formula with the Bernoulli numbers. What can we say about the divisibility of the obtained number by two, three, and four?
4. Find the connections between the alternating sum of elements of the nth row of Euler's triangle and the nth *Genocchi number* $G_n = 2(1-2^n)B_n$. Check that the first few Genocchi numbers are $0, 1, -1, 0, 1, 0, -3, 0, 17, 0, \ldots$ (sequence A226158 in the OEIS). Prove that $E(n, 0) - E(n, 1) + \cdots (-1)^n E(n, n) = -\frac{2^n G_n}{n+1}$. What can we say about the divisibility of the alternating sum of the elements of the nth row of Euler's triangle by powers of two if it is known that G_n are positive integers?
5. For $n = 4, 5, 6, 7, 8$, find the alternating sum of the values $\frac{E(n,k)}{\binom{n}{k}}$ obtained through sequential division of the elements of the nth row of Euler's triangle by the corresponding elements of the nth row of Pascal's triangle, directly and using the Bernoulli numbers. Compare the results.
6. For $n = 4, 5, 6, 7, 8$, find the alternating sum of values $\frac{E(n,k)}{\binom{n-1}{k}}$, obtained through the sequential division of the elements of the

nth row of Euler's triangle by the corresponding elements of the $(n-1)$th row of Pascal's triangle, directly. Check that all sums are equal to zero.

7. For $n = 4, 5, 6, 7, 8$ and $k = 1, 2, 3, \ldots, n-1$, check that

$$E(n, n-k)\binom{n-k}{n-k} + E(n, n-k+1)\binom{n-k+1}{n-k} + \cdots + E(n,n)\binom{n}{n-k} = k!S(n,k).$$

Formulate a rule to find the Stirling numbers of the second kind if we can use Euler's triangle and Pascal's triangle.

3.6. Polynomials Related to Eulerian Numbers of the First Order

Eulerian polynomials

3.6.1. In this chapter, we have already considered the *Eulerian polynomials* $A_n(x)$, $n = 0, 1, 2, \ldots$, defined as

$$A_n(x) = E(n,0) + E(n,1)x + E(n,2)x^2 + \cdots + E(n,n)x^n.$$

Using the triangle of Eulerian numbers of the first order, we can easily get these polynomials for small n:

$$A_0(x) \equiv 1; \quad A_1(x) \equiv 1;$$
$$A_2(x) = 1 + x; \quad A_3(x) = 1 + 4x + x^2;$$
$$A_4(x) = 1 + 11x + 11x^2 + x^3;$$
$$A_5(x) = 1 + 26x + 66x^2 + 26x^3 + x^4, \ldots.$$

3.6.2. These polynomials first appeared in 1755, when Euler considered in his book, *Institutiones Calculi Differentialis* (*Foundations of*

Differential Calculus), the polynomials

$$\alpha = \frac{1}{1 \cdot (p-1)}, \quad \beta = \frac{p+1}{1 \cdot 2 \cdot (p-1)^2},$$
$$\gamma = \frac{pp+4p+1}{1 \cdot 2 \cdot 3 \cdot (p-1)^3}, \quad \delta = \frac{p^3+11p^2+11p+1}{1 \cdot 2 \cdot 3 \cdot 4 \cdot (p-1)^4}, \ldots,$$
$$\eta = \frac{p^6+120p^5+1191p^4+2416p^3+1191p^2+120p+1}{1 \cdot 2 \cdot 3 \cdot 4 \cdot 5 \cdot 6 \cdot 7 \cdot (p-1)^7},$$

already encountered in the solution of combinatorial problems related to the Eulerian numbers and having in our notation the form $\frac{A_n(p)}{n!(p-1)^n}$, $n = 1, 2, 3, 4, 5, 6, 7$.

Euler introduced these polynomials in an attempt to evaluate the *Dirichlet eta function* $\eta(s)$, defined as

$$\eta(s) = \sum_{k=1}^{\infty} \frac{(-1)^{n-1}}{k^s} = 1 - \frac{1}{2^s} + \frac{1}{3^s} - \frac{1}{4^s} + \cdots, \qquad Re\ s > 0,$$

at $s = -1, -2, -3, \ldots$, or, in other words, to obtain information about the behaviour of the alternating sums $1^n - 2^n + 3^n - 4^n + \cdots$ of positive integer powers k^n of positive integers $k = 1, 2, 3, \ldots$.. This led him to the functional equation of the eta function, which immediately implies the functional equation of the *Riemann zeta function*

$$\zeta(s) = \sum_{l=1}^{\infty} \frac{1}{k^s} = 1 + \frac{1}{2^s} + \frac{1}{3^s} + \frac{1}{4^s} + \cdots, Re \quad s > 1.$$

In these terms, we can state that

$$A_n(-1) = E(n,0) - E(n,1) + \cdots + (-1)^{n-1} E(n, n-1)$$
$$= (2^{n+1} - 4^{n+1})\,\zeta(-n), \quad n \geq 0,$$

as $\zeta(-n) = -\frac{B^+_{n+1}}{n+1}$, where B^+_{n+1} is the $(n+1)$th second Bernoulli number (thus, $B^+_1 = \frac{1}{2}$, and $\zeta(0) = -\frac{1}{2}$).

So, the historical origin of the Eulerian polynomials is closely connected with the formula

$$\sum_{k=0}^{\infty}(k+1)^n x^k = \frac{A_n(x)}{(1-x)^{n+1}},$$

which is called the *Carlitz identity.* It was considered in Section 3.2 in connection with the Eulerian fractions:

$$1+x+x^2+\cdots+x^k+\cdots = \frac{A_0(x)}{1-x} = \frac{1}{1-x};$$

$$1+2x+3x^2+\cdots+(k+1)x^k+\cdots = \frac{A_1(x)}{(1-x)^2} = \frac{1}{(1-x)^2};$$

$$1+4x+9x^2+\cdots+(k+1)^2x^k+\cdots = \frac{A_2(x)}{(1-x)^3} = \frac{1+x}{(1-x)^3};$$

$$1+8x+27x^2+\cdots+(k+1)^3x^k+\cdots = \frac{A_3(x)}{(1-x)^3} = \frac{1+4x+x^2}{(1-x)^3};\ldots.$$

3.6.3. From the properties of the Eulerian numbers of the first order follow several simple properties of the polynomials $A_n(x)$.

- *The polynomial $A_n(x)$, $n \geq 1$, is a polynomial of degree $n-1$ with the coefficient at the highest term x^{n-1} and the coefficient at the term x^0 equal to unities; the sum of coefficients of the Euler polynomial $A_n(x)$ is equal to $n!$.*

□ By definition, $A_n(x) = \sum_{k=0}^{n} E(n,k)x^k$, and for $n = 0$, we get the polynomial $A_0(x) \equiv 1$. For $n \geq 1$, $E(n,n) = 0$ holds, and $A_n(x) = \sum_{k=0}^{n} E(n,k)x^k = \sum_{k=0}^{n-1} E(n,k)x^k$. The coefficient at the highest term x^{n-1} is $E(n,n-1) = 1$; the coefficient at the term x^0 is $E(n,0) = 1$; at last, the sum of the coefficients is $E(n,0)+E(n,1)+\cdots+e(n,n-1) = n!$. □

3.6.4. We have already proved that:

- *the Eulerian polynomials can be computed using the recurrence*

$$A_n(x) = x(1-x)A'_{n-1}(x) + A_{n-1}(x)\cdot(1+(n-1)\,x), \quad A_0(x) \equiv 1.$$

However, there exist several other recurrences for the sequence $A_n(x)$, $n = 0, 1, 2, \ldots$.

• *For any $n > 0$, the following quadratic polynomial recurrent relation holds*:

$$A_n(x) = A_{n-1}(x) + x\sum_{i=0}^{n-2}\binom{n-1}{i}A_i(x)A_{n-1-i}(x), \quad A_0(x) \equiv 1.$$

□ In fact, it is easy to see that

$$A_1(x) = A_0(x) \equiv 1, \quad A_2(x) = 1 + x = A_1(x) + x\binom{1}{0}A_0(x)A_1(x),$$

$$A_3(x) = 1 + 4x + x^2 = (1 + x) + x((1 + x) + 2)$$
$$= A_2(x) + x\left(\binom{2}{0}A_0(x)A_2(x) + \binom{2}{1}A_1(x)A_1(x)\right).$$

In general, note, that $A_n(x)$ is the generating function of ascents in the permutations of the set $\{1, 2, \ldots, n\}$: the coefficient at x^k in $A_n(x)$ is equal to the number of such permutations with exactly k ascents, $k = 0, 1, 2, \ldots, n-1$. We represent a permutation $\pi = (\pi_1, \pi_2, \ldots, \pi_n)$ of the set $\{1, 2, \ldots, n\}$ in the form $\rho n \delta$, where ρ and δ are permutations of some (possibly empty) sets X_1 and X_2, such that $X_1 \cap X_2 = \emptyset$, and $X_1 \cup X_2 = \{1, 2, \ldots, n-1\}$. If $|X_2| = n-1$, i.e., $X_1 = \emptyset$, $X_2 = \{1, 2, \ldots, n-1\}$, and $\pi = n\delta$, where δ is a permutation of $\{1, 2, \ldots, n-1\}$, then the number of ascents in the permutation π is equal to the number of ascents in the permutation δ, which gives as a result $A_{n-1}(x)$ in $A_n(x)$. If $|X_2| = i$, $0 \le i \le n-2$, then $X_1 \ne \emptyset$, $|X_1| = n - 1 - i$, and the number of ascents in the permutation $\pi = \rho n \delta$ is one more than the sum of the ascents in ρ and in δ. Then, $xA_i(x)A_{n-1-i}(x)$ is the generating function for the ascents of permutations with the same elements to the left and to the right of n, i.e., with fixed sets X_1 and X_2 of sizes $n-1-i$ and i, respectively. Now, choosing i elements of X_2 by $\binom{n-1}{i}$ ways and then summing over all possible i, we get the equality

$$A_n(x) = A_{n-1}(x) + \sum_{i=0}^{n-2}\binom{n-1}{i}xA_i(x)A_{n-1-i}(x). \qquad \square$$

3.6.5. There exists another formula which can be considered a recurrent relation for the Eulerian polynomials.

• *For any $n \geq 1$, it holds that*

$$A_n(x) = \sum_{k=0}^{n-1} \binom{n}{k} A_k(x) \cdot (x-1)^{n-1-k}, \quad A_0(x) \equiv 1.$$

□ In fact, we get that

$$\begin{aligned}
A_1(x) &= \binom{1}{0} A_0(x)(x-1)^0 \equiv 1;\\
A_2(x) &= \binom{2}{0} A_0(x)(x-1)^1 + \binom{2}{1} A_1(x)(x-1)^0\\
&= (x-1) + 2 = x+1;\\
A_3(x) &= \binom{3}{0} A_0(x)(x-1)^2 + \binom{3}{1} A_1(x)(x-1)^1 + \binom{3}{2} A_2(x)(x-1)^0\\
&= (x-1)^2 + 3(x-1) + 3(x+1) = 1 + 4x + x^2;\\
A_4(x) &= \binom{4}{0} A_0(x)(x-1)^3 + \binom{4}{1} A_1(x)(x-1)^2 + \binom{4}{2} A_2(x)(x-1)^1\\
&\quad + \binom{4}{3} A_3(x)(x-1)^0 = (x-1)^3 + 4(x-1)^2\\
&\quad + 6(x+1)(x-1)^1 + 4(1+4x+x^2)\\
&= 1 + 11x + 11x^2 + x^3.
\end{aligned}$$

In general, we try to prove this formula using the identity $\frac{xA_n(x)}{(1-x)^{n+1}} = \sum_{k=0}^{\infty} k^n x^k$. It is easy to see that

$$\begin{aligned}
&\sum_{k=0}^{n-1} \binom{n}{k} A_k(x)(x-1)^{n-1-k}\\
&= \frac{1}{x} \sum_{k=0}^{n-1} \binom{n}{k} \frac{xA_k(x)}{(1-x)^{k+1}} (1-x)^{k+1}(x-1)^{n-1-k}\\
&= \frac{1}{x} \sum_{k=0}^{n-1} (-1)^{n-1-k} \binom{n}{k} \frac{xA_k(x)}{(1-x)^{k+1}} (1-x)^n
\end{aligned}$$

$$
\begin{aligned}
&= \frac{(1-x)^n}{x} \sum_{k=0}^{n-1} (-1)^{n-1-k} \binom{n}{k} \frac{xA_k(x)}{(1-x)^{k+1}} \\
&= \frac{(1-x)^n}{x} \sum_{k=0}^{n-1} (-1)^{n-1-k} \binom{n}{k} \left(\sum_{t=1}^{\infty} t^k x^t \right) \\
&= \frac{(1-x)^n}{x} \sum_{t=1}^{\infty} x^t \sum_{k=0}^{n-1} (-1)^{n-1-k} \binom{n}{k} t^k \\
&= \frac{(1-x)^n}{x} \sum_{t=1}^{\infty} x^t \left(\binom{n}{n-1} t^{n-1} - \binom{n}{n-2} t^{n-2} \right. \\
&\quad \left. + \cdots + (-1)^{n-1} \binom{n}{0} t^0 \right) \\
&= \frac{(1-x)^n}{x} \sum_{t=1}^{\infty} x^t \left(-\binom{n}{n} t^n + \binom{n}{n-1} t^{n-1} - \binom{n}{n-2} t^{n-2} \right. \\
&\quad \left. + \cdots + (-1)^{n-1} \binom{n}{0} t^0 + \binom{n}{n} t^n \right) \\
&= \frac{(1-x)^n}{x} \sum_{t=1}^{\infty} x^t \left(t^n - (t-1)^n \right) \\
&= \frac{(1-x)^n}{x} \left(\sum_{t=1}^{\infty} t^n x^t - \sum_{t=1}^{\infty} (t-1)^n x^t \right) \\
&= \frac{(1-x)^n}{x} \left(\sum_{t=0}^{\infty} t^n x^t - \sum_{m=0}^{\infty} m^n x^{m+1} \right) \\
&= \frac{(1-x)^n}{x} \left(\sum_{t=0}^{\infty} t^n x^t - x \sum_{t=0}^{\infty} t^n x^t \right) \\
&= \frac{(1-x)^{n+1}}{x} \cdot \sum_{t=0}^{\infty} t^n x^t \\
&= \frac{(1-x)^{n+1}}{x} \cdot \frac{xA_n(x)}{(1-x)^{n+1}} \\
&= A_n(x). \qquad \square
\end{aligned}
$$

3.6.6. There also exists a formula, proved by Frobenius ([Frob10]), which allows us to calculate the Eulerian polynomials using the Stirling numbers of the second kind.

- *For the Eulerian polynomial $A_n(x)$, $n \geq 1$, it holds that*

$$A_n(x) = 1!S(n,1)(x-1)^{n-1} + 2!S(n,2)(x-1)^{n-2}$$
$$+ \cdots + n!S(n,n)(x-1)^0.$$

☐ Let us prove the property $A_n(x) = \sum_{k=1}^{n} k!\ \mathrm{S}(n,k)(x-1)^{n-k}$ by induction.

In fact, $A_1(x) \equiv 1 = 1!S(1,1)(x-1)^0$. Furthermore, $A_2(x) = 1 + x = (x-1) + 2 = 1!S(1,1)(x-1)^1 + 2!S(2,2)(1-x)^0$. Similarly, we have

$$A_3(x) = 1 + 4x + x^2 = (x-1)^2 + 6(x-1) + 6$$
$$= 1!S(3,1)(x-1)^2 + 2!S(3,2)(x-1)^1 + 3!S(3,3)(1-x)^0.$$

Suppose now that the property holds for a positive integer n: $A_n(x) = \sum_{k=1}^{n} k!S(n,k)(x-1)^{n-k}$. Then,

$$A^{'}(x) = \sum_{k=1}^{n-1} k!S(n,k)(n-k)(x-1)^{n-1-k}.$$

Using the recurrent relation $A_{n+1}(x) = x(1-x)A^{'}(x) + (1+nx)A_n(x)$, we get that

$A_{n+1}(x)$

$$= x(1-x)\sum_{k=1}^{n-1} k!S(n,k)(n-k)(x-1)^{n-1-k}$$
$$+ (1+nx)\sum_{k=1}^{n} k!S(n,k)(x-1)^{n-k}$$
$$= \sum_{k=1}^{n-1}(k!S(n,k)(1+nx)(x-1)^{n-k} - k!S(n,k)x(n-k)(x-1)^{n-k})$$
$$+ (1+nx)S(n,n)$$
$$= \sum_{k=1}^{n-1}(k!S(n,k)(x-1)^{n-k}(xk - nx + nx + 1) + (1+nx)S(n,n)$$

$$
\begin{aligned}
&= \sum_{k=1}^{n-1} (k!S(n,k)(x-1)^{n-k}(xk+1) + (1+nx)S(n,n) \\
&= \sum_{k=1}^{n-1} (k!S(n,k)(x-1)^{n-k}((x-1)k + (k+1)) + (1+nx)S(n,n) \\
&= \sum_{k=1}^{n-1} (k!kS(n,k)(x-1)^{n+1-k} \\
&\quad + \sum_{k=1}^{n-1} ((k+1)!S(n,k)(x-1)^{(n+1)-(k+1)} + (1+nx)S(n,n) \\
&= \sum_{k=1}^{n-1} (k!kS(n,k)(x-1)^{n+1-k} + \sum_{m=2}^{n} m!S(n,m-1)(x-1)^{n+1-m} \\
&\quad + (1+nx)S(n,n) \\
&= 1!S(n,1)(x-1)^n + \sum_{k=2}^{n-1} k!(kS(n,k) + S(n,k-1)(x-1)^{n+1-k} \\
&\quad + n!S(n,n-1)(x-1) + (1+nx)S(n,n).
\end{aligned}
$$

As $S(n,1) = S(n,n) = 1$, then $1!S(n,1)(x-1)^n = 1!S(n+1,1)(x-1)^n$, and

$$
\begin{aligned}
&n!S(n,n-1)(x-1) + (1+nx)S(n,n) \\
&\quad = n!S(n+1,n)(x-1) + (n+1)!S(n+1,n+1).
\end{aligned}
$$

Moreover, for $2 \leq k \leq n-2$, it holds that $kS(n,k) + S(n,k-1) = S(n+1,k)$. So,

$$
\begin{aligned}
A_{n+1}(x) = 1!S(n,1)(x-1)^n &+ \sum_{k=2}^{n-1} k!(kS(n,k) \\
&+ S(n,k-1)(x-1)^{n+1-k} + n!S(n,n-1)(x-1) \\
&+ (1+nx)S(n,n)
\end{aligned}
$$

$$= 1!S(n+1,1)(x-1)^n + \sum_{k=2}^{n-1} k!S(n+1,k)(x-1)^{n+1-k}$$

$$+ n!S(n+1,n)(x-1) + (n+1)!S(n+1,n+1)(x-1)^0$$

$$= \sum_{k=1}^{n+1} k!S(n+1,k)(x-1)^{n+1-k}. \qquad \square$$

3.6.7. Using for a polynomial $P_n(x) = a_0 + a_1x + \cdots + a_nx^n$ the classic notation $[x^k]P_n(x) = a_k$, $k = 0, 1, 2, \ldots, n$, we can represent Worpitzky's identity in terms of the Eulerian polynomials.

- *There exists the identity*

$$x^n = \sum_{k=0}^{n-1} \binom{x+k}{n} [x^k]A_n(x).$$

□ As $[x^k]A_n(x)$ denotes the coefficient at x^k in $A_n(x)$, then $[x^k]A_n(x) = E(n,k)$, and we immediately obtain the classic form of Worpitzky's identity, which was proven in Section 3.2 of this chapter. □

3.6.8. We can obtain another property of the Euler polynomials by working with special sums containing the Eulerian numbers of the first order.

- *For any real $x \neq 0$, it holds that*

$$A_n(x^{-1}) = x^{1-n}A_n(x).$$

□ It is easy to check the following:

$$A_n(x^{-1})$$

$$= E(n,0) + E(n,1)x^{-1} + E(n,2)x^{-2} + \cdots + E(n,n-1)x^{-(n-1)}$$

$$= x^{-(n-1)} \left(E(n,0)x^{n-1} + E(n,1)x^{n-2} + E(n,2)x^{n-3}\right.$$

$$\left. + \cdots + E(n,n-1)\right)$$

$$= x^{-n+1}(E(n,n-1)x^{n-1}$$

$$+E(n,n-2)x^{n-2} + E(n,n-3)x^{n-3} + \cdots + E(n,0))$$

$$= x^{-n+1}A_n(x). \qquad \square$$

3.6.9. One more interesting property of the Euler polynomials has the following form:

- *For the Eulerian polynomial $A_n(x)$, it holds that*

$$\frac{e}{1-ex} = \sum_{n=0}^{\infty} \frac{A_n(x)}{n!(1-x)^{n+1}}.$$

□ In fact, as $\frac{1-x}{e^{(x-1)y}-x} = \sum_{n=0}^{\infty} A_n(x)\frac{y^n}{n!}$, then, for $y = \frac{1}{1-x}$, we have

$$\sum_{n=0}^{\infty} \frac{A_n(x)}{n!(1-x)^{n+1}} = \frac{1}{1-x}\sum_{n=0}^{\infty} A_n(x)\frac{\frac{1}{(1-x)^n}}{n!}$$

$$= \frac{1}{1-x} \cdot \frac{1-x}{e^{(x-1)\cdot\frac{1}{1-x}} - x} = \frac{1}{e^{-1}-x} = \frac{e}{1-ex}. \qquad \square$$

3.6.10. Consider now the properties of the roots of Eulerian polynomials. It turns out that all roots of $A_n(x)$, $n \geq 2$, are distinct real negative numbers. Moreover, the polynomials $A_n(x)$, $n = 2, 3, \ldots$, form a *Sturm sequence*, that is, the roots of $A_{n+1}(x)$ are separated by the roots of $A_n(x)$. (This fact can be used to prove that the sequence $E(n, k)$, $k = 0, 1, \ldots, n-1$ is unimodal.)

As was proved earlier, the number -1 is the root of any polynomial $A_n(x)$ with even index n:

$$A_{2k}(-1) = 0.$$

Furthermore, we have proven that $A_n(x^{-1}) = x^{1-n}A_n(x)$; that is, if a is a root of $A_n(x)$, then $\frac{1}{a}$ also is a root of $A_n(x)$.

- *Let $x_1, \ldots, x_{n-1}$ be the roots of the Eulerian polynomial $A_n(x)$, $n \geq 2$, and $y_1, \ldots, y_n$ br the roots of the Eulerian polynomial $A_{n+1}(x)$. Then, the following properties hold:*
  - *the numbers $x_1, \ldots, x_{n-1}$, as well as the numbers $y_1, \ldots, y_n$, are distinct negative real numbers;*
  - *the sequence $y_1, \ldots, y_n$ is separated by the sequence $x_1, \ldots, x_{n-1}$, i.e.,*

$$y_1 < x_1 < y_2 < x_2 < \cdots < y_{n-1} < x_{n-1} < y_n;$$

– *the set* $\{y_1, x_1, y_2, x_2, \dots, y_{n-1}, x_{n-1}, y_n\}$ *contains the number* -1;
– *for any* x_i, $i = 1, 2, \dots, n-1$, *there exist* x_j, $j = 1, 2, \dots, n-1$, *such that* $x_i x_j = 1$.

□ In fact, the polynomial $A_2(x) = 1 + x$ has the only (negative and simple) root -1.

The polynomial $A_3(x) = 1 + 4x + x^2$ has the two (negative and simple) roots $-2 - \sqrt{3}$ and $-2 + \sqrt{3}$, separated by -1; moreover, $(-2 - \sqrt{3})(-2 + \sqrt{3}) = 1$.

The polynomial $A_4(x) = 1 + 11x + 11x^2 + x^3$ has the three (negative and simple) roots $-5 - 2\sqrt{6}$, -1, and $-5 + 2\sqrt{6}$, separated by $-2 - \sqrt{3}$ and $-2 + \sqrt{3}$; moreover, $(-5 - 2\sqrt{6})(-5 + 2\sqrt{6}) = 1$.

A general proof, a result due to Frobenius (see [Frob10] and [CKSS72]), can de obtained using the recurrent relation

$$\alpha_{n+1}(x) = \frac{A_{n+1}(x)}{(1-x)^{n+2}} - (x\alpha_n(x))' - \left(\frac{xA_n(x)}{(1-x)^{n+1}}\right)'. \qquad \square$$

3.6.11. Let us now list some special values of the Eulerian polynomials.

In fact, $A_n(0) = 1$, $n \geq 0$. So, we get the simplest sequence $1, 1, 1, 1, 1, 1, 1, 1, 1, 1, \dots$ (sequence A000012 in the OEIS).

For $A_n(1) = n!$, $n \geq 0$, and we get the sequence 1, 1, 2, 6, 24, 120, 720, 5040, 40320, $362880, \dots$ (sequence A000142 in the OEIS).

$A_n(2) = \sum_{k=0}^{n} k!S(n,k)$. The numbers $OB(n) = \sum_{k=0}^{n} k!S(n,k)$ are called the *Fubini numbers* (or *ordered Bell numbers*). The number $OB(n)$ gives the number of ordered partitions of an n-set. This sequence starts with the elements $1, 1, 3, 13, 75, 541, 4683, 47293, 545835, 7087261, \dots$ (sequence A000670 in the OEIS).

The sequence $A_n(3)$, $n \geq 0$, starts with the elements $1, 1, 4, 22, 160, 1456, 15904, 202672, 2951680, 48361216, \dots$ (sequence A122704 in the OEIS).

The sequence $A_n(-1)$, $n \geq 0$, has the form $1, 1, 0, -2, 0, 16, 0, -272, 0, 7936, \dots$ (sequence A155585 in the OEIS). Its entries are $\frac{2^{n+1}(2^{n+1}-1)B_{n+1}^{+}}{n+1}$, where B_{n+1}^{+} are the second Bernoulli numbers.

The sequence $A_n(-2)$, $n \geq 0$, starts with the numbers $1, 1, -1, -3, 15, 21, -441, 477, 19935, -101979, \ldots$ (sequence A087674 in the OEIS).

The OEIS also features the sequences $2^n A_n(x)$, $n \geq 0$, for $x = \frac{1}{2}$, $\frac{3}{2}$, and $-\frac{1}{2}$.

For $x = \frac{1}{2}$, the first few elements of the sequence are 1, 2, 6, 26, 150, 1082, 9366, 94586, 1091670, 14174522, ... (sequence A000629 in the OEIS). This gives the number of necklaces for the partitions of $n+1$ labelled beads.

For $x = \frac{3}{2}$, the first few elements of the sequence are $1, 2, 10, 74, 730, 9002, 133210, 2299754, 45375130, 1007179562, \ldots$ (sequence A004123 in the OEIS). They are the numbers of generalized weak orders on n points.

For $x = -\frac{1}{2}$, the first few elements of the sequence are $1, 2, 2, -6, -30, 42, 882, 954, -39870, -203958, \ldots$ (sequence A179929 in the OEIS).

Relatives of Eulerian polynomials

3.6.12. Similarly to the construction of polynomials $A_n(x) = \sum_{k=0}^{n} E(n,k)x^k$, corresponding to the Eulerian numbers $E(n,k)$, $k = 0, 1, 2, \ldots, n$, of the first order, we can obtain the polynomials $U_n(x) = \sum_{k=0}^{n} E^2(n,k)x^k$, corresponding to the Eulerian numbers $E^2(n,k)$, $k = 0, 1, 2, \ldots, n$, of the second order, which we will consider later.

In this context, it is convenient to use the term *Eulerian polynomials of the first order* for polynomials $A_n(x)$, and the term *Eulerian polynomials of the second order* for polynomials $U_n(x)$.

In addition, during the study of Euler numbers E_n, the so-called *Euler polynomials* $E_n(x)$ will appear. They are close relatives of the *Bernoulli polynomials.* These polynomials have a number of interesting and useful properties, some of which we will consider when discussing Euler numbers.

The Eulerian polynomials $A_n(x)$ and the Euler polynomials $E_n(x)$ have a sequence of values in common (up to a binary shift). If $B_n(x)$ are Bernoulli polynomials and $\zeta(s)$ is the Riemann Zeta function, we can state that, for all $n \geq 0$, the following holds:

$$A_n(-1) = 2^n E_n(1) = (2^{n+1} - 4^{n+1})\zeta(-n) = \frac{(4^{n+1} - 2^{n+1})}{(n+1)} B_{n+1}(1).$$

Exercises

1. Using Euler's triangle, find the Euler polynomials $A_n(x)$, $n = 5, 6, 7, 8, 9, 10, 11, 12$.
2. Find the Eulerian polynomials $A_n(x)$, $n = 5, 6, 7, 8, 9, 10, 11, 12$ using the recurrent relation $A_n(x) = x\,(1 - x)\,A_{n-1}^{'}(x) + A_{n-1}(x)\,(1 + (n-1)\,x)$, with the initial condition $A_0(x) \equiv 1$.
3. Find the Eulerian polynomials $A_n(x)$, $n = 5, 6, 7, 8, 9, 10, 11, 12$ using the formula

$$A_n(x) = A_{n-1}(x) + x \sum_{i=0}^{n-2} \binom{n-1}{i} A_i(x) A_{n-1-i}(x).$$

4. Find the Eulerian polynomials $A_n(x)$, $n = 5, 6, 7, 8, 9, 10, 11, 12$ using the formula

$$A_n(x) = \sum_{k=0}^{n-1} \binom{n}{k} A_k(x)(x-1)^{n-1-k}.$$

5. Find the Eulerian polynomials $A_n(x)$, $n = 5, 6, 7, 8, 9, 10, 11, 12$ using the formula

$$A_n(x) = \sum_{k=1}^{n} k!S(n,k)(x-1)^{n-k}.$$

6. Find $A_n(x)$ if $n = 0, 1, 2, 3, 4, 5$, $x = 2, 3, 4$. Find $A_n(x)$ if $n = 0, 1, 2, 3, 4, 5$, $x = \frac{1}{2}, \frac{1}{3}, \frac{1}{4}$.
7. Prove that $1 + \sum_{n=1}^{\infty} A_n(x)\frac{t^n}{n!} = \frac{1-x}{e^{t(x-1)}-x}$, $\sum_{n=0}^{\infty} xA_n(x)\frac{t^n}{n!} = \frac{1-x}{1-xe^{t(1-x)}}$, $\sum_{n=0}^{\infty} \frac{A_n(x)}{(x-1)^n}\frac{t^n}{n!} = \frac{1-x}{e^t-x}$.
8. Prove that the value of the polynomial $(xA_n(x))^{'}$ at $x = 1$ is equal to $\frac{1}{2}(n+1)!$. Prove that the value of the polynomial $(xA_n(x))^{''}$ at $x = 1$ is equal to $\frac{1}{12}(3n-2)(n+1)!$

3.7. Eulerian Numbers of the First Order in the Family of Special Numbers

Eulerian numbers of the first order and elements of the Pascal's triangle

3.7.1. We considered many connections of Eulerian numbers of the first order with other classes of special numbers In particular, the construction of Euler's triangle is similar to the construction of Pascal's triangle, and the Eulerian numbers of the first order have many common features with the binomial coefficients.

First of all, these two triangular arrays are defined by a recurrent relation, giving a considering element of the nth row as a weighted sum of two elements of the previous row, placing above the constructing element:

$$\binom{n}{k} = \binom{n-1}{k} + \binom{n-1}{k-1}, \quad \text{and}$$

$$E(n,k) = (k+1)E(n-1,k) + (n-k)E(n-1,k-1).$$

Furthermore, both triangles have several common properties:

- the elements $\binom{n}{k}$, $k = 0, 1, \ldots, n$, of the nth row of Pascal's triangle, as well as the elements $E(n,k)$, $k = 0, 1, \ldots, n$, of Euler's triangle, form *unimodal sequences*, i.e., the sequences with the only maximum achieved at one or at two consecutive values of k;
- the elements $\binom{n}{k}$, $k = 0, 1, \ldots, n$, of the nth row of Pascal's triangle, as well as the elements $E(n,k)$, $k = 0, 1, \ldots, n$, of Euler's triangle, satisfy the strong logarithmic concavity property:

$$\binom{n}{k}^2 > \binom{n-1}{k}\binom{n+1}{k}, \quad \text{and}$$

$$(E(n,k))^2 > E(n-1,k)E(n+1,k)$$

for all acceptable values of k;

- there exist special divisibility properties for the elements of the triangles' rows with prime indices p: all infernal elements of the pth row of Pascal's triangle are divisible by p; all non-zero elements of the pth row of Euler's triangle are congruent to 1 modulo p.

The list can be continued.

Besides, many properties of the Eulerian numbers of the first order are formulated using the binomial coefficients, including the decomposition

$$x^n = E(n,0)\binom{x}{n} + E(n,1)\binom{x+1}{n} + \cdots + E(n,n-1)\binom{x+n-1}{n}$$

and the closed formula

$$E(n,k) = \binom{n+1}{0}(k+1)^n - \binom{n+1}{1}k^n + \cdots + (-1)^k\binom{n+1}{k}1^n.$$

Eulerian numbers of the first order and factorial numbers

3.7.2. Any binomial coefficient $\binom{n}{m}$ can be calculated through the classical formula using factorials:

$$\binom{n}{m} = \frac{n!}{m!(n-m)!}.$$

Therefore, the Eulerian numbers of the first order, connected with the binomial coefficients, are also closely related to the *factorial numbers*.

Recall that the *factorial* of a positive integer n, denoted by $n!$, is the product of all positive integers less than or equal to n:

$$n! = 1 \cdot 2 \cdot \cdots \cdot (n-1) \cdot n, \quad n \in \mathbb{N}.$$

The value of $0!$ is 1, according to the convention for an empty product.

The sequence of factorials starts with the elements $1, 1, 2, 6, 24, 120, 720, 5040, 40320, 362880, \ldots$ (sequence A000142 in the OEIS).

As the number of all permutations of the set $\{1, 2, \ldots, n\}$ is equal to $n!$, we can state that:

- *the sum of the elements of the nth row of Euler's triangle is equal to the nth factorial number:*

$$E(n,0)+E(n,1)+\cdots+E(n,n)=n!,\quad n\geq 0.$$

In particular, $E(0,0)=1=0!$; $E(1,0)+E(1,1)=1+0=1!$; $E(2,0)+E(2,1)+E(2,2)=1+1+0=2!$, $E(3,0)+E(3,1)+E(3,2)+E(3,3)=1+4+1+0=3!$.

In this sense, the factorial numbers can be considered relatives of the powers of two 2^n, which give the sums of the elements of the nth row of Pascal's triangle, and the Bell numbers $B(n)$, which give the sums of the elements of the nth row of the triangle of Stirling numbers of the second kind. As for the triangle of unsigned Stirling numbers of the first kind, the corresponding sum is exactly $n!$:

$$\binom{n}{0}+\binom{n}{1}+\cdots+\binom{n}{n}=2^n,\quad n\geq 0;$$

$$S(n,0)+S(n,1)+\cdots+S(n,n)=B(n),\quad n\geq 0;$$

$$|s(n,0)|+|s(n,1)|+\cdots+|s(n,n)|=n!,\quad n\geq 0.$$

Eulerian numbers of the first order and Stirling numbers

3.7.3. Similar connections exist between Euler's triangle and the triangles of Stirling numbers of the second kind and (unsigned) Stirling numbers of the first kind, including the form of the recurrent relations used for the construction of these triangles, the unimodality property, and the strong logarithmic concavity property.

Besides, these three classes of numbers, as well as the set of Lah numbers $L(n,k)$, can be considered as the coefficients of some transformations among the system $x^0, x^1, \ldots, x^n, \ldots$ of ordinary powers, the system $x^{\underline{0}}, x^{\underline{1}}, \ldots, x^{\underline{n}}, \ldots$ of falling powers, the system $x^{\overline{0}}, x^{\overline{1}}, \ldots, x^{\overline{n}}, \ldots$ of rising powers, and the system $\binom{x}{0}, \binom{x}{1}, \ldots, \binom{x}{n}, \ldots$ of generalised binomial coefficients.

Moreover, we have proven several formulas, including simultaneously the Eulerian numbers of the second order, the Stirling numbers of the second kind, and other numerical constructions. For example,

$$S(n,k) = \frac{1}{k!}\left(E(n,n-k)\binom{n-k}{n-k} + E(n,n-k+1)\binom{n-k+1}{n-k} + \cdots + E(n,n-1)\binom{n-1}{n-k}\right),$$

On the other hand,

$$\begin{aligned}&E(n,n-1) + E(n,n-2)x + \cdots + E(n,0)x^{n-1}\\ &\quad = n!S(n,n) + (n-1)!S(n,n-1)(x-1)\\ &\qquad + \cdots + 1!S(n,1)(x-1)^{n-1}.\end{aligned}$$

Eulerian numbers of the first order and Bernoulli numbers

3.7.4. There exist deep connections between the Eulerian numbers of the first order and the Bernoulli numbers. They are based on the fundamental properties of the Eulerian polynomials, the Bernoulli polynomials, the Euler polynomials, and other analytic functions, including the Riemann zeta function.

For example, as $\zeta(-n) = -\frac{B_{n+1}}{n+1}$, we can state that:

- *the alternating sum of the elements of the nth row of Euler's triangle can be expressed by the value of the Riemann zeta function at the argument $-n$:*

$$\begin{aligned}&E(n,0) - E(n,1) - \cdots + (-1)^{n-1}\\ &\quad = (2^{n+1} - 4^{n+1})\frac{B^{+}_{n+1}}{n+1} = (4^{n+1} - 2^{n+1})\zeta(-n), n \geq 1.\end{aligned}$$

Similarly, it holds that

$$\begin{aligned}&\frac{E(n,0)}{\binom{n}{0}} - \frac{E(n,1)}{\binom{n}{1}} + \cdots + (-1)^n\frac{E(n,n)}{\binom{n}{n}}\\ &\quad = (n+1)B_{n+1} = -(n+1)^2\zeta(-n).\end{aligned}$$

Eulerian numbers of the first order and Euler numbers

3.7.5. The Eulerian numbers of the first and second orders and the Euler numbers are connected not only by the name Euler.

First of all, these three classes of numbers, as well as the Euler zig-zag numbers, count some special sets of permutations:

- The Eulerian number $E(n,k)$ of the first order gives the number of permutations $\pi = (\pi_1, \ldots, \pi_n)$ of the set $\{1, 2, \ldots, n\}$ with exactly k ascents i: $\pi_i < \pi_{i+1}$, $1 \leq i \leq n-1$.
- The Eulerian number $E^2(n,k)$ of the second order gives the number of Stirling permutations $\pi = (\pi_1, \ldots, \pi_{2n})$ of the multiset $\{1, 1, 2, 2, \ldots, n, n\}$ with exactly k ascents i: $\pi_i < \pi_{i+1}$, $1 \leq i \leq 2n-1$.
- The absolute value $|E_n|$ of the Euler number E_n gives the number of proper up-down permutations of the set $\{1, 2, \ldots, n\}$, i.e., the permutations $\pi = (\pi_1, \ldots, \pi_n)$ with $\pi_1 < \pi_2 > \pi_3 < \pi_4 > \cdots < \pi_n$; therefore, $E_{2k-1} = 0$, and $E_{2k} = (-1)^k |E_{2k}|$.
- The Euler zig-zag number A_n gives the number of all up-down permutations of the set $\{1, 2, \ldots, n\}$, i.e., the permutations $\pi = (\pi_1, \ldots, \pi_n)$ with $\pi_1 < \pi_2 > \pi_3 < \pi_4 > \cdots < \pi_n$, or with $\pi_1 < \pi_2 > \pi_3 < \pi_4 > \cdots < \pi_{n-1} > \pi_n$; therefore, $A_{2n} = |E_{2n}|$.

The sequence of Euler zig-zag numbers A_n starts with the elements $1, 1, 1, 2, 5, 16, 61, 272, 1385, 7936, 50521, \ldots$ (sequence A000111 in the OEIS).

It can be proven that the exponential generating function $A(x) = \sum_{n=0}^{\infty} A_n \frac{x^n}{n!}$ for the sequence A_n, $n = 0, 1, 2, \ldots$, has the form

$$A(x) = \sec x + \tan x,$$

i.e., $A(x)$ is the sum of the secant and tangent functions. This result is known as *André's theorem* (see [Stan10]).

3.7.6. Since the secant function is even and the tangent function is odd, we obtain two additional classes of numbers.

The numbers A_{2n} with even indices are called the *secant numbers* (or *zig numbers*). The first few values are 1, 1, 5, 61, 1385, 50521,

2702765, 199360981, 19391512145, 2404879675441, ... (sequence A000364 in the OEIS).

By definition, the secant numbers give the absolute values of the Euler numbers with even indices:

$$A_{2n} = |E_{2n}|.$$

As $E_{2n} = (-1)^n|E_{2n}| = (-1)^n A_{2n}$ and $E_{2n+1} = 0$, the first few elements of the sequence E_n, $n = 0, 1, 2, \ldots$, of Euler numbers are $1, 0, -1, 0, 5, 0, -61, 0, 1385, 0, \ldots$ (sequence A122045 in the OEIS).

The numbers A_{2n+1} with odd indices are called the *tangent numbers* (or *zag numbers*). The first few values are 1, 2, 16, 272, 7936, 353792, 22368256, 1903757312, 209865342976, 29088885112832, ... (sequence A000182 in the OEIS). The tangent numbers are closely related to the *Bernoulli numbers*. The relation is given by the formula

$$B_{2n} = (-1)^{n-1}\frac{2n}{4^{2n} - 2^{2n}}A_{2n-1}, \quad n \geq 1.$$

In the context of our consideration, it means that the sums of the non-zero alternating rows of Euler's triangle can be represented by the tangent numbers:

$$E(2n+1,0) - E(2n+1,1) + \cdots - E(2n+1,2n) = (-1)^n A_{2n+1}, n \geq 0.$$

Eulerian numbers of the first order and Mersenne numbers

3.7.7. There exists a simply but elegant connection between the Eulerian numbers of the first order and the Mersenne numbers.

Recall that a *Mersenne number* is a positive integer of the form

$$M_n = 2^n - 1, \quad n \in \mathbb{N}.$$

The first few Mersenne numbers are 1, 3, 7, 15, 31, 63, 127, 255, 511, 1023, 2047, ... (sequence A000225 in OEIS).

This numerical set is well known due to the problem of primality of its members. For $2^n - 1$ to be prime, it is necessary that n itself be prime. By this reason, many authors require that the exponent n in the definition above be a prime. The Mersenne numbers $2^p - 1$

with prime p form the sequence 3, 7, 31, 127, 2047, 8191, 131071, 524287, 8388607, 536870911, ... (sequence A001348 in OEIS). Moreover, the exponents n which give the Mersenne primes are 2, 3, 5, 7, 13, 17, 19, 31, ... (sequence A000043 in the OEIS), and the resulting Mersenne primes are 3, 7, 31, 127, 8191, 131071, 524287, 2147483647, ... (sequence A000668 in the OEIS).

In order to find a connection between the Mersenne numbers and the Eulerian numbers of the first order, it is enough to remember that the elements $E(n, 1)$ of the first column of Euler's triangle have the form $E(n, 1) = 2^n - n - 1$. In other words,

$$E(n, 1) = M_n - n, \quad \text{or} \quad M_n - E(n, 1) = n.$$

In particular, the following property holds:

- *Any Mersenne prime M_p can be represented as the sum of its prime index p and the number $E(p, 1)$, which is equal to the number of p-permutations with exactly one ascent.*

Eulerian numbers and Catalan numbers

3.7.8. In combinatorial mathematics, the *Catalan numbers* form a sequence of positive integers that occur in various counting problems, often involving recursively defined objects.

A *Catalan number* is a member of the sequence $C_0, C_1, C_2, \dots, C_n, \dots$, built according to the following recurrent law:

$$C_0 = 1, \quad \text{and} \quad C_n = C_0C_{n-1} + C_1C_{n-2} + \cdots + C_{n-1}C_0 \quad \text{for } n \geq 1.$$

The Catalan sequence was described by Euler, who was interested in the number of different ways of dividing a polygon into triangles.

Using this simple algorithm, it is easy to find the first few terms of the sequence of Catalan numbers: 1, 1, 2, 5, 14, 42, 132, 429, 1430, 4862, ... (sequence A000108 in the OEIS).

There are many counting problems leading to the Catalan numbers. In particular, similar to the Eulerian numbers of the first order, the Catalan numbers count some types of permutations (in fact, some *pattern-avoiding permutations*), as well as some types of trees.

3.7.9. A *pattern* is a smaller permutation embedded in a larger permutation. More precisely, a permutation σ of an n-set contains a permutation α of a k-set, $k \leq n$, as a pattern if we can find k digits of σ that appear in the same relative order as the digits of α. Consider, for example, the permutation $\sigma = 18274653$. If, instead of looking at all eight digits, we focus on the digits 8, 4, and 6, we would observe that the largest of these three digits comes first, the smallest of these digits comes second, and the middle of these three digits comes last, just as in the permutation $\alpha = 312$. So, the digits 846 are called a 312-*pattern* inside σ. It is easy to see that the permutation σ also contains the patterns 123, 132, 231, and 321. On the other hand, it does not contain the pattern 213. In this case, we say that the permutation σ *avoids* the pattern 213.

Pattern-avoiding permutations have provided many counting problems (see [Deza25]). In particular:

- *the number of permutations of an n-set that avoid the pattern* 213 *is equal to the nth Catalan number C_n, $n \geq 0$.*

3.7.10. It is widely known that the Catalan numbers appear quite often when counting the number of some classes of trees.

Recall that a *full binary tree* is a tree whose internal vertices all have a degree of 3. A tree is called a *rooted tree* if it has a selected vertex, called the *root.* If the tree is placed on the plane without self-intersecting edges, it is called a *plane.*

Consider the class of all (full) plane binary trees with n internal vertices. A direct check indicates that there is exactly one plane binary tree with one internal vertex, exactly two plane binary trees with two internal vertices, and exactly five plane binary trees with three internal vertices. The plane binary trees with n internal vertices for $n = 1, 2, 3$ are presented in the following figure (one tree for $n = 1$; two trees for $n = 2$; five trees for $n = 3$):

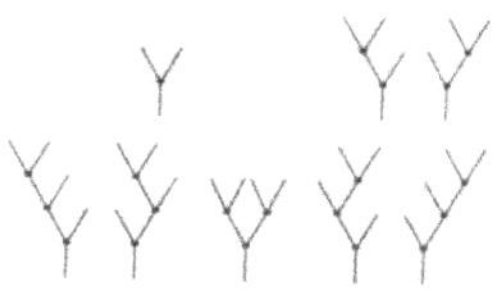

Note that the plane binary trees shown in the figure can be called *planted*, as any such tree has a *trunk*; thus, the graph is drawn to simulate a tree growing up from the ground. The lowest leaf can be considered the root of this planted tree.

Cayley proved the following fact:

- *The number of plane binary trees with n internal vertices is equal to the nth Catalan number C_n, $n \geq 0$.*

3.7.11. Consider now the set of all plane rooted trees. It is easy to check that there is exactly one plane rooted tree with zero edges, exactly one plane rooted tree with one edge, exactly two plane rooted trees with two edges, exactly five plane rooted trees with three edges, and exactly 14 plane rooted trees with four edges.

Note that any plane rooted tree can be presented as a *planted tree*, i.e., drawn to simulate a tree growing up from the root in the ground. In general, we can prove the following result:

- *The number of plane rooted trees with n edges is equal to the nth Catalan number C_n, $n \geq 0$.*

3.7.12. Other definitions of the Catalan numbers can be found in the literature. In the widely known work of John H. Conway and Richard K. Guy. *The Book of Numbers* ([CoGu96]), the elements $C_0, C_1, C_2, \ldots, C_n, \ldots$ of the sequence of Catalan numbers are defined using *Pascal's triangle*:

$$C_n = \frac{1}{n+1} \cdot \binom{2n}{n},$$

where $\binom{2n}{n} = \frac{(2n)!}{n!n!}$, $n \geq 0$, is a *central binomial coefficient.*

Thus, we can obtain the sequence $1, 1, 2, 5, 14, \ldots$ of Catalan numbers by writing out the central elements $1, 2, 6, 20, 70, \ldots$ of Pascal's triangle and dividing them by the numbers $1, 2, 3, 4, 5, \ldots,$ respectively.

$$
\begin{array}{c}
\mathbf{1}\\
1 \quad 1\\
1 \quad \mathbf{2} \quad 1\\
1 \quad 3 \quad 3 \quad 1\\
1 \quad 4 \quad \mathbf{6} \quad 4 \quad 1\\
1 \quad 5 \quad 10 \quad 10 \quad 5 \quad 1\\
1 \quad 6 \quad 15 \quad \mathbf{20} \quad 15 \quad 6 \quad 1\\
1 \quad 7 \quad 21 \quad 35 \quad 35 \quad 21 \quad 7 \quad 1\\
1 \quad 8 \quad 28 \quad 56 \quad \mathbf{70} \quad 56 \quad 28 \quad 8 \quad 1\\
\vdots \quad \vdots \quad \vdots \quad \vdots \quad \vdots \quad \vdots \quad \vdots \quad \vdots \quad \vdots \quad \vdots \quad \vdots \quad \vdots
\end{array}
$$

3.7.13. This approach allows us to construct a new class of numbers, called the *Eulerian Catalan numbers*, applying the considered algorithm to Euler's triangle.

The sequence of Eulerian Catalan numbers EC_n, $n = 0, 1, 2, \ldots$, is defined as

$$EC_n = \frac{1}{n+1} \cdot E(2n+1, n),$$

where $E(2n+1, n)\, n \geq 0$, are the *central Eulerian numbers.*

Thus, we can obtain the sequence of Eulerian Catalan numbers by writing out the central elements 1, 4, 66, 2416, 156190, 15724248, 2275172004, 447538817472, 114890380658550, 37307713155613000, ... (sequence A025585 in the OEIS) of the isosceles Euler's triangle and dividing them by the numbers $1, 2, 3, 4, 5, \ldots$, respectively.

$$
\begin{array}{c}
\mathbf{1}\\
1 \quad 1\\
1 \quad \mathbf{4} \quad 1\\
1 \quad 11 \quad 11 \quad 1\\
1 \quad 26 \quad \mathbf{66} \quad 26 \quad 1\\
1 \quad 57 \quad 302 \quad 302 \quad 57 \quad 1\\
1 \quad 120 \quad 1191 \quad \mathbf{2416} \quad 1191 \quad 120 \quad 1\\
\vdots \quad \vdots \quad \vdots \quad \vdots \quad \vdots \quad \vdots \quad \vdots \quad \vdots \quad \vdots \quad \vdots \quad \vdots \quad \vdots \quad \vdots
\end{array}
$$

The first few Eulerian Catalan numbers, beginning with $EC_0 = 1$, are 1, 2, 22, 604, 31238, 2620708, 325024572, 55942352184, 12765597850950, 3730771315561300, ... (sequence in the OEIS).

It is easy to prove that:

- *the nth Eulerian Catalan number EC_n, $n \geq 1$ is twice the number $E(2n, n)$:*

$$EC_n = 2E(2n, n), \quad n \geq 0.$$

□ In fact, due to the recurrent relation $E(n,k) = (n+1)E(n-1,k) + (n-k)E(n-1,k-1)$ and the property $E(n,k) = E(n, n-1-k)$, we have that

$$\begin{aligned}(n+1)EC_n & \\ &= E(2n+1, n) = (n+1)E(2n,n) + (2n+1-n)E(2n, n-1) \\ &= (n+1)(E(2n,n) + E(2n, 2n-1-(n-1))) = 2(n+1)E(2n,n).\end{aligned}$$

So, $EC_n = 2E(2n, n)$. □

The first few elements of the sequence $E(2n, n)$, $n = 0, 1, 2, \ldots$, of permutations of the set $\{1, 2, \ldots, 2n\}$ with n ascents have the form 1, 1, 11, 302, 15619, 1310354, 162512286, 27971176092, 6382798925475, 1865385657780650, ... (sequence A180056 in the OEIS). The Eulerian Catalan numbers can be obtained as doubled elements of this sequence.

From the last formula, one important result holds:

- *All Eulerian Catalan numbers are positive integers; moreover, all Euler Catalan numbers with positive integer indices are even numbers.*

The Eulerian Catalan numbers have a simple combinatorial interpretation.

- *The nth Eulerian Catalan number EC_n is the number of permutations $\pi = (\pi_1, \pi_2, \ldots, \pi_{2n+1})$ of the set $\{1, 2, \ldots, 2n+1\}$ with n ascents, such that every left segment π_1, $\pi_1\pi_2$, $\pi_1\pi_2\pi_3, \ldots$, $\pi_1\pi_2\cdots\pi_i\cdots$, $\pi_1\pi_2\cdots\pi_{2n+1}$ of π has at least as many ascents as descents.*

□ For $n = 1$, we count the two permutations $(1, 3, 2)$ and $(2, 3, 1)$. Both permutations have one ascent; their left segments (1, 13, 132 and 2, 23, 231, correspondingly) have zero ascents and zero descants, $0 \geq 0$; one ascent and zero descents, $1 \geq 0$; and one ascent and one descent, $1 \geq 1$. The permutations $(2, 1, 3)$ and $(3, 2, 1)$ with one ascent do not satisfy the condition, having the segments 21 and 32 with one descent and zero ascents.

The general proof (see, for example, [Deza25] and [BiSi11]) can be obtained using a variant of the *method of images*, which, using the circular permutations of the set $\{1, 2, \ldots, 2n + 1\}$, allows us to construct the equivalence classes of the permutations of $\{1, 2, \ldots, 2n+1\}$, having exactly n ascents (and, hence, exactly n descents) with the following properties: each class has $n+1$ elements and contains only one permutation $(\pi_1, \pi_2, \ldots, \pi_{2n+1})$, such that each element of its left segment $\pi_1\pi_2 \ldots \pi_i$, $i = 1, 2, \ldots, 2n + 1$, has at least as many ascents as descents. So, the number of permutations satisfying the specified conditions can be calculated as $\frac{E(2n+1,n)}{n+1}$. □

This approach leads to connections of the Eulerian numbers of the first order with the *Narayana numbers* $N(n, k)$ ([Deza25]), possessing many of the same properties as $E(n, k)$ and having many similar interpretations. In particular:

- *the Narayana number $N(n, k)$ gives the number of permutations of an n-set, which avoid the pattern* 132 *and have $k - 1$ descents.*

Eulerian numbers of the first order with negative indices

3.7.14. We built Euler's triangle by looking at its elements $E(n, k)$ for all non-negative integers n and all non-negative integers k, $k \leq n$.

A possible expansion of the set of indices is completely natural. Exactly, let us assume that, for any fixed non-negative integer n, the values $E(n, k)$ are defined for all integers values of k: they are equal to zero for all negative integers k and for all positive integers

k, $k > n$:

$$E(n, k) = 0,$$

$$n = 0, 1, 2, 3, \ldots, k = \ldots, -3, -2, -1, n + 1, n + 2, n + 3, \ldots.$$

This is fully consistent with the combinatorial interpretation of the elements of Euler's triangle and with the recurrent scheme of the construction of the triangle. Placing the unity in the position $E(0, 0)$ and using the recurrent relation $E(n, k) = (k+1)E(n-1, k)+(n-k)$ $E(n - 1, k - 1)$, we can now build the elements

$$E(n, k), \quad n = 0, 1, 2, \ldots, \quad k = 0, 1, 2, \ldots, n.$$

So, we have complete information about the behaviour of the numbers $E(n, k)$ for non-negative values of n and for all integer values of k.

Let us now see what happens with negative values of n if we assume that, for negative n and negative k, it holds that $E(n, k) = 0$:

$$E(n, k) = 0, \quad n = -1, -2, -3, \ldots, \quad k = -1, -2, -3, \ldots.$$

So, for the integers n and k, we know the behaviour of $E(n, k)$ for non-negative integers n and non-negative integers k (in fact, for $0 \leq k \leq n$), and we would like to see the values of the Eulerian numbers of the first order for negative integers n and non-negative integers k if the same recurrent relation were to be used.

Let us consider $k = 0$. If $n = 0$, we have $E(0, 0) = 0$. If $n \geq 1$, we have $E(n, 0) = 0$.

But what happens if $n \leq -1$?

- *For $n \geq 1$, it holds that $E(-n, 0) = 1$.*

□ In fact, $E(n, k) = (k + 1)E(n - 1, k) + (n - k)E(n - 1, k - 1)$. So,

$$E(n - 1, k) = \frac{1}{k + 1}\left(E(n, k) + (k - n)E(n - 1, k - 1)\right).$$

If $k = 0$, we get, as $E(n, -1) = 0$ for any integer n, the recurrent relation

$$E(n - 1, 0) = \frac{1}{0 + 1}\left(E(n, 0) + (0 - n)E(n - 1, -1)\right) = E(n, 0).$$

Then, $E(-1,0) = E(0,0) = 1$, $E(-2,0) = E(-1,0) = 1, \ldots, E(-n,0) = E(-n+1) = 1, \ldots$. □

- *For $k \geq 0$, it holds that $E(-1,k) = 1$.*

□ Using the recurrent relation $E(n-1,k) = \frac{1}{k+1}(E(n,k) + (k-n)E(n-1,k-1))$, we get for the pair $n = 0, k = 0$ the following result:

$$E(-1,0) = \frac{1}{0+1}E(0,0) = 1.$$

For $n = 0, k = 1$, we have

$$E(-1,1) = \frac{1}{2}(E(0,1) + (1-0)E(-1,0)) = \frac{1}{2}(0 + 1 \cdot 1) = \frac{1}{2}.$$

Similarly, for $n = 0, k = 2$, we have

$$E(-1,2) = \frac{1}{3}(E(0,2) + (2-0)E(-1,1)) = \frac{1}{3}\left(0 + 2 \cdot \frac{1}{2}\right) = \frac{1}{3}.$$

Suppose that $E(-1,k-1) = \frac{1}{k}$. Then, for $n = 0, k$, we have

$$\begin{aligned} E(-1,k) &= \frac{1}{k+1}(E(0,k) + (k-0)E(-1,k-1) \\ &= \frac{1}{k+1}\left(0 + k \cdot \frac{1}{k}\right) = \frac{1}{k+1}. \end{aligned}$$ □

- *For $k \geq 0$, it holds that*

$$E(-2,k) = H_{k+1}^{(2)} = 1 + \frac{1}{2^2} + \frac{1}{3^2} + \cdots + \frac{1}{(k+1)^2}.$$

□ Using the recurrent relation

$$E(n-1,k) = \frac{1}{k+1}(E(n,k) + (k-n)E(n-1,k-1),$$

we get for the pair $n = -1, k = 0$ the following result:

$$E(-2,0) = E(0,0) = 1 = H_1^{(2)}.$$

For $n = -1, k = 1$, we have

$$E(-2,1) = \frac{1}{2}\left(E(-1,1) + (1+1)E(-2,0)\right)$$
$$= \frac{1}{2}\left(\frac{1}{2} + 2 \cdot H_1^2\right) = \frac{1}{2^2} + H_1^{(2)} = H_2^{(2)}.$$

Similarly, for $n = -1, k = 2$, we have

$$E(-2,2) = \frac{1}{3}\left(E(-1,2) + (2+1)E(-2,1)\right)$$
$$= \frac{1}{3}\left(\frac{1}{3} + 3 \cdot H_2^{(2)}\right) = \frac{1}{3^2} + H_2^{(2)} = H_3^{(2)}.$$

Suppose that $E(-2, k-1) = H_k^{(2)}$. Then, for $n = -1, k$, we have

$$E(-2,k) = \frac{1}{k+1}\left(E(-1,k) + (k+1)E(-2,k-1)\right)$$
$$= \frac{1}{k+1}\left(\frac{1}{k+1} + (k+1) \cdot H_k^{(2)}\right)$$
$$= \frac{1}{(k+1)^2} + H_k^{(2)} = H_{k+1}^{(2)}. \qquad \square$$

In general, for any integer n, it holds that

$$E(n,k) = \sum_{j=0}^{k} \binom{n+1}{j} (k+1-j)^n.$$

This result coincides with our previous considerations. For $n = -1$, we have that

$$E(-1,k) = \binom{0}{0}(k+1)^0 - \binom{0}{1}k^0 + \cdots + (-1)^k \binom{0}{k} 1^0 = 1;$$

For $n = -2$, we get that

$$E(-2,k) = \binom{-1}{0}(k+1)^{-1} - \binom{-1}{1}k^{-1} + \cdots + (-1)^k \binom{-1}{k} 1^1$$
$$= \frac{1}{k+1} - \frac{(-1)(-2)}{1 \cdot 2}\frac{1}{k} + \cdots + (-1)^k \frac{(-1)(-2)\cdots(-k)}{1 \cdot 2 \cdot \cdots \cdot k}\frac{1}{1}$$
$$= 1 + \frac{1}{2} + \cdots + \frac{1}{k+1}.$$

Euler–Frobenius numbers

3.7.15. For a given real number ρ, $0 \le \rho \le 1$, and for any non-negative integers n and k, we define the *Euler–Frobenius numbers* $E(n, k, \rho)$ by the recurrent relation

$$E(n,k,\rho) = (k+\rho)E(n-1,k,\rho) + (n-k+1-\rho)E(n-1,k-1,\rho),$$

with the initial conditions $E(0, 0, \rho) = 1$,and $E(n, k, \rho) = 0$ for all positive integers $k > n$ and for all negative integers k (see [Jans13] and [GaNe13]).

So, starting from $E(0, 0, \rho) = 1$, we get that

$$\begin{aligned}
E(1,0,\rho) &= \rho E(0,0) + (2-\rho)E(0,-1) = \rho \cdot 1 + (2-\rho) \cdot 0 = \rho,\\
E(1,1,\rho) &= (1+\rho)E(0,1,\rho) + (1-\rho)E(0,0,\rho)\\
&= (1+\rho) \cdot 0 + (1-\rho) \cdot 1 = 1-\rho,\\
E(2,0,\rho) &= \rho E(1,0,\rho) + (3-\rho)E(1,-1,\rho) = \rho \cdot \rho + (3-\rho) \cdot 0 = \rho^2,\\
E(2,1,\rho) &= (1+\rho)E(1,1) + (2-\rho)E(1,0)\\
&= (1+\rho) \cdot (1-\rho) + (2-\rho) \cdot \rho = 1 + 2\rho - 2\rho^2,\\
E(2,2,\rho) &= (2+\rho)E(1,2,\rho) + (1-\rho)E(1,1,\rho)\\
&= (2+\rho) \cdot 0 + (1-\rho) \cdot (1-\rho) = 1 - 2\rho + \rho^2.
\end{aligned}$$

In the following table, the values of $E(n, k, \rho)$ for small n, k are given:

$n\backslash k$	0	1	2	3
0	1			
1	ρ	$1-\rho$		
2	ρ^2	$1+2\rho-\rho^2$	$(1-\rho)^2$	
3	ρ^3	$1+3\rho+3\rho^2-3\rho^3$	$4-6\rho^2+3\rho^3$	$(1-\rho)^3$

3.7.16. The special cases of $\rho = 0, 1$ yield the Eulerian numbers of the first order.

In fact, for $\rho = 0$, we have that $E(n, k, 0) = kE(n-1, k, 0) + (n-k+1)E(n-1, k-1)$. So, $E(n, k, 0) = E(n, k-1)$, forming a triangle with the elements $1, 0, 1, 0, 1, 1, 0, 1, 4, 1, \ldots$ (sequence A123125 in the OEIS).

For $\rho = 1$, we have that $E(n, k, 1) = (k+1)E(n-1, k, 1) + (n-k)$ $E(n-1, k-1)$, So, $E(n, k, 1) = E(n, k)$, forming a triangle with the elements $1, 1, 0, 1, 1, 0, 1, 4, 1, 0, \ldots$ (sequence A173018 in the OEIS).

3.7.17. The case of $\rho = \frac{1}{2}$ also occurs in some applications. In this case, it is more convenient to consider the numbers $B(n, k) = 2^n E(n, k, \frac{1}{2})$, which are called the *Eulerian numbers of type B*. These numbers are integers and satisfy the recurrent relation

$$B(n, k) = (2k+1)B(n-1, k) + (2n-2k+1)B(n-1, k-1),$$

with the initial conditions $B(0, 0) = 1$ and $B(n, k) = 0$ for positive integers $k > n$ and for negative integers k.

Moreover, the following closed formula holds:

$$B(n, k) = \sum_{i=0}^{k} \binom{n+1}{i} (-1)^i (2(k-i)+1)^n.$$

These numbers were considered already by Euler ([Eule49]) in connection with the study of the alternating sums

$$\sum_{k=1}^{\infty} (-1)^k (2k-1)^n = 1^n - 3^n + 5^n - 7^n + \cdots$$

of positive integer powers of odd positive integers $1, 3, 5, \ldots$.

The numbers $B(n, k)$ give the numbers of descents in the so-called *signed permutations* on the set $\{1, 2, \ldots, n\}$, i.e., the functions

$$\pi^{\pm} = \begin{pmatrix} 1 & 2 & \cdots & n \\ \epsilon_1\pi_1 & \epsilon_2\pi_2 & \cdots & \epsilon_n\pi_n \end{pmatrix} = (\epsilon_1\pi_1, \epsilon_2\pi_2, \ldots, \epsilon_n\pi_n),$$

where π is a permutation of $\{1, 2, \ldots, n\}$, and $\epsilon_i = \pm 1$.

The numbers $B(n, k) = 2^n E(n, k, \frac{1}{2})$ for small n, k are give in the following table (sequence A060187 in the OEIS). As the number of signed permutations of the set $\{1, 2, \ldots, n\}$ is $2^n n!$, the sums of the rows of the constructed triangle are equal to $2^n n!$, $n = 0, 1, 2, \ldots$.

3.7.18. The special case of $\rho = \frac{1}{N}$, where $N > 1$ is an integer, leads to a new set of special integers of the form $D(n, k, N) =$

$n\backslash k$	0	1	2	3	4	5	6	7	8
0	1								
1	1	1							
2	1	6	1						
3	1	23	23	1					
4	1	76	230	76	1				
5	1	237	1682	1682	237	1			
6	1	722	10543	23548	10543	722	1		
7	1	2179	60657	259723	259723	60657	2179	1	
8	1	6552	33161	2485288	4675014	22485288	33161	6552	1

$N^n E(n, k, \frac{1}{N})$. The numbers $D(n, k, N)$ satisfy a simple recurrent relation:

$$D(n, k, N) = (Nk + 1)D(n-1, k, N) + (N(n-k) + (N-1))D(n-1, k-1, N).$$

Moreover, the following closed formula holds:

$$D(n, k, N) = \sum_{i=0}^{k} \binom{n+1}{i} (-1)^i (N(k-i) + 1)^n.$$

These numbers are connected with the enumeration of the so-called *indexed permutations*.

For a given positive integer r, the *indexed permutation* (or *r-signed permutation*) of the set $\{1, 2, \ldots, n\}$ with the index set $Z_r = \{0, 1, \ldots, r-1\}$ is a function

$$\pi^r = \begin{pmatrix} 1 & 2 & \cdots & n \\ \pi_1^{ind_1} & \pi_2^{ind_2} & \cdots & \pi_n^{ind_n} \end{pmatrix} = (\pi_1^{ind_1}, \pi_2^{ind_2}, \ldots, \pi_n^{ind_n}),$$

where $\pi = (\pi_1, \pi_2, \ldots, \pi_n)$ is an ordinary permutation of the set $\{1, 2, \ldots, n\}$ and $ind_1, ind_2, \ldots, ind_n \in Z_r$. The set S_n^r of all the signed permutations of $\{1, 2, \ldots, n\}$ with the index set Z_r can be considered the Cartesian product of the set S_n of all the ordinary permutations of $\{1, 2, \ldots, n\}$ and the set $\{(ind_1, ind_2, \ldots, ind_n) \mid ind_i \in Z_r\}$ of all possible index vectors. For example, the indexed permutation $(3^1, 2^2, 4^0, 1^0, 5^1) \in S_n^3$ can be represented as the product of the permutation $(3,2,4,1,5) \in S_5$ and the index vector $(1,2,0,0,1)$ with values from the set $Z_3 = \{0,1,2\}$.

In particular, any ordinary permutation can be considered an indexed permutation with $r = 1$; the index set contains only one element: 0 or, say, 1. Similarly, any signed permutation can be considered as an indexed permutation with $r = 2$; the index set has two elements: $\{0,1\}$ or, say, $\{-1,1\}$.

For a given indexed permutation $(\pi_1^{ind_1}, \pi_2^{ind_2}, \ldots, \pi_n^{ind_n}) \in S_n^r$, we call the position i, $1 \le i \le n-1$, an ascent if $ind_i < ind_{i+1}$, or if $ind_i = ind_{i+1}$, and $\pi_i < \pi_{i+1}$. Similarly, we call the position i, $1 \le i \le n-1$, a descent if $ind_i > ind_{i+1}$, or if $ind_i = ind_{i+1}$, and $\pi_i > \pi_{i+1}$.

For example, the permutation $(3^0, 2^1, 4^0, 1^2, 5^1)$ has ascents at the positions 1 and 4 and has descents at the positions 2 and 3.

It turns out that:

- *the numbers* $D(n,k,N) = N^n E(n,k,\frac{1}{N})$ *enumerate the indexed permutations of the set* $\{1,2,\ldots,n\}$ *with* k *descents.*

So, we get a generalisation of the cases $N = 1$ (ordinary permutations, the Eulerian numbers of the first order) and $N = 2$ (signed permutations, the Eulerian numbers of type B).

As the number of all indexed permutations belonging to the set S_n^r is equal to $r^n n!$, we can state that

$$D(n,0,N) + D(n,1,N) + \cdots + D(n,n,N) = N^n \cdot n!.$$

3.7.19. Some other particular cases of the numbers $E(n,k,\rho)$ can be found in the special literature. For example, the numbers $E(n,k,\rho)$ are special case of the *generalised Eulerian numbers* $A(r,s|\alpha,\beta)$, introduced in [CaSc74]:

$$E(n,k,\rho) = A(n-k,k|\rho,1-\rho).$$

On the other hand, the *generalised Eulerian numbers* from [XTH12] $A_{n,k}(a,d)$ coincide with the numbers $d^n E(n,k+1,1-\frac{a}{d})$.

Eulerian numbers of the mth order

3.7.20. The Eulerian numbers of the mth order $E^m(n,k)$, $m = 1, 2, 3, \ldots$, give a natural generalisation of both classes of the Eulerian numbers, considered in detail in this book.

In order to define these numbers (see, for example, [He23]), let us consider the set of all permutations of the multiset $Q_n^m = \{1, \ldots, 1, 2, \ldots, 2, \ldots, n, \ldots, n\}$, with the multiplicity of m for each integer. The number of such permutations is equal to $P_{(m,m,\ldots,m)} = \frac{(mn)!}{(m!)^n}$. We are interested in the so-called *m-Stirling permutations*, i.e., permutations of the set such that Q_n^m and that for each value i appearing in the permutation, any values between the two copies of i are larger than i.

Define the *Eulerian number of the mth order* $E^m(n,k)$ as the number of m-Stirling permutations $\pi = (\pi_1, \pi_2, \ldots, \pi_{mn})$ with exactly k ascents, i.e., the positions i, $1 \le i \le mn - 1$, with $\pi_i < \pi_{i+1}$. Then, we can see that, by definition, $E^1(n,k) = E(n,k)$ are the Eulerian numbers of the first order, considered in this chapter. As the numbers $E^2(n,k)$ will be studied in Chapter 4, we will discuss the numbers $E^m(n,k)$ as a common generalisation of the classical Eulerian numbers at the end of Chapter 4.

Exercises

1. Define the *refined Eulerian number* $A_{n,k,m}$ as the number of permutations of the set $\{1, 2, \ldots, n\}$ with k descents, ending with the element m. Prove that

$$A_{n+1,k,n-m+1} = (k+1)A_{n,k,n-m} + (n-k)A_{n,k-1,n-m},$$

$$A_{n+1,k,n-m+1} = kA_{n,k,n-m+1} + (n-k+1)A_{n,k-1,n-m+1}.$$

2. Prove that there exists the following closed formula for $A_{n,k,m}$:

$$A_{n,k,m} = \sum_{i=0}^{k} \binom{m}{i} (-1)^i (k-i)^{n-m} (k-i+1)^{m-1}.$$

3. Let $[0,1]^n$ be the n-dimensional unit cube. Prove that the number $\frac{E(n,k,\rho)}{n!}$ gives the volume of the slice $\{(x_1,\dots,x_n) \in [0,1]^n \mid k+\rho-1 \leq x_1+\cdots+x_n \leq k+\rho\}$.
4. Define the *Euler–Frobenius polynomial* $A_{n,\rho}(x)$ by the equality $A_{n,\rho}(x) = \sum_{k=0}^{n} E(n,k,\rho)x^k$, where $E(n,k,\rho)$ are the Euler–Frobenius numbers. Prove that

$$\frac{A_{n,\rho}(x)}{(1-x)^{n+1}} = \left(\rho + x\frac{d}{dx}\right)^n \frac{1}{1-x} = \sum_{k=0}^{\infty}(k+\rho)^n x^k.$$

5. Prove that the Euler–Frobenius polynomial $A_{n,\rho}(x)$ satisfies the recurrent relation

$$A_{n,\rho}(x) = (nx+\rho(1-x))A_{n-1,\rho}(x) + x(1-x)A'_{n-1,\rho}(x),$$
$$A_{0,\rho}(x) \equiv 1.$$

   Prove that $A_{n,\rho}(x)$ is a polynomial in x of degree at most n.
6. Let $x_1,\dots,x_n$ be independent random variables uniformly distributed on $[0,1]$. Prove that, for any $n > 1$ and for any $\rho \in [0,1]$, the value $\frac{E(n,k,1-\rho)}{n!}$ gives the probability of the event $\lfloor x_1+\cdots+x_n+\rho \rfloor = k$.

References

[AbSt72], [Abra74], [Adam97], [Apos86], [BeSl95], [Bhar00], [BiSi11], [Bron01], [BuGr94], [BHM23], [Carl60], [Carl75], [Carl78], [CKSS72], [CaRi53], [CaSc74], [Cata44], [ChMo75], [ClKi08], [CoGu96], [Comt74], [CYY20], [Dede63], [DeZa80], [DeDe12], [Deza17], [Deza18], [Deza21], [Deza24], [Dick05], [Eule36], [Eule48], [Eule49], [Eule55], [Foat10], [FoSh70], [FoSh70a], [Frob10], [Gard88], [GaNe13], [Gelf59], [Geno52], [GrCh17], [GKP94], [Goul85], [He23], [HCD19], [Jord65], [Kord95], [Kurt72], [KPP94], [Lah54], [MacM16], [Pasc54], [Pete15], [QiBa17], [Sach97], [Sand78], [Sloa25], [SlPl95], [Stan97], [Stan15], [Stan86], [Stei94], [Uspe76], [WaXu10], [Weis25], [Went99], [Wiki25], [Worp83].

Chapter 4

Eulerian Numbers of the Second Order

4.1. Construction of Eulerian Numbers of the Second Order

A history of the question

4.1.1. The Eulerian numbers of the second order were investigated by Carlitz [Carl65] and Riordan [Rior76], using different approach. Carlitz obtained their formal properties and searched for a combinatorial interpretation. Riordan found such an interpretation but not in terms of ascents of some type of permutations.

The first combinatorial consideration concerning the interpretation of the Eulerian numbers of the second order as the number of special permutations, was proposed by Ira Gessel and Richard P. Stanley in 1978 [GeSt78]. They noted that these second-order Eulerian numbers can be represented as *Stirling permutations* of a given size with a given number of ascents. More precisely, if we form permutations of the multiset $\{1, 1, 2, 2, \ldots, n, n\}$ with the special property that all numbers between the two occurrences of m are greater than m, then $E^2(n, k)$ is the number of such permutations that have exactly k ascents.

4.1.2. The second-order Eulerian numbers are important chiefly because of their connection with the Stirling numbers (see [GKP94], [Gins28], [Gins65], and [Deza24]). Recently, several new interesting results related to the second-order Eulerian numbers were obtained. For example, in [Fu21], several identities related to the second-order Eulerian numbers were proved. In [He23], the mth-order Eulerian numbers, a common generalisation of the Eulerian numbers of the first and second orders, were proposed, along with numerous properties and applications. We cite many more works in this chapter.

In the literature, many different notations for the Eulerian numbers of the second order are used. The most common ones are $\left\langle\left\langle {n \atop k} \right\rangle\right\rangle$, $B_{n,k}$, $T_{n,k}$, etc.

In this book, we use for the Eulerian numbers of the second order the notation $E^2(n,k)$; it is the number of Stirling permutations of the multiset $\{1,1,2,2,\ldots,n,n\}$ having exactly k ascents.

Combinatorial definition of Eulerian numbers of the second order

4.1.3. Consider a *multiset* $\{1,1,2,2,\ldots,n,n\}$, in which every element $1,2,3,\ldots,n$ is repeated twice. The number of all possible permutations of the multiset $\{1,1,2,2,\ldots,n,n\}$ can be calculated using the formula

$$P_{(2,2,\ldots,2)} = \frac{(2n)!}{2!2!\cdots 2!} = \frac{(2n)!}{2^n}.$$

For $n=1$, there is the only permutation of the multiset $\{1,1\}$: $(1,1)$. For $n=2$, there are $\frac{4!}{2^2}=6$ permutations of the multiset $\{1,1,2,2\}$: $(1,1,2,2)$, $(1,2,1,2)$, $(1,2,2,1)$, $(2,2,1,1)$, $(2,1,2,1)$, and $(2,1,1,2)$. For $n=3$, there are $\frac{6!}{2^3}=90$ permutations of the multiset: $\{1,1,2,2,3,3\}$, etc.

Consider now all permutations of the multiset $\{1,1,2,2,\ldots,n,n\}$, such that for each m, $m=1,2,\ldots,n$, all numbers located between two m will be greater than m. Among the six permutations of the multiset $\{1,1,2,2\}$, there are exactly three permutations with this property: $(1,1,2,2)$, $(1,2,2,1)$, and $(2,2,1,1)$. For $n=3$, among the

90 permutations of multiset $\{1, 1, 2, 2, 3, 3\}$, exactly 15 permutations have this property. They are:

$$(3,3,2,2,1,1),\ (2,2,1,1,3,3),\ (2,2,1,3,3,1),\ (2,2,3,3,1,1),\ (2,3,3,2,1,1),$$
$$(1,1,3,3,2,2),\ (1,3,3,2,2,1),\ (3,3,1,1,2,2),\ (3,3,1,2,2,1),\ (1,1,2,2,3,3),$$
$$(1,2,2,1,3,3),\ (1,1,2,3,3,2),\ (1,2,3,3,2,1),\ (1,3,3,1,2,2),\ (1,2,2,3,3,1).$$

In general, the following statement holds:

- *The number of permutations of a multiset* $\{1, 1, 2, 2, \ldots, n, n\}$ *such that for each* m, $m = 1, 2, \ldots, n$, *all numbers located between two* m *will be greater than* m *can be calculated using the formula*

$$(2n-1)!! = 1 \cdot 3 \cdot \cdots \cdot (2n-1).$$

□ This fact can be easily proven by induction. In fact, the statement is true for $n = 1$:

$$(2 \cdot 1 - 1)!! = 1!! = 1,$$

and only the permutation $(1, 1)$ of the multiset $\{1, 1\}$ has the considered property. The statement also holds for $n = 2$:

$$(2 \cdot 2 - 1)!! = 3!! = 1 \cdot 3 = 3,$$

and there are exactly three permutations of the multiset $\{1, 1, 2, 2\}$ with the considered property. For $n = 3$, we have that

$$(2 \cdot 3 - 1)!! = 5!! = 1 \cdot 3 \cdot 5 = 15,$$

and there are exactly 15 permutations of the multiset $\{1, 1, 2, 2, 3, 3\}$ with the considered property.

Let us suppose that the statement holds for a positive integer $n - 1$: there are exactly

$$(2(n-1) - 1)!! = (2n-3)!! = 1 \cdot 3 \cdot \cdots \cdot (2n-3)$$

permutations of the multiset $\{1, 1, 2, 2, 3, 3, \ldots, n-1, n-1\}$ with the considered property. Any "good" permutation of the multiset $\{1, 1, 2, 2, 3, 3, \ldots, n-1, n-1, n, n\}$ can be obtained from a "good" permutation of the multiset $\{1, 1, 2, 2, 3, 3, \ldots, n-1, n-1\}$ by properly placing the two elements n and n. As by condition there can be no elements between the two elements, nn should be placed together at any possible place in the fixed "good" permutation of $\{1, 1, 2, 2, \ldots, n-1, n-1\}$: before the first element, between the first and second elements, ..., between the $(2n-3)$th and $(2n-2)$th (the last) elements, after the last element; altogether, there are $(2n-3)+2 = 2n-1$ possible places for nn. So, from any "good" permutation of the multiset $\{1, 1, 2, 2, \ldots, n-1, n-1\}$, we get exactly $2n-1$ "good" permutations of the multiset $\{1, 1, 2, 2, \ldots, n-1, n-1, n, n\}$. As there are $(2n-3)!!$ "good" permutations of the multiset $\{1, 1, 2, 2, \ldots, n-1, n-1\}$, we have proven that there are exactly $(2n-3)!! \cdot (2n-1) = (2n-1)!!$ "good" permutations of the multiset $\{1, 1, 2, 2, \ldots, n-1, n-1, n, n\}$. □

The permutations of the multiset $\{1, 1, 2, 2, \ldots, n, n\}$ which have the property that for each k, $k = 1, 2, \ldots, n$, all the numbers appearing between the two occurrences of k in the permutation are greater than k are called the *Stirling permutations.*

So, we have proven the following property:

- *The Stirling permutations of the multiset* $\{1, 1, 2, 2, \ldots, n, n\}$ *are counted by the double factorial number* $(2n-1)!!$.

The sequence of *double factorials* (or *semifactorials*) for odd $n = 1, 3, 5, 7, 9, \ldots$ starts with 1, 1, 3, 15, 105, 945, 10395, 135135, 2027025, 34459425, ... (sequence A001147 in the OEIS) The term *odd factorial* is sometimes used for the double factorial of an odd number.

For a given permutation $\pi = (\pi_1, \pi_2, \ldots, \pi_{2n})$ of the multiset $\{1, 1, 2, 2, \ldots, n, n\}$, we define an ascent as the position i, $1 \leq i \leq 2n-1$, such that $\pi_i < \pi_{i+1}$. Similarly, we define a descent as the position i, $1 \leq i \leq 2n-1$, such that $\pi_i > \pi_{i+1}$.

4.1.4. Now, we can define the *Eulerian number of the second order* $E^2(n, k)$ as the number of Stirling permutations of the multiset $\{1, 1, 2, 2, \ldots, n, n\}$ which have exactly k ascents.

For $n = 1$, only the permutation $(1, 1)$ of $\{1, 1\}$ has no ascents. So, $E^2(1, 0) = 1$.

For $n = 2$, among the three Stirling permutations $(1, 1, 2, 2)$, $(1, 2, 2, 1)$, and $(2, 2, 1, 1)$ of the multiset $\{1, 1, 2, 2\}$, the first two have one ascent each, and the third has no ascents. Thus, $E^2(2,0) = 1$ and $E^2(2,1) = 2$.

For $n = 3$, among the 15 Stirling permutations of the multiset $\{1,1,2,2,3,3\}$ one has no ascents, eight have one ascent, and six have two ascents. Thus, $E^2(3,0) = 1$, $E^2(3,1) = 8$, and $E^2(3,2) = 6$.

It is easy to see that $E^2(n,0) = 1$ for any $n \geq 1$.

- *There is exactly one permutation of the multiset* $\{1,1,2,2,\ldots, n,n\}$, $n \geq 1$, *without ascents: it is the Stirling permutation* $(n,n,n-1,n-1,\ldots,2,2,1,1)$.

□ In fact, this can be checked by induction. $\{1,1\}$ is the only permutation that has no ascents. There is only one permutation of $\{1,1,2,2\}$ without ascents: it is $(2,2,1,1)$. In order to get a permutation of $\{1,1,2,2,\ldots,n-1,n-1,n,n\}$ without ascents, we should use a permutation of $\{1,1,2,2,\ldots,n-1,n-1\}$ without ascents, i.e., the permutation $(n-1,n-1,\ldots,2,2,1,1)$; the only position of n,n which gives no ascents is the first position. So, the only permutation of $\{1,1,2,2,\ldots,n-1,n-1,n,n\}$ without ascents is the permutation $(n,n,n-1,n-1,\ldots,2,2,1,1)$. Of course, it is a Stirling permutation. □

On the other hand, $E^2(n,n) = 0$ for any $n \geq 1$.

- *There are no Stirling permutations of the multiset* $\{1,1,2,2,\ldots, n,n\}$ *having* n *ascents.*

□ In fact, $\{1,1\}$ is the only permutation that has no, i.e., zero, ascents. The Stirling permutations of the set $\{1,1,2,2\}$ have zero or one ascent. The Stirling permutations of the set $\{1,1,2,2,3,3\}$ have zero, one, or two ascents. So, for $n = 1,2,3$, it holds that $E^2(n,n) = 0$; moreover, $E^2(n,n+k) = 0$ for any $k \geq 0$. Going from n to $n+1$, we simply put the pair $n+1, n+1$ in a possible position. So, the number of descents in the new permutation will remain the same as in the old permutation, i.e., at most $n-1$, or it will increase by unity, i.e., at most n. It means that the number of ascents on any Stirling permutations of the multiset $\{1,1,2,2,\ldots,n+1,n+1\}$ is less than ore equal to n. □

So, for any positive integer n, it holds that

$$E^2(n,n)=0,\quad E^2(n,n+k)=0,\quad k\geq 0.$$

But for $n=0$, as always, we use the classic condition

$$E^2(0,0)=1$$

(convention for the empty product), while

$$E^2(0,k)=0,\quad k\geq 1.$$

Recurrent relation for Eulerian numbers of the second order

4.1.5. The Eulerian numbers of the second order satisfy the recurrence relation that follows directly from the above considerations.

- *The following is the recurrent relation for the Eulerian numbers of the second order:*

  $$E^2(n,k)=(k+1)E^2(n-1,k)+(2n-k-1)E^2(n-1,k-1)$$

  if we use for $n=0$ the values $E^2(0,0)=1$ and $E^2(0,k)=0$ for any positive integer k.

□ In fact, each Stirling permutation $\rho=(\rho_1,\rho_2,\ldots,\rho_{2n-2})$ of elements $\{1,1,2,2,,\ldots,n-1,n-1\}$ leads to $2n-1$ Stirling permutations of elements $\{1,1,2,2,\ldots,n-1,n-1,n,n\}$ if we insert the pair of new elements n,n in all possible ways: between two elements ρ_i,ρ_{i+1}, $i=1,2,3,\ldots,2n-3$ ($2n-3$ positions); before the first element ρ_1 (one position); and after the last element ρ_{2n-2} (one position). Upon inserting n,n in some position, we get three possibilities:

- a permutation $\pi=(n,n,\rho_1,\rho_2,\ldots,\rho_{2n-2})$;
- a permutation $\pi=(\rho_1,\rho_2,\ldots,\rho_i,n,n,\rho_{i+1},\ldots,\rho_{2n-2})$ in the cases $i=1,2,\ldots,2n-3$;
- a permutation $\pi=(\rho_1,\rho_2,\ldots,\rho_{2n-2},n,n)$.

The number of ascents in the permutation π is equal to the number of ascents in the permutation ρ for the first possibility

$\pi = (n, n, \rho_1, \rho_2, \ldots, \rho_{2n-2})$, or for the second possibility $\pi = (\rho_1, \rho_2, \ldots, \rho_i, n, n, \rho_{i+1}, \ldots, \rho_{2n-2})$, if $\rho_i < \rho_{i+1}$. The number of ascents in π is greater than the number of ascents in the permutation ρ for the third possibility $\pi = (\rho_1, \rho_2, \ldots, \rho_{2n-2}, n, n)$, or for the second possibility $\pi = (\rho_1, \rho_2, \ldots, \rho_i, n, n, \rho_{i+1}, \ldots, \rho_{2n-2})$, if $\rho_i \geq \rho_{i+1}$. Therefore, in order to obtain a permutation π of $\{1, 1, 2, 2, \ldots, n-1, n-1, n, n\}$ with k ascents, we can:

- use any permutation ρ of $\{1, 1, 2, 2, \ldots, n-1, n-1\}$ with k ascents, putting the elements n, n in the first position, or between two elements, forming an ascent, altogether giving $k+1$ possibilities;
- use any permutation ρ of $\{1, 1, 2, 2, \ldots, n-1, n-1\}$ with $k-1$ ascents (and, hence, with $(2n-3)-(k-1) = 2n-k-2$ non-ascents), putting the elements n, n in the last position, or between two elements, forming a non-ascent, altogether giving $(2n-k-2)+1 = 2n-k-1$ possibilities.

In other words, there are $(k+1)E^2(n-1, k)$ ways to construct a permutation π of $\{1, 1, 2, 2, \ldots, n-1, n-1, n, n\}$ with k ascents from the permutations ρ of $\{1, 1, 2, 2, \ldots, n-1, n-1\}$ with k ascents and $(2n-1)E^2(n-1, k-1)$ ways to construct a permutation π of $\{1, 1, 2, 2, \ldots, n-1, n-1, n, n\}$ from the permutations ρ of $\{1, 1, 2, 2, \ldots, n-1, n-1\}$ with $k-1$ ascents. The sum of all possibilities gives the recurrent formula

$$E^2(n, k) = (k+1)E^2(n-1, k) + (2n-k-1)E^2(n-1, k-1). \quad \square$$

For example, if $n = 2$, $k = 1$, we have

$$(1,1,2,2) \begin{matrix} \to (1,1,3,3,2,2) \\ \to (3,3,1,1,2,2) \end{matrix}, \quad (1,2,2,1) \begin{matrix} \to (1,3,3,2,2,1) \\ \to (3,3,1,2,2,1) \end{matrix},$$

$$(2,2,1,1) \begin{matrix} \to (2,2,1,1,3,3) \\ \to (2,2,1,3,3,1) \\ \to (2,2,3,3,1,1) \\ \to (2,3,3,2,1,1) \end{matrix}.$$

So, $E^2(3, 1) = 2E^2(2, 1) + 4E^2(2, 0) = 2 \cdot 2 + 4 \cdot 1 = 8$.

Using the initial conditions $E^2(0,0) = 1$, $E^2(n,0) = 0$, and $E^2(n,n) = 0, n \geq 1$, we can obtain all values of the Eulerian numbers of the second order.

Starting from $E^2(1,0) = 1$ and $E^2(1,1) = 0$, we obtain that

$$E^2(2,1) = 2 \cdot E^2(1,1) + 2 \cdot E^2(1,0) = 2 \cdot 1 + 2 \cdot 0 = 2.$$

Starting from $E^2(2,0) = 1$ and $E^2(2,1) = 2$, we obtain that

$$E^2(3,1) = 2 \cdot E^2(2,1) + 4E^2(2,0) = 2 \cdot 2 + 4 \cdot 1 = 8.$$

Similarly, as $E^2(2,1) = 1$ and $E^2(2,2) = 0$, we obtain that

$$E^2(3,2) = 3 \cdot E^2(2,2) + 3 \cdot E^2(2,1) = 3 \cdot 0 + 3 \cdot 2 = 6.$$

Starting from $E^2(3,0) = 1$ and $E^2(3,1) = 8$, we obtain that

$$E^2(4,1) = 2 \cdot E^2(3,1) + 6E^2(3,0) = 2 \cdot 8 + 6 \cdot 1 = 22.$$

Similarly, as $E^2(3,1) = 8$ and $E^2(3,2) = 6$, we obtain that

$$E^2(4,2) = 3 \cdot E^2(3,2) + 5 \cdot E^2(3,1) = 3 \cdot 6 + 5 \cdot 8 = 58.$$

As $E^2(3,2) = 6$ and $E^2(3,3) = 0$, we obtain that

$$E^2(4,3) = 4 \cdot E^2(3,3) + 4 \cdot E^2(3,2) = 4 \cdot 0 + 4 \cdot 6 = 24.$$

Triangle of Eulerian numbers of the second order

4.1.6. Using the relation obtained earlier, we can build the *triangle of Eulerian numbers of the second order*: $E^2(n,k)$, $n \geq 0$, $0 \leq k \leq n$. It can be considered as another edge-weighted version of the Pascal's triangle (sequence A007318 in the OEIS), which shares some common characteristics with this and other similar arithmetic constructions.

Formally, we can define the *triangle of Eulerian numbers of the second order* as an arithmetic triangle consisting of the numbers $E^2(n,k)$, $n,k \in \mathbb{Z}, n \geq 0, 0 \leq k \leq n$, constructed according to the following law: the top (zero) row of the triangle contains only one unity:

$$E^2(0,0) = 1.$$

The next (first) row contains one (left) unity and one (right) zero:

$$E^2(1,0) = 1, \quad E^2(1,1) = 0.$$

In the second row, the two extreme elements are again equal to unity (extreme left) and zero (extreme right):

$$E^2(2,0) = 1, \quad E^2(2,2) = 0.$$

The internal element of the second row is obtained as the sum of the two elements of the first row, located above it, up and up-left, with weights 2 and 1, correspondingly:

$$E^2(2,1) = 2 \cdot E^2(1,1) + 2 \cdot E^2(1,0) = 2 \cdot 0 + 2 \cdot 1 = 1.$$

In general, the extreme left elements of the nth row of the triangle, $n \in \mathbb{N}$, are equal to 1:

$$E^2(n,0) = 1, \quad n \in \mathbb{N}.$$

The extreme right elements of the nth row of the triangle, $n \in \mathbb{N}$, are equal to 0:

$$E^2(n,n) = 0, \quad n \in \mathbb{N}.$$

Any internal element $E^2(n,k)$ is obtained as the sum of the two elements located in the previous row above it (up and up-left), taken with their weights: the weight for the element $E^2(n-1,k)$ located in the same column k, as the calculated element, is equal to the number $k+1$ of the next column; the weight for the element $E^2(n-1,k-1)$ located up-left, i.e., in the previous column $k-1$, is equal to $2n-k-1$:

$$E^2(n,k) = (k+1) \cdot E^2(n-1,k) + (2n-k-1) \cdot E^2(n-1,k-1).$$

So, one can say that the triangle of Eulerian numbers of the second order represents a sort of a generalisation of Pascal's triangle, as well as a relative of Euler's triangle and the triangles of Stirling numbers, where the recurrent relation used is the weighted sum of the two neighbours of the element being calculated, and the weights are simple functions of the corresponding row's and column's numbers.

By sequential calculations, we get, for $0 \leq n \leq 9$ and $0 \leq k \leq n$, the following construction:

$n \backslash k$	0	1	2	3	4	5	6	7	8	9
0	1									
1	1	0								
2	1	2	0							
3	1	8	6	0						
4	1	22	58	24	0					
5	1	52	328	444	120	0				
6	1	114	1452	4400	3708	720	0			
7	1	240	5610	32120	58140	33984	5040	0		
8	1	494	19950	195800	644020	785304	341136	40320	0	
9	1	1004	67260	1062500	5765500	12440064	11026296	3733920	362880	0

This version of the triangle contains the Eulerian numbers $E^2(n,k)$ with $n, k \in \mathbb{Z}, n \geq 0, 0 \leq k \leq n$. This approach coincides with the constructions represented in [GKP94]. This triangle, read by rows, gives the sequence $1, 1, 0, 1, 2, 0, 1, 8, 6, 0, \ldots$ (sequence A201637 in the OEIS).

4.1.7. If we delete the vertex $E^2(0,0) = 1$ and the zeros $E^2(n,n) = 0$, $n \geq 1$, forming the right-hand side of the triangle, we can replace the remaining elements in the form of an isosceles triangle, i.e., we get the following arrangement:

$$
\begin{array}{c}
1 \\
1 \quad 2 \\
1 \quad 8 \quad 6 \\
1 \quad 22 \quad 58 \quad 24 \\
1 \quad 52 \quad 328 \quad 444 \quad 120 \\
1 \quad 114 \quad 1452 \quad 4400 \quad 3708 \quad 720 \\
1 \quad 240 \quad 5610 \quad 32120 \quad 58140 \quad 33984 \quad 5040 \\
\vdots \quad \vdots \quad \vdots \quad \vdots \quad \vdots \quad \vdots \quad \vdots
\end{array}
$$

This triangle, read by rows, gives the sequence 1, 1, 2, 1, 8, 6, 1, 22, 58, 24, , ... (sequence A008517 in the OEIS). This version of the triangle contains the Eulerian numbers $E^2(n,k)$ with $n,k \in \mathbb{Z}, n \geq 1, 0 \leq k \leq n-1$. In fact, in this sequence, the numbers $A^2(n,k) = E^2(n,k-1)$ with $n,k \in \mathbb{Z}, n \geq 1, 1 \leq k \leq n$, are given (see [Rior80], [Comt74], etc.). The sequence A008517 is considered the main entry for the Eulerian numbers of the second order in the OEIS. Note that the row-reversed triangle, starting with the numbers $1, 2, 1, 6, 8, 1, 24, 58, 22, 1, \ldots$, is represented by the sequence A112007 in the OEIS.

4.1.8. The triangle of Eulerian numbers of the second order, represented by the sequence $1, 0, 1, 0, 1, 2, 0, 1, 8, 6, \ldots$ (sequence A340556 in the OEIS), also does not quite agree with the usual version. The kth element of the nth row of this triangle gives the number of Stirling permutations of order n with exactly k ascents if the first index is considered to be an ascent. So, any permutation has at least one ascent, and the zeroth column contains only zeros (except the vertex of the triangle). On the other hand, the kth column, $1 \leq k \leq n$, contains the numbers $E^2(n,k-1)$:

$n\backslash k$	0	1	2	3	4	5	6	7
0	1							
1	0	1						
2	0	1	2					
3	0	1	8	6				
4	0	1	22	58	24			
5	0	1	52	328	444	120		
6	0	1	114	1452	4400	3708	720	
7	0	1	240	5610	32120	58140	33984	5040

The reader can obtain the complete description of the structure and properties of these modifications of the triangle of Eulerian numbers of the second order by themselves.

4.1.9. Consider some simple properties of the constructed triangle:

- *The sum of elements of the nth row of the triangle of Eulerian numbers of the second order is equal to* $(2n-1)!!$.

□ This statement is a direct consequence of the fact that this sum gives all the Stirling permutations of the set $\{1, 1, 2, 2, \ldots, n, n\}$, and there are exactly $(2n-1)!!$ such permutations. □

4.1.10. Direct checking shows that the left-hand side of the triangle is formed by unities (sequence A000012 in the OEIS):

$$E^2(n, 0) = 1, \quad n \geq 0.$$

The right-hand side of the triangle consists of zeros, except the vertex $E^2(0, 0)$ of the triangle:

$$E^2(n, n) = 0, \quad n \geq 1.$$

Moreover, it can be proven that:

- *the first right descending diagonal consists of the numbers $n!$, $n = 1, 2, 3, \ldots$:*

$$E^2(n, n-1) = n!, \quad n \geq 1.$$

The first elements are $1, 1, 2, 6, 24, 120, 720, 5040, 40320, 362880, \ldots$ (sequence A000142 in the OEIS).

□ This fact can be easily proven by induction. As $E^2(n, n) = 0$, $n \geq 1$,

$$\begin{aligned} E^2(n+1, n) &= (n+1)E^2(n, n) + (2(n+1) - n - 1)E^2(n, n-1) \\ &= (n+1) \cdot 0 + (n+1)E^2(n, n-1). \end{aligned}$$

In other words, for the numbers $E^2(n, n-1)$, the following recurrent relation holds:

$$E^2(n+1, n) = (n+1)E^2(n, n-1), \quad E^2(1, 0) = 1.$$

Starting with $E^2(1, 0) = 1$, we get that $E^2(2, 1) = 2E^2(1, 0) = 2 \cdot 1 = 2!$, $E^2(3, 2) = 3 \cdot E^2(2, 1) = 3 \cdot 2! = 3!, \ldots, E^2(n+1, n) = (n+1)E^2(n, n-1) = (n+1) \cdot n! = (n+1)!$. □

4.1.11. The small special cases of $E^2(n, 1)$, $n \geq 1$, $E^2(n, 2)$, $n \geq 2$, and $E^2(n, 3)$, $n \geq 3$, are given by the sequences 0, 2, 8, 22, 52, 114, 240, 494, ... (see sequence A005803 in the OEIS), 0, 6, 58, 328, 1452, 5610, 19950, 67260, 218848, 695038, ... (sequence A004301 in the OEIS), and 0, 24, 444, 4400, 32120, 195800, 1062500, 5326160, 25243904, 114876376, ... (sequence A006260 in the OEIS).

More exactly, we have the following properties:

- *For the numbers $E^2(n,1)$, the following recurrent relation holds:*

$$E^2(n+1,1) = 2E^2(n,1) + 2n, \quad E^2(1,1) = 0.$$

□ In fact, using the recurrent relation for the Eulerian numbers of the second order and the values $E^2(n,0) = 1$, we get that

$$E^2(n+1,1) = 2E^2(n,1) + 2nE^2(n,0) = 2E^2(n,1) + 2n.$$

So, starting from $E^2(1,1) = 0$, we have that $E^2(2,1) = 2E^2(1,1) + 2 = 2$, $E^2(3,1) = 2E^2(2,1) + 4 = 8$, $E^2(4,1) = 2E^2(3,1) + 6 = 22$, etc. Using these considerations, it is easy to obtain the following closed formula:

$$E^2(n,1) = 2^{n+1} - 2(n+1).$$

It can be proved by induction. For $n = 1$, we have that $E^2(1,1) = 2^2 - 2 \cdot 2 = 0$. For $n = 2$, it holds that $E^2(2,1) = 2^3 - 6 = 2$. In general, supposing $E^2(n,1) = 2^{n+1} - 2(n+1)$, we get that

$$\begin{aligned} E^2(n+1,1) &= 2E^2(n,1) + 2n = 2\left(2^{n+1} - 2(n+1)\right) + 2n \\ &= 2^{n+2} - 2(n+2). \end{aligned}$$ □

- *For the numbers $E^2(n,2)$, $n = 2,3,\ldots$, we have the recurrent relation*

$$E^2(n+1,2) = 3E^2(n,2) + (2n-1)2^{n+1} - (2n+2)(2n-1),$$
$$E^2(2,2) = 0.$$

□ In fact, using the recurrent relation for the Eulerian numbers of the second order and the values $E^2(n,1) = 2^{n+1} - 2(n+1)$, we get that

$$\begin{aligned} E^2(n+1,2) &= 3E^2(n,2) + (2n-1)E^2(n,1) \\ &= 3E^2(n,2) + (2n-1)(2^{n+1} - 2(n+1)) \\ &= 3E^2(n,2) + (2n-1)2^{n+1} - (2n+2)(2n-1). \end{aligned}$$

So, starting from $E^2(2,2) = 0$, we have that $E^2(3,2) = 3E^2(2,2) + 3 \cdot 2^3 - 6 \cdot 3 = 6$, $E^2(4,2) = 3E^2(3,2) + 5 \cdot 16 - 8 \cdot 5 = 58$, $E^2(5,2) = 3E^2(4,2) + 7 \cdot 32 - 10 \cdot 7 = 328$, etc. □

4.1.12. Considering the triangle of Eulerian numbers of the second order modulo two (using black points for odd numbers and white points for even numbers), we can obtain several simple properties.

- *For the triangle of Eulerian numbers of the second order modulo two, we have the following statements*:
  - the left-hand side of the triangle consists of black points: $E^2(n,0) = 1$, $n \geq 0$;
  - the right-hand side of the triangle consists of white points, except the vertex of the triangle: $E^2(0,0) = 1$ and $E^2(n,n) = 0$, $n \geq 1$;
  - all the internal elements of the triangle are white points: $E^2(n,k)$ is even for all $1 \leq k \leq n$.

In fact, we only need to check the last statement:

- *The Eulerian numbers of the second order $E^2(n,k)$ are even for all $1 \leq k \leq n$.*

□ Using the recurrent relation

$$E^2(n,k) = (k+1)E^2(n-1,k) + (2n-k-1)E^2(n-1,k-1)$$

and checking this fact for several first rows of the triangle, we can in fact state that in the nth row, all elements $E^2(n,k)$, $2 \leq k \leq n-2$, are even: for such k, the numbers $E^2(n-1,k)$ and $E^2(n-1,k-1)$ are even, as they are the internal elements of the $(n-1)$th row, and $E^2(n,k)$ are even, as they are the sums of two even numbers. On the other hand, for $k = 1$, we have that $E^2(n,1) = 2^{n+1} - 2(n+1)$ is even; similarly, for $k = n-1$, the number $E^2(n,n-1) = n!$ is even for all $n \geq 2$. The statement is proven. □

Considering the isosceles triangle of Eulerian numbers of the second order, we can now get the following symmetrical arrangement:

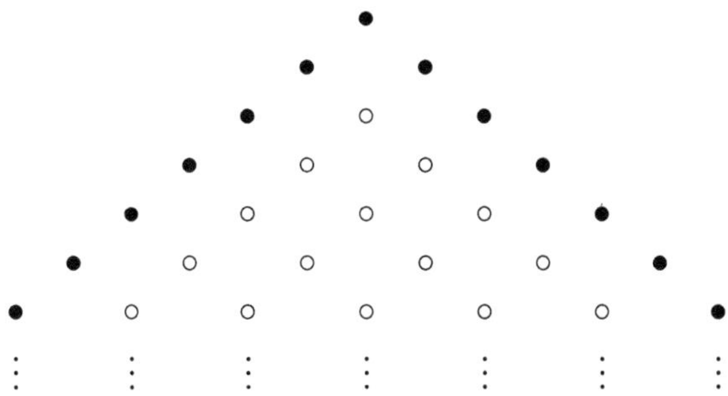

Exercises

1. Find the values of $(2k-1)!! = 1 \cdot 3 \cdot \cdots \cdot (2k-1)$ for $k = 1, 2, 3, 4, 5$. Check that the sequence $n!!$ starts with the elements $1, 3, 15, 105, 945, 10395, 135135, \ldots$ (sequence A001147 in the OEIS). Find the values of $(2k)!! = 2 \cdot 4 \cdot \cdots \cdot 2k$ for $k = 1, 2, 3, 4, 5$. Check that the sequence $n!!$ starts with the elements 1, 1, 2, 3, 8, 15, 48, 105, 384, 945, 3840. (sequence A000165 in the OEIS).
2. Prove that $(2k)!! = 2^k \cdot k!$, $(2k+1)!! = \frac{(2k+1)!}{2^k \cdot k!}$; $(2k)!! = \frac{(2n+1)!}{(k+1)!!}$; $(2k+1)!! = \frac{(2k+1)!}{(2k)!!}$, $(2k-1)!! = \frac{(2k)^{\underline{k}}}{2^k}$.
3. Find all permutations of the multiset $\{1, 1, 2, 2, \ldots, n, n\}$ for $n = 1, 2, 3, 4$. Check that the number of such permutations is equal to $\frac{(2n)!}{2^n}$. Find all permutations of the multiset $\{1, 1, 1, 2, 2, 2, \ldots, n, n, n\}$ for $n = 1, 2, 3, 4$. Check that the number of such permutations is equal to $\frac{(3n)!}{6^n}$.
4. For $n = 1, 2, 3, 4$, find all permutations of the multiset $\{1, 1, 2, 2, \ldots, n, n\}$ such that for each m, $m = 1, 2, \ldots, n$, all numbers between two m are greater than m. Check that the number of such permutations is equal to $(2n-1)!!$.
5. Using the recurrent relation $E^2(n, k) = (k+1)E^2(n-1, k) + (2n-k-1)E^2(n-1, k-1)$, build the 10th and 11th rows of the triangle of Eulerian numbers of the second order.

6. Construct the 8th and 9th rows of the isosceles triangle of Eulerian numbers of the second order. What properties does it have?

4.2. Problems Related to Eulerian Numbers of the Second Order

4.2.1. The Eulerian numbers of the second order arise in many combinatorial problems.

The most important is the connection of Eulerian numbers of the second order $E^2(n,k)$, $n=0,1,2,\ldots$, $k=0,1,\ldots,n$, with representations of the Stirling numbers.

These representations are similar to others connecting the ordinary powers, falling and rising powers, and generalised binomial coefficients. The following main decompositions of such forms were listen in Chapter 3:

- using the Stirling numbers of the second kind, we can express the ordinary powers x^n of x through the falling powers $x^{\underline{k}}$ of x:

$$x^n = S(n,0)x^{\underline{0}} + S(n,1)x^{\underline{1}} + S(n,2)x^{\underline{2}} + \cdots + S(n,3)x^{\underline{3}} \\ + \cdots + S(n,n)x^{\underline{n}};$$

- using the Stirling numbers of the second kind, we can express the ordinary powers x^n of x through the rising powers $x^{\overline{k}}$ of x:

$$x^n = (-1)^n \left(S(n,0)x^{\overline{0}} - S(n,1)x^{\overline{1}} + S(n,2)x^{\overline{2}} - S(n,3)x^{\overline{3}} \right. \\ \left. + \cdots + (-1)^n S(n,n)x^{\overline{n}}\right);$$

- using the Stirling numbers of the first kind, we can express the falling powers $x^{\underline{n}}$ of x through the ordinary powers $x^0, x^1, \ldots, x^n$ of x:

$$x^{\underline{n}} = s(n,0) + s(n,1)x + s(n,2)x^2 + \cdots + s(n,n-1)x^{n-1} \\ + s(n,n)x^n;$$

- using the Stirling numbers of the first kind, we can express the rising powers $x^{\overline{n}}$ of x through the ordinary powers

$x^0, x^1, \ldots, x^n$ of x:

$$x^{\overline{n}} = (-1)^n \left(s(n,0) - s(n,1)x + s(n,2)x^2 - \cdots + (-1)^{n-1}s(n,n-1)x^{n-1} + (-1)^n s(n,n)x^n\right);$$

- using the Lah numbers (the Stirling numbers of the third kind), we can express the rising powers $x^{\overline{n}}$ of x through the falling powers $x^{\underline{n}}$ of x:

$$x^{\overline{n}} = L(n,0)x^{\underline{0}} + L(n,1)x^{\underline{1}} + L(n,2)x^{\underline{2}} + \cdots L(n,3)x^{\underline{3}} + \cdots + L(n,n)x^{\underline{n}};$$

- using the Lah numbers, we can express the falling powers $x^{\underline{n}}$ of x through the rising powers $x^{\overline{n}}$:

$$x^{\underline{n}} = (-1)^n \left(L(n,0)x^{\overline{0}}L(n,1)x^{\overline{1}} + L(n,2)x^{\overline{2}} - \cdots L(n,3)x^{\overline{3}} + \cdots + L(n,n)x^{\overline{n}}\right);$$

- using the Eulerian numbers of the first order, we can express the ordinary powers x^n of x through the generalised binomial coefficients $\binom{x+m}{n} = \frac{(x+m)(x+m-1)\cdots(b+m-n+1)}{1\cdot 2\cdot\cdots\cdot n} = \frac{(x+m)^{\underline{n}}}{n!}$:

$$x^n = E(n,0)\binom{x+n-1}{n} + E(n,1)\binom{x+n-2}{n} + E(n,2)\binom{x+1}{n} + \cdots + E(n,n-1)\binom{x}{n}.$$

In particular, using the Eulerian numbers of the first order, we have that

$$x^1 = E(1,0)\binom{x}{1};$$

$$x^2 = E(2,0)\binom{x+1}{2} + E(2,1)\binom{x}{2}$$

$$= E(2,0)\binom{x}{2} + E(2,1)\binom{x+1}{2};$$

$$x^3 = E(3,0)\binom{x+2}{3} + E(3,1)\binom{x+1}{3} + E(3,2)\binom{x}{3}$$
$$= E(3,0)\binom{x}{3} + E(3,1)\binom{x+1}{3} + E(3,2)\binom{x+2}{3};$$
$$x^4 = E(4,0)\binom{x+3}{4} + E(4,1)\binom{x+2}{4} + E(4,2)\binom{x+1}{4}$$
$$+ E(4,3)\binom{x}{4}$$
$$= E(4,0)\binom{x}{4} + E(4,1)\binom{x+1}{4} + E(4,2)\binom{x+2}{4}$$
$$+ E(4,3)\binom{x+3}{4}.$$

4.2.2. On the other hand, the Eulerian numbers of the second order allow us to represent the Stirling numbers $S(x, x-n)$ of the second kind and the Stirling numbers of the first kind $s(x, x-n)$ as linear combinations of generalised binomial coefficients $\binom{x+m}{2n} = \frac{(x+m)(x+m-1)\cdots(x+m-2n+1)}{1\cdot 2\cdot\cdots\cdot 2n} = \frac{(x+m)^{\underline{2n}}}{(2n)!}$.

- *For the Stirling numbers of the second kind, the following identity holds:*

$$S(x, x-n) = E^2(n,0)\binom{x+n-1}{2n} + E^2(n,1)\binom{x+n-2}{2n}$$
$$+ \cdots + E^2(n, n-1)\binom{x}{2n}.$$

In particular, it holds that

$$S(x, x-1) = E^2(1,0)\binom{x}{2};$$
$$S(x, x-2) = E^2(2,0)\binom{x+1}{4} + E^2(2,1)\binom{x}{4};$$
$$S(x, x-3) = E^2(3,0)\binom{x+2}{6} + E^2(3,1)\binom{x+1}{6} + E^2(3,2)\binom{x}{6};$$

$$S(x,x-4) = E^2(4,0)\binom{x+3}{8} + E^2(4,1)\binom{x+2}{8} + E^2(4,2)\binom{x+1}{8} + E^2(4,3)\binom{x}{8}.$$

Using the triangle of Eulerian numbers of the second order, we get that

$$S(x,x-1) = 1\cdot\binom{x}{2};$$

$$S(x,x-2) = 1\cdot\binom{x+1}{4} + 2\cdot\binom{x}{4};$$

$$S(x,x-3) = 1\cdot\binom{x+2}{6} + 8\cdot\binom{x+1}{6} + 6\cdot\binom{x}{6};$$

$$S(x,x-4) = 1\cdot\binom{x+3}{8} + 22\cdot\binom{x+2}{8} + 58\cdot\binom{x+1}{8} + 24\cdot\binom{x}{8}.$$

□ We can easy check this equalities for small values of x: for example,

$$S(4,2) = 7 = 1\cdot 5 + 2\cdot 1 = 1\cdot\binom{5}{4} + 2\cdot\binom{4}{4}.$$

Moreover, the result $S(x,x-1) = \binom{x}{2}$ is well known for any integer $x \geq 1$.

In general, there are many possible proofs of this fact. Here, we try to get a proof by induction.

We prove the identity, for integers $x \geq 0$, in the form of

$$\begin{aligned} S(x,x-n) &= E^2(n,0)\binom{x+n-1}{2n} + E^2(n,1)\binom{x+n-2}{2n} \\ &\quad + E^2(n,2)\binom{x+n-3}{2n} + \cdots + E^2(n,n-1)\binom{x}{2n} \\ &\quad + E^2(n,n)\binom{x-1}{2n} \\ &= \sum_{k=0}^{n} E^2(n,k)\binom{x+n-k-1}{2n}. \end{aligned}$$

For $n = 0$, it holds that

$$S(x, x) = 1 = E^2(0, 0)\binom{x-1}{0}.$$

For $n = 1$, it holds that

$$\begin{aligned} S(x, x-1) &= \binom{x}{2} = 1 \cdot \binom{x}{2} + 0 \cdot \binom{x-1}{2} \\ &= E^2(1, 0)\binom{x}{2} + E^2(1, 1)\binom{x-1}{2}. \end{aligned}$$

Note that $E^2(n, n) = 0$ for positive integers n, so we can use for $n \geq 1$ the shorter version of the formula:

$$\begin{aligned} S(x, x-n) &= \sum_{k=0}^{n-1} E^2(n, k)\binom{x+n-k-1}{2n} \\ &= E^2(n, 0)\binom{x+n-1}{2n} + E^2(n, 1)\binom{x+n-2}{2n} \\ &\quad + E^2(n, 2)\binom{x+n-3}{2n} + \cdots + E^2(n, n-1)\binom{x}{2n}. \end{aligned}$$

For $n = 2$, we should check that

$$S(x, x-2) = E^2(2, 0) \cdot \binom{x+1}{4} + E^2(2, 1) \cdot \binom{x}{4}.$$

Let us check this formula for small values of x.

For $x = 2$, it holds that

$$\begin{aligned} S(x, x-2) = S(2, 0) = 0 = 1 \cdot 0 + 1 \cdot 0 = E^2(2, 0) \cdot \binom{3}{4} \\ + E^2(2, 1) \cdot \binom{2}{4}. \end{aligned}$$

For $x = 3$, it holds that

$$\begin{aligned} S(x, x-2) = S(3, 1) = 1 = 1 \cdot 1 + 1 \cdot 0 = E^2(2, 0) \cdot \binom{4}{4} \\ + E^2(2, 1) \cdot \binom{3}{4}. \end{aligned}$$

For $x = 4$, it holds that

$$S(x, x-2) = S(4, 2) = 7 = 1 \cdot 5 + 2 \cdot 1 = E^2(2, 0) \cdot \binom{5}{4} + E^2(2, 1) \cdot \binom{4}{4}.$$

Now, we prove the identity

$$S(x, x-2) = E^2(2, 0) \cdot \binom{x+1}{4} + E^2(2, 1) \cdot \binom{x}{4}$$

for a positive integer x, supposing that it holds for a positive integer $x - 1$:

$$S(x-1, (x-1)-2) = E^2(2, 0) \cdot \binom{(x-1)+1}{4} + E^2(2, 1) \cdot \binom{x-1}{4},$$

or, in other words,

$$S(x-1, x-3) = E^2(2, 0) \cdot \binom{x}{4} + E^2(2, 1) \cdot \binom{x-1}{4}.$$

Recall that $S(x, x-2) = (x-2)S(x-1, x-2) + S(x-1, x-3)$. As (we have supposed) $S(x-1, x-3) = E^2(2, 0) \cdot \binom{x}{4} + E^2(2, 1) \cdot \binom{x-1}{4}$ and (see the proof for $n = 1$) $S(x-1, x-2) = S(x-1, (x-1)-1) = E^2(1, 0)\binom{x-1}{2}$, we get that

$$\begin{aligned} S(x, x-2) &= (x-2)S(x-1, x-2) + S(x-1, x-3) \\ &= (x-2)E^2(1, 0)\binom{x-1}{2} + E^2(2, 0)\binom{x}{4} \\ &\quad + E^2(2, 1)\binom{x-1}{4}. \end{aligned}$$

On the other hand, considering the sum $E^2(2,0)\binom{x+1}{4} + E^2(2,1)\binom{x}{4}$. As

$$(x-2)E^2(1,0)\binom{x-1}{2} + E^2(2,0)\binom{x}{4} + E^2(2,1)\binom{x-1}{4}$$
$$= E^2(2,0)\binom{x+1}{4} + E^2(2,1)\binom{x}{4} \Leftrightarrow (x-2)E^2(1,0)\binom{x-1}{2}$$
$$+ E^2(2,0)\binom{x}{4} + E^2(2,1)\binom{x-1}{4}$$
$$= E^2(2,0)\left(\binom{x}{4} + \binom{x}{3}\right) + E^2(2,1)\left(\binom{x-1}{4} + \binom{x-1}{3}\right)$$
$$\Leftrightarrow (x-2)E^2(1,0)\binom{x-1}{2} = E^2(2,0)\binom{x}{3} + E^2(2,1)\binom{x-1}{3}$$
$$\Leftrightarrow (x-2)\cdot 1 \cdot \binom{x-1}{2} = 1 \cdot \binom{x}{3} + 1 \cdot \binom{x-1}{3}$$
$$\Leftrightarrow \frac{x-2}{x-3} = \frac{x}{3(x-3)} + \frac{2}{3} \Leftrightarrow 3x - 6 = x + 2x - 6,$$

we have proven that, for any $x \geq 2$, the following holds:

$$S(x, x-2) = E^2(2,0) \cdot \binom{x+1}{4} + E^2(2,1) \cdot \binom{x}{4}.$$

Note that we have proven the considered property for all Stirling numbers of the second kind from the first three descending diagonals of the triangle of Stirling numbers of the second kind:

$$S(0,0), S(1,1), \ldots S(x,x), \ldots; \quad S(1,0), S(2,1), \ldots S(x, x-1), \ldots;$$
$$S(2,0), S(2,1), \ldots S(x, x-2), \ldots.$$

Suppose now that the property holds for the elements

$$S(n-1, 0), S(n-1, 1), \ldots, S(x, x-(n-1)), \ldots$$

of the $(n-1)$th diagonal, and we prove it for the elements

$$S(1,0), S(2,1), \ldots, S(x, x-1), \ldots$$

of the nth diagonal. In other words, suppose that

$$\begin{aligned}
S(x, x-(n-1)) & \\
&= \sum_{k=0}^{n-2} E^2(n-1,k)\binom{x+(n-1)-k-1}{2(n-1)} \\
&= E^2(n-1,0)\binom{x+n-2}{2n-2} + E^2(n-1,1)\binom{x+n-3}{2n-2} \\
&\quad + E^2(n-1,2)\binom{x+n-4}{2n-2} + \cdots + E^2(n-1,n-2)\binom{x}{2n-2},
\end{aligned}$$

and try to prove that

$$\begin{aligned}
S(x, x-n) &= \sum_{k=0}^{n-1} E^2(n,k)\binom{x+n-k-1}{2n} \\
&= E^2(n,0)\binom{x+n-1}{2n} \\
&\quad + E^2(n,1)\binom{x+n-2}{2n} + E^2(n,2)\binom{x+n-3}{2n} + \cdots \\
&\quad + E^2(n,n-1)\binom{x}{2n}.
\end{aligned}$$

For small x, the statement is true. In fact, for $x = n$, it holds that

$$\begin{aligned}
S(x,0) = 0 = E^2(n,0)\binom{2n-1}{2n} + E^2(n,1)\binom{2n-2}{2n} \\
+ E^2(n,2)\binom{2n-3}{2n} + \cdots + E^2(n,n-1)\binom{n}{2n}.
\end{aligned}$$

For $x = n+1$, it holds that

$$\begin{aligned}
S(x,1) = 1 = 1\cdot 1 + 0 = E^2(n,0)\binom{2n}{2n} + E^2(n,1)\binom{2n-1}{2n} \\
+ E^2(n,2)\binom{2n-2}{2n} + \cdots + E^2(n,n-1)\binom{n+1}{2n}.
\end{aligned}$$

Suppose that the statement is true for a positive integer $x-1$:

$$\begin{aligned}
&S(x-1,(x-1)-n)=S(x-1,x-n-1)\\
&\quad=E^2(n,0)\binom{(x-1)+n-1}{2n}+E^2(n,1)\binom{(x-1)+n-2}{2n}\\
&\qquad+\cdots+E^2(n,n-1)\binom{x-1}{2n}.
\end{aligned}$$

Prove the statement for a positive integer x:

$$\begin{aligned}
S(x,x-n)&=E^2(n,0)\binom{x+n-1}{2n}+E^2(n,1)\binom{x+n-2}{2n}\\
&\quad+\cdots+E^2(n,n-1)\binom{x}{2n}.
\end{aligned}$$

Recall that $S(x,x-n)=(x-n)S(x-1,x-n)+S(x-1,x-n-1)$. As by the induction's assertion, it holds that

$$\begin{aligned}
&S(x-1,x-n)\\
&\quad=S(x-1,(x-1)-(n-1))\\
&\quad=\sum_{k=0}^{n-2}E^2(n-1,k)\binom{(x-1)+(n-1)-k-1}{2(n-1)}\\
&\quad=E^2(n-1,0)\binom{x+n-3}{2n-2}+E^2(n-1,1)\binom{x+n-4}{2n-2}\\
&\qquad+\cdots+E^2(n-1,n-2)\binom{x-1}{2n-2},
\end{aligned}$$

and

$$\begin{aligned}
&S(x-1,x-n-1)\\
&\quad=\sum_{k=0}^{n-1}E^2(n,k)\binom{(x-1)+n-k-1}{2n}=E^2(n,0)\binom{x+n-2}{2n}\\
&\qquad+E^2(n,1)\binom{x+n-3}{2n}+E^2(n,2)\binom{x+n-4}{2n}\\
&\qquad+\cdots+E^2(n,n-1)\binom{x-1}{2n},
\end{aligned}$$

then

$$(x-n)S(x-1,x-n)+S(x-1,x-n-1)$$
$$= (x-n)\left(E^2(n-1,0)\binom{x+n-3}{2n-2}+E^2(n-1,1)\binom{x+n-4}{2n-2}\right.$$
$$\left.+\cdots+E^2(n-1,n-3)\binom{x}{2n-2}+E^2(n-1,n-2)\binom{x-1}{2n-2}\right)$$
$$+\left(E^2(n,0)\binom{x+n-2}{2n}+E^2(n,1)\binom{x+n-3}{2n}\right.$$
$$\left.+\cdots+E^2(n,n-1)\binom{x-1}{2n}\right).$$

On the other hand,

$$E^2(n,0)\binom{x+n-1}{2n}+E^2(n,1)\binom{x+n-2}{2n}+E^2(n,2)\binom{x+n-3}{2n}$$
$$+\cdots+E^2(n,n-1)\binom{x}{2n}$$
$$=E^2(n,0)\left(\binom{x+n-2}{2n}\right.$$
$$\left.+\binom{x+n-2}{2n-1}\right)+E^2(n,1)\left(\binom{x+n-3}{2n}+\binom{x+n-3}{2n-1}\right)$$
$$+E^2(n,2)\left(\binom{x+n-4}{2n}+\binom{x+n-4}{2n-1}\right)$$
$$+\cdots+E^2(n,n-1)\left(\binom{x-1}{2n}+\binom{x-1}{2n-1}\right)$$
$$=\left(E^2(n,0)\binom{x+n-2}{2n}+E^2(n,1)\binom{x+n-3}{2n}\right.$$
$$\left.+\cdots+E^2(n,n-1)\binom{x-1}{2n}\right)+\left(E^2(n,0)\binom{x+n-2}{2n-1}\right.$$
$$\left.+E^2(n,1)\binom{x+n-3}{2n-1}+\cdots+E^2(n,n-1)\binom{x-1}{2n-1}\right).$$

Using the recurrent relation

$$E^2(x,n)=(n+1)E^2(x-1,n)+(2x-n-1)E^2(x-1,n-1)$$

and the properties $E^2(x,0) = 1$, $E^2(x,x) = 0$, and $x \geq 1$, we can rewrite the sum

$$E^2(n,0)\binom{x+n-2}{2n-1} + E^2(n,1)\binom{x+n-3}{2n-1}$$
$$+ E^2(n,2)\binom{x+n-4}{2n-1} + \cdots + E^2(n,n-1)\binom{x-1}{2n-1}$$

in the following form:

$$E^2(n,0)\binom{x+n-2}{2n-1} + E^2(n,1)\binom{x+n-3}{2n-1} + E^2(n,2)\binom{x+n-4}{2n-1}$$
$$+ \cdots + E^2(n,n-1)\binom{x-1}{2n-1}$$
$$= E^2(n-1,0)\binom{x+n-2}{2n-1}$$
$$+ \left((2n-2)E^2(n-1,0) + 2E^2(n-1,1)\right)\binom{x+n-3}{2n-1}$$
$$+ \left((2n-3)E^2(n-1,1) + 3E^2(n-1,2)\right)\binom{x+n-4}{2n-1}$$
$$+ \cdots + \left(nE^2(n-1,n-2) + nE^2(n-1,n-1)\right)\binom{x-1}{2n-1}$$
$$= E^2(n-1,0)\left(\binom{x+n-2}{2n-1} + (2n-2)\binom{x+n-3}{2n-1}\right)$$
$$= E^2(n-1,1)\left(2\binom{x+n-3}{2n-1} + (2n-3)\binom{x+n-4}{2n-1}\right)$$
$$+ \cdots + E^2(n-1,n-3)\left((n-2)\binom{x+1}{2n-1} + (n+1)\binom{x}{2n-1}\right)$$
$$+ E^2(n-1,n-2)\left((n-1)\binom{x}{2n-1} + n\binom{x-1}{2n-1}\right).$$

Therefore, we have that

$$S(x,x-n) = (x-n)S(x-1,x-n) + S(x-1,x-n-1)$$
$$= (x-n)\left(E^2(n-1,0)\binom{x+n-3}{2n-2} + E^2(n-1,1)\binom{x+n-4}{2n-2}\right.$$

$$
\begin{aligned}
&+\cdots+E^2(n-1,n-3)\binom{x}{2n-2}+E^2(n-1,n-2)\binom{x-1}{2n-2}\Bigg)\\
&+\left(E^2(n,0)\binom{x+n-2}{2n}+E^2(n,1)\binom{x+n-3}{2n}\right.\\
&\left.+\cdots+E^2(n,n-1)\binom{x-1}{2n}\right)\\
=\;&E^2(n,0)\binom{x+n-1}{2n}\\
&+E^2(n,1)\binom{x+n-2}{2n}+E^2(n,2)\binom{x+n-3}{2n}\\
&+\cdots+E^2(n,n-1)\binom{x}{2n}\Leftrightarrow E^2(n-1,0)(x-n)\binom{x+n-3}{2n-2}\\
&+E^2(n-1,1)(x-n)\binom{x+n-4}{2n-2}\\
&+\cdots+E^2(n-1,n-3)(x-n)\binom{x}{2n-2}\\
&+E^2(n-1,n-2)(x-n)\binom{x-1}{2n-2}\\
=\;&E^2(n-1,0)\left(\binom{x+n-2}{2n-1}+(2n-2)\binom{x+n-3}{2n-1}\right)\\
&+E^2(n-1,1)\left(2\binom{x+n-3}{2n-1}+(2n-3)\binom{x+n-4}{2n-1}\right)\\
&+\cdots+E^2(n-1,n-3)\left((n-2)\binom{x+1}{2n-1}+(n+1)\binom{x}{2n-1}\right)\\
&+E^2(n-1,n-2)\left((n-1)\binom{x}{2n-1}+n\binom{x-1}{2n-1}\right).
\end{aligned}
$$

In other words, we should prove that

$$
\begin{aligned}
(x-n)\binom{x+n-t-3}{2n-2}=(t+1)&\binom{x+n-t-2}{2n-1}\\
&+(2n-t-2)\binom{x+n-t-3}{2n-1}
\end{aligned}
$$

for any $0 \leq t \leq n-2$. In fact,

$$(x-n)\binom{x+n-t-3}{2n-2} = (t+1)\binom{x+n-t-2}{2n-1}$$

$$+ (2n-t-2)\binom{x+n-t-3}{2n-1} \Leftrightarrow \frac{(x-n)(x_n-t-3)!}{(2n-2)!(x-n-t-1)!}$$

$$= \frac{(t+1)(x_n-t-2)!}{(2n-1)!(x-n-t-1)!} + \frac{(2n-t-2)(x_n-t-3)!}{(2n-1)!(x-n-t-2)!}$$

$$\Leftrightarrow \frac{(x-n)(2n-1)}{x-n-t-1} = \frac{(t+1)(x_n-t-2)}{(2n-1)(x-n-t-1)} + \frac{(2n-t-2)}{2n-1}.$$

The last equality can be checked directly:

$$\begin{aligned}(x-n)(2n-1) &= (t+1)(x+n-t-2) - (t+1) + (2n-t-2)\\ &\quad \times (x-n-t-1) \Leftrightarrow (t+1)(x+n-t-2)\\ &\quad - (t+1) + (2n-t-2)(x-n-t-1)\\ &= (t+1)(x-n-t-1) + (t+1)(2n-1)\\ &\quad + (2n-1-(t+1))(x-n-t-1)\\ &= (t+1)(x-n-t-1) + (t+1)(2n-1)\\ &\quad + (2n-1)(x-n) - (2n-1)(t+1)\\ &\quad - (t+1)(x-n-t-1)\\ &= (x-n)(2n-1).\end{aligned}$$

So, we have proven that

$$\sum_{k=0}^{n-1} E^2(n,k)\binom{x+n-k-1}{2n}$$

$$= \sum_{k=0}^{n-1} E^2(n,k)\left(\binom{x+n-k-2}{2n}\right.$$

$$\left.+\binom{x+n-k-2}{2n-1}\right) = S(x-1, x-1-n) + \sum_{k=0}^{n-1} E^2(n,k)$$

$$\times \binom{x+n-k-2}{2n-1} = S(x-1, x-n-1) + \sum_{k=0}^{n-1}((k+1)E^2$$

$$\times (n-1,k) + (2n-k-1)E^2(n-1,k-1))\binom{x+n-k-2}{2n-1}$$

$$= S(x-1,x-n-1) + E^2(n,0)\binom{x+n-2}{2n-1} + \sum_{k=1}^{n-1}((k+1)E^2$$

$$\times (n-1,k) + (2n-k-1)E^2(n-1,k-1))\binom{x+n-k-2}{2n-1}$$

$$= S(x-1,x-n-1) + E^2(n-1,0)\binom{x+n-2}{2n-1} + \sum_{k=1}^{n-1} E^2(n-1,k)$$

$$\times \left((k+1)\binom{x+n-t-2}{2n-1} + (2n-k-2)\binom{x+n-k-3}{2n-1}\right)$$

$$= S(x-1,x-n-1) + (x-n)\sum_{k=0}^{n-1} E^2(n-1,k)\binom{x+n-k-3}{2n-2}$$

$$= S(x-1,x-n-1) + (x-n)S(x-1,x-n)$$

$$= S(x,x-n). \qquad \square$$

4.2.3. There exists a similar representation for the unsigned Stirling numbers of the first kind.

- *For the unsigned Stirling numbers of the first kind, the following identity holds:*

$$|s(x,x-n)|$$

$$= E^2(n,0)\binom{x}{2n} + E^2(n,1)\binom{x+1}{2n} + E^2(n,2)\binom{x+2}{2n}$$

$$+ \cdots + E^2(n,n-1)\binom{x+(n-1)}{2n}.$$

In particular, it holds that

$$|s(x,x-1)| = E^2(1,0)\binom{x}{2};$$

$$|s(x,x-2)| = E^2(2,0)\binom{x}{4} + E^2(2,1)\binom{x+1}{4};$$

$$|s(x, x-3)| = E^2(3,0)\binom{x}{6} + E^2(3,1)\binom{x+1}{6} + E^2(3,2)\binom{x+2}{6};$$

$$|s(x, x-4)| = E^2(4,0)\binom{x}{8} + E^2(4,1)\binom{x+1}{8} + E^2(4,2)\binom{x+2}{8} + E^2(4,3)\binom{x+3}{8}.$$

Using the triangle of Eulerian numbers of the second order, we get that

$$|s(x, x-1)| = 1 \cdot \binom{x}{2};$$

$$|s(x, x-2)| = 1 \cdot \binom{x}{4} + 2 \cdot \binom{x+1}{4};$$

$$|s(x, x-3)| = 1 \cdot \binom{x}{6} + 8 \cdot \binom{x+1}{6} + 6 \cdot \binom{x+2}{6};$$

$$|s(x, x-4)| = 1 \cdot \binom{x}{8} + 22 \cdot \binom{x+1}{8} + 58 \cdot \binom{x+2}{8} + 24 \cdot \binom{x+3}{8}.$$

□ We can easily check this equality for small values of x: for example,

$$|s(4,2)| = 11 = 1 \cdot 1 + 2 \cdot 5 = 1 \cdot \binom{4}{4} + 2 \cdot \binom{5}{4}.$$

Moreover, the result $|s(x, x-1)| = \binom{x}{2}$ is well known for any integer $x \geq 1$.

In general, the result

$$|s(x, x-n)| = \sum_{k=0}^{n} E^2(n,k)\binom{x+k}{2n}$$

can be proven by induction, following the algorithm, considered earlier. However, there exist other approaches to possible proofs. The reader can find them later. □

Note that the identity giving the decomposition of $|s(x, x-n)|$ as a linear combination of binomial coefficients with weights $E^2(n, k)$ is simply the mirror image of the corresponding identity giving the decomposition of x^n as a linear combination of binomial coefficients with weights $E(n,k)$.

4.2.4. As, by definition, $s(x,x-n) = (-1)^{x+x-n}|s(x,x-n)| = (-1)^n|s(x,x-n)|$, we obtain the formulas representing the Stirling numbers of the first kind as linear combinations of of binomial coefficients with weights $E^2(n,k)$.

- *For the Stirling numbers of the first kind, the following identity holds:*

$$s(x,x-n) = (-1)^n \left(E^2(n,0)\binom{x}{2n} + E^2(n,1)\binom{x+1}{2n} + \cdots + E^2(n,n-1)\binom{x+(n-1)}{2n} \right).$$

In particular, we have that

$$s(x,x-1) = -E^2(1,0)\binom{x}{2};$$

$$s(x,x-2) = E^2(2,0)\binom{x}{4} + E^2(2,1)\binom{x+1}{4};$$

$$s(x,x-3) = -\left(E^2(3,0)\binom{x}{6} + E^2(3,1)\binom{x+1}{6} + E^2(3,2)\binom{x+2}{6} \right);$$

$$s(x,x-4) = E^2(4,0)\binom{x}{8} + E^2(4,1)\binom{x+1}{8} + E^2(4,2)\binom{x+2}{8} + E^2(4,3)\binom{x+3}{8}.$$

Using the triangle of Eulerian numbers of the second order, we get that

$$s(x,x-1) = -1\cdot\binom{x}{2};$$

$$s(x,x-2) = 1\cdot\binom{x}{4} + 2\cdot\binom{x+1}{4};$$

$$s(x, x-3) = -\left(1 \cdot \binom{x}{6} + 8 \cdot \binom{x+1}{6} + 6 \cdot \binom{x+2}{6}\right);$$

$$s(x, x-4) = 1 \cdot \binom{x}{8} + 22 \cdot \binom{x+1}{8} + 58 \cdot \binom{x+2}{8} + 24 \cdot \binom{x+3}{8}.$$

4.2.5. As $\binom{x+k}{n} = \frac{(x+k)(x+k-1)\cdots(x+k-n+1)}{n!} = \frac{x^{\underline{n}}}{n!}$, we get several expressions that allow us to represent the Stirling numbers using the Eulerian numbers of the second order and the "shifted" falling powers.

- *For the Stirling numbers of the second kind, the following identity holds:*

$$S(x, x-n) = \frac{1}{(2n)!}\left(E^2(n,0)(x+n-1)^{\underline{2n}} + E^2(n,1)(x+n-2)^{\underline{2n}}\right.$$
$$\left. + E^2(n,2)(x+b-3)^{\underline{2n}} + \cdots + E^2(n,n-1)x^{\underline{2n}}\right).$$

In particular, we have that

$$S(x, x-1) = \frac{1}{2!}E^2(1,0)x^{\underline{2}};$$

$$S(x, x-2) = \frac{1}{4!}\left(E^2(2,0)(x+1)^{\underline{4}} + E^2(n,1)x^{\underline{4}}\right);$$

$$S(x, x-3) = \frac{1}{6!}\left(E^2(3,0)(x+2)^{\underline{6}} + E^2(3,1)(x+1)^{\underline{6}} + E^2(3,2)x^{\underline{6}}\right);$$

$$S(x, x-4) = \frac{1}{8!}\left(E^2(4,0)(x+3)^{\underline{8}} + E^2(4,1)(x+2)^{\underline{8}}\right.$$
$$\left. + E^2(4,2)(x+1)^{\underline{8}} + E^2(4,3)x^{\underline{8}}\right).$$

In other words, it holds that

$$S(x, x-1) = \frac{1}{2!} \cdot 1 \cdot x^{\underline{2}};$$

$$S(x, x-2) = \frac{1}{4!}\left(1 \cdot (x+1)^{\underline{4}} + 2 \cdot x^{\underline{4}}\right);$$

$$S(x, x-3) = \frac{1}{6!}\left(1 \cdot (x+2)^{\underline{6}} + 8 \cdot (x+1)^{\underline{6}} + 6 \cdot x^{\underline{6}}\right);$$

$$S(x, x-4) = \frac{1}{8!}\left(1 \cdot (x+3)^{\underline{8}} + 22 \cdot (x+2)^{\underline{8}} + 58 \cdot (x+1)^{\underline{8}} + 24 \cdot x^{\underline{8}}\right).$$

4.2.6. As $(x+n-1-k)^{\underline{2n}} = (x+n-k-1)(x+n-k-2)\cdots(x-n-k+1) = (x-n-k)^{\overline{2n}}$, we can rewrite the last result in the following form.

- *For the Stirling numbers of the second kind, the following identity holds:*

$$S(x, x-n) = \frac{1}{(2n)!}\Big(E^2(n,0)(x-n)^{\overline{2n}} + E^2(n,1)(x-n-1)^{\overline{2n}} + E^2(n,2)(x-n-2)^{\overline{2n}} + \cdots + E^2(n,n-1)(x-2n+1)^{\overline{2n}}\Big).$$

4.2.7. Similar decompositions can be obtained for the Stirling numbers of the first kind.

- *For the unsigned Stirling numbers of the first kind, the following identity holds:*

$$|s(x, x-n)| = \frac{1}{(2n)!}\left(E^2(n,0)x^{\underline{2n}} + E^2(n,1)(x+1)^{\underline{2n}} + E^2(n,2)(x+2)^{\underline{2n}} + \cdots + E^2(n,n-1)(x+n-1)^{\underline{2n}}\right).$$

In particular, we have that

$$|s(x, x-1)| = \frac{1}{2!}E^2(1,0)x^{\underline{2}};$$

$$|s(x, x-2)| = \frac{1}{4!}\left(E^2(2,0)x^{\underline{4}} + E^2(n,1)(x+1)^{\underline{4}}\right);$$

$$|s(x,x-3)| = \frac{1}{6!}\left(E^2(3,0)x^{\underline{6}} + E^2(3,1)(x+1)^{\underline{6}} + E^2(3,2)(x+2)^{\underline{6}}\right);$$

$$|s(x,x-4)| = \frac{1}{8!}\left(E^2(4,0)x^{\underline{8}} + E^2(4,1)(x+1)^{\underline{8}} + E^2(4,2)(x+2)^{\underline{8}}\right.$$
$$\left.+E^2(4,3)(x+3)^{\underline{8}}\right).$$

In other words, it holds that

$$|s(x,x-1)| = \frac{1}{2!}1 \cdot x^{\underline{2}};$$

$$|s(x,x-2)| = \frac{1}{4!}\left(1 \cdot x^{\underline{4}} + 2 \cdot (x+1)^{\underline{4}}\right);$$

$$|s(x,x-3)| = \frac{1}{6!}\left(1 \cdot x^{\underline{6}} + 8 \cdot (x+1)^{\underline{6}} + 6 \cdot (x+2)^{\underline{6}}\right);$$

$$|s(x,x-4)| = \frac{1}{8!}\left(1 \cdot x^{\underline{8}} + 22 \cdot (x+1)^{\underline{8}} + 58 \cdot (x+2)^{\underline{8}}\right.$$
$$\left.+24 \cdot (x+3)^{\underline{8}}\right).$$

- *For the Stirling numbers of the first kind, the following identity holds*:

$$s(x,x-n) = \frac{(-1)^n}{(2n)!}\left(E^2(n,0)x^{\underline{2n}} + E^2(n,1)(x+1)^{\underline{2n}}\right.$$
$$\left.+\cdots+E^2(n,n-1)(x+n-1)^{\underline{2n}}\right).$$

In particular, we have that

$$s(x,x-1) = -\frac{1}{2!}E^2(1,0)x^{\underline{2}};$$

$$s(x,x-2) = \frac{1}{4!}\left(E^2(2,0)x^{\underline{4}} + E^2(n,1)(x+1)^{\underline{4}}\right);$$

$$s(x,x-3) = -\frac{1}{6!}\left(E^2(3,0)x^{\underline{6}} + E^2(3,1)(x+1)^{\underline{6}} + E^2(3,2)(x+2)^{\underline{6}}\right);$$

$$s(x,x-4) = \frac{1}{8!}\left(E^2(4,0)x^{\underline{8}} + E^2(4,1)(x+1)^{\underline{8}} + E^2(4,2)(x+2)^{\underline{8}}\right.$$
$$\left.+E^2(4,3)(x+3)^{\underline{8}}\right).$$

In other words, it holds that

$$s(x, x-1) = -\frac{1}{2!} 1 \cdot x^{\underline{2}};$$

$$s(x, x-2) = \frac{1}{4!} \left(1 \cdot x^{\underline{4}} + 2 \cdot (x+1)^{\underline{4}}\right);$$

$$s(x, x-3) = -\frac{1}{6!} \left(1 \cdot x^{\underline{6}} + 8 \cdot (x+1)^{\underline{6}} + 6 \cdot (x+2)^{\underline{6}}\right);$$

$$s(x, x-4) = \frac{1}{8!} \left(1 \cdot x^{\underline{8}} + 22 \cdot (x+1)^{\underline{8}} + 58 \cdot (x+2)^{\underline{8}} + 24 \cdot (x+3)^{\underline{8}}\right).$$

4.2.8. Going from the falling powers to the rising powers, we get the following result:

- *For the unsigned Stirling numbers of the first kind, the following identity holds:*

$$\begin{aligned} &|s(x, x-n)| \\ &\quad = \frac{1}{(2n)!} \Big(E^2(n,0)(x-2n+1)^{\overline{2n}} + E^2(n,1)(x-2n+2)^{\overline{2n}} \\ &\qquad + E^2(n,2)(x+2)^{\overline{2n}} + \cdots + E^2(n,n-1)(x-n)^{\overline{2n}}\Big). \end{aligned}$$

- *For the Stirling numbers of the first kind, the following identity holds:*

$$\begin{aligned} &s(x, x-n) \\ &\quad = \frac{(-1)^n}{(2n)!} \Big(E^2(n,0)(x-2n+1)^{\overline{2n}} + E^2(n,1)(x-2n+2)^{\underline{2n}} \\ &\qquad + E^2(n,2)(x+2)^{\underline{2n}} + \cdots + E^2(n,n-1)(x-n)^{\underline{2n}}\Big). \end{aligned}$$

4.2.9. The representations of $S(x, x-n)$ in terms of the "falling" binomial coefficients $\binom{x+n-1}{2n}$, $\binom{x+n-2}{2n}, \ldots, \binom{x}{2n}$ can be rewritten in terms of the "rising" binomial coefficients $\binom{x}{2n}$, $\binom{x+1}{2n}$, $\ldots$, $\binom{x+n-1}{2n}$ if we use the rows of the reversal triangle of Eulerian numbers of the second order (see sequence A112007 in the OEIS), i.e., the rows of the triangle, going "from right to left".

- *For the Stirling numbers of the second kind, the following identity holds:*

$$S(x, x-n) = E^2(n, n-1)\binom{x}{2n} + E^2(n, n-2)\binom{x-1}{2n} + \cdots + E^2(n, 0)\binom{x+n-1}{2n}.$$

□ As was noted above, the property $S(x, x-n) = \sum_{k=1}^{n} E^2(n, n-k)\binom{x+k-1}{2n}$ is a simple consequence of the previous result. □

In particular, we have that

$$S(x, x-1) = E^2(1, 0)\binom{x}{2};$$

$$S(x, x-2) = E^2(2, 1)\binom{x}{4} + E^2(2, 0)\binom{x+1}{4};$$

$$S(x, x-3) = E^2(3, 2)\binom{x}{6} + E^2(3, 1)\binom{x+1}{6} + E^2(3, 0)\binom{x+2}{6};$$

$$S(x, x-4) = E^2(4, 3)\binom{x}{8} + E^2(4, 2)\binom{x+1}{8} + E^2(4, 1)\binom{x+2}{8} + E^2(4, 0)\binom{x+3}{8}.$$

Using the triangle of Eulerian numbers of the second order, we get that

$$S(x, x-1) = 1 \cdot \binom{x}{2};$$

$$S(x, x-2) = 2 \cdot \binom{x}{4} + 1 \cdot \binom{x+1}{4};$$

$$S(x, x-3) = 6 \cdot \binom{x}{6} + 8 \cdot \binom{x+1}{6} + 1 \cdot \binom{x+2}{6};$$

$$S(x, x-4) = 22 \cdot \binom{x}{8} + 24 \cdot \binom{x+1}{8} + 58 \cdot \binom{x+2}{8} + 1 \cdot \binom{x+3}{8}.$$

In other words, we can always represent the Stirling numbers $S(x, x-n)$ and $s(x, x-n)$ in the form of linear combinations of the same binomial coefficients placed in the same order.

4.2.10. Our considerations allow us to expand the table, showing the significant changes among the systems x^n, $x^{\overline{n}}$, and $x^{\overline{n}}$, adding to the list the systems $S(x, x-n)$, $|s(x, x-n)|$, and $s(x, x-n)$.

Change of basis	Coefficients
$x^n \to x^{\underline{k}}$	$S(n, k)$
$x^n \to x^{\overline{k}}$	$(-1)^{n+k} S(n, k)$
$x^{\underline{n}} \to x^k$	$s(n, k)$
$x^{\overline{n}} \to x^k$	$(-1)^{n+k} s(n, k)$
$x^{\overline{n}} \to x^{\underline{k}}$	$L(n, k)$
$x^{\underline{n}} \to x^{\overline{k}}$	$(-1)^{n+k} L(n, k)$
$x^n \to \binom{x+k}{n}$	$E(n, k)$
$S(x, x-n) \to \binom{x+n-1-k}{2n}$	$E^2(n, k)$
$s(x, x-n) \to \binom{x+k}{2n}$	$(-1)^n E^2(n, k)$

4.2.11. In Chapter 3, we used the decomposition $x^n = \sum_{k=0}^{n} E(n, k) \binom{x+k}{n}$ to obtain the equality

$$\sum_{k=0}^{\infty} k^n \cdot x^k = \frac{x A_n(x)}{(1-x)^{n+1}}, n \geq 1,$$

where $A(x) = E(n, 0) + E(n, 1)x + \cdots + E(n, n)x^n$ is the *Eulerian polynomial of the first order.*

It turns out that a similar result can be obtained for the Eulerian numbers of the second order:

- *For the Eulerian numbers of the second order, there exists the identity*

$$\sum_{k=0}^{\infty} S(k+n, k) \cdot x^k$$

$$= \frac{E^2(n, 0)x + E^2(n, 1)x^2 + \cdots + E^2(n, n-1)x^n}{(1-x)^{2n+1}}, \quad n \geq 1.$$

□ Define the nth *Eulerian polynomial of the second order* $U_n(x)$, $0, 1, 2, \ldots$, as

$$U_n(x) = E^2(n,0) + E^2(n,1)x + \cdots + E^2(n,n)x^n.$$

Then, it is easy to see that

$$U_0(x) \equiv 1, \quad U_1(x) \equiv 1, \quad U_2(x) = 1 + 2x, \quad U_3(x) = 1 + 8x + 6x^2.$$

In these terms, the property takes the form

$$\sum_{k=0}^{\infty} S(k+n,k) \cdot x^k = \frac{xU_n(x)}{(1-x)^{2n+1}}, \qquad n \geq 1.$$

We prove this fact by induction.

Note that, for $n = 0$, there is a small difference (compared with the decomposition of $\sum_{k=1}^{\infty} k^n x^k$ for the Eulerian numbers of the first order). In fact, in this case, we should use $S(0,0) = 1$, which will provide an "additional" term in the sum $\sum_{k=0}^{\infty} S(k+n,k) \cdot x^k$. More exactly, for $n = 0$, we have

$$\sum_{k=0}^{\infty} S(k+0,k)x^k = S(0,0) + S(1,1)x + \cdots = 1 + x + x^2 + \cdots$$

$$= \frac{1}{1-x} \neq \frac{xU_0(x)}{1-x}.$$

But if we will ignore the value $S(0,0) = 1$, we get that

$$x + x^2 + \cdots = x(1 + x + \cdots) = \frac{x \cdot 1}{1-x} = \frac{xU_0(x)}{1-x}.$$

In other words, the identity

$$\sum_{k=1}^{\infty} S(k+n,k) \cdot x^k = \frac{xU_n(x)}{(1-x)^{2n+1}}$$

holds for any $n \geq 0$. This observation will be used later.

For $n = 1$, we have

$$\sum_{k=0}^{\infty} S(k+1,k)x^k = S(1,0)x^0 + S(2,1)x^1 + \cdots = 0 + \binom{2}{2}x + \binom{3}{2}x^2$$
$$+ \cdots \binom{k+1}{2}x^k + \cdots = \frac{2 \cdot 1}{2}x + \frac{3 \cdot 2}{2}x^2 + \cdots$$
$$+ \frac{(k+1)k}{2}x^k + \cdots .$$

On the other hand, it is easy to see that

$$1 + x + x^2 + \cdots + x^k + \cdots = \frac{1}{1-x},$$
$$(1 + x + x^2 + \cdots + x^k + \cdots)'$$
$$= 1 + 2x + 3x^2 + \cdots + (k+1)x^k + \cdots = \frac{1}{(1-x)^2},$$
$$(1 + 2x + 3x^2 + \cdots + (k+1)x^k + \cdots)'$$
$$= 2 \cdot 1 + 3 \cdot 2x + \cdots + (k+1)kx^k + \cdots = \frac{2}{(1-x)^3}.$$

Therefore, it holds that

$$\sum_{k=0}^{\infty} S(k+1,k)x^k = \frac{2 \cdot 1}{2}x + \frac{3 \cdot 2}{2}x^2 + \cdots + \frac{(k+1)k}{2}x^k + \cdots$$
$$= \frac{x}{(1-x)^3} = \frac{x \cdot U_1(x)}{(1-X)^{2 \cdot 1+1}}.$$

Suppose that, for a given positive integer n, the following holds:

$$\sum_{k=0}^{\infty} S(k+n,k) \cdot x^k = \frac{xU_n(x)}{(1-x)^{2n+1}}, \quad U_n(x) = \sum_{k=0}^{n} E^2(n,k)x^k.$$

Prove the formula for $n+1$:

$$\sum_{k=0}^{\infty} S(k+(n+1),k) \cdot x^k = \frac{xU_{n+1}(x)}{(1-x)^{2(n+1)+1}},$$
$$U_{n+1}(x) = \sum_{k=0}^{n+1} E^2(n+1,k)x^k.$$

In fact, it holds that

$$\left(\sum_{k=0}^{\infty} S(k+n,k)\cdot x^k\right)' = \sum_{k=1}^{\infty} kS(n+k,k)x^{k-1}$$

$$= 1\cdot S(n+1,1) + 2\cdot S(n+2,2)x$$

$$+ \cdots + (k+1)\cdot S(n+k+1,k+1)x^k + \cdots,$$

and we have that

$$x\left(\sum_{k=1}^{\infty} S(k+n,k)\cdot x^k\right)' = \sum_{k=0}^{\infty} kS(n+k,k)x^k$$

$$= 0\cdot S(n,0) + 1\cdot S(n+1,1)x^1 + 2\cdot S(n+2,2)x^2 + \cdots.$$

On the other hand, it holds that

$$\left(\frac{xU_n(x)}{(1-x)^{2n+1}}\right)' = \frac{x(1-x)U_n'(x) + (1+2nx)U_n(x)}{(1-x)^{2n+2}}.$$

As $U_n(x) = \sum_{k=0}^{n} E^2(n,k)x^k$, $U_n'(x) = \sum_{k=0}^{n} kE^2(n,k)x^{k-1}$, and the polynomial

$$x\,(1-x)\,U_n'(x) + U_n(x)\,(1+2n\,x)$$

takes the form

$$x(1-x)\sum_{k=0}^{n} kE^2(n,k)x^{k-1} + (1+2nx)\sum_{k=0}^{n} E^2(n,k)x^k$$

$$= \sum_{k=0}^{n} kE^2(n,k)x^k - \sum_{k=0}^{n} kE^2(n,k)x^{k+1} + \sum_{k=0}^{n} E^2(n,k)x^k$$

$$+ 2n\sum_{k=0}^{n} E^2(n,k)x^{k+1}$$

$$= \sum_{k=0}^{n}(k+1)E^2(n,k)$$

$$+ \sum_{k=0}^{n}((2n-k)E^2(n,k)x^{k+1}$$

$$= E^2(n,0) + \sum_{k=1}^{n}(k+1)E^2(n,k)x^k$$

$$+\sum_{k=1}^{n+1} E^2(2n+1-k)E^2(n,k-1)x^k$$

$$= E^2(n,0)$$

$$+\sum_{k=1}^{n}(k+1)E^2(n,k)x^k + \sum_{k=1}^{n}(n+1-k)E^2(n,k-1)x^k$$

$$= E^2(n,0) + \sum_{k=1}^{n}((k+1)E^2(n,k)$$

$$+ (2(n+1)-k-1)E^2(n,k-1))x^k$$

$$= E^2(n+1,0) + \sum_{k=1}^{n} E^2(n+1,k)x^k$$

$$= \sum_{k=0}^{n+1} E^2(n+1,k)x^k$$

$$= U_{n+1}(x).$$

Therefore, we get that

$$\left(\frac{xU_n(x)}{(1-x)^{2n+1}}\right)' = \frac{U_{n+1}(x)}{(1-x)^{2n+2}},$$

and hence, it holds that

$$\sum_{k=0}^{\infty} kS(n+k,k)x^k = \frac{xU_{n+1}(x)}{(1-x)^{2n+2}}.$$

Furthermore, we can state that

$$\frac{1}{1-x}\cdot\sum_{k=0}^{\infty} kS(n+k,k)x^k = \frac{xU_{k+1}(x)}{(1-x)^{2n+3}}.$$

Using the decomposition $\frac{1}{1-x} = 1+x+x^2+\cdots+x^k+\cdots$, we obtain that

$$\frac{1}{1-x}\cdot\sum_{k=0}^{\infty} kS(n+k,k)x^k$$

$$= (1+x+x^2+\cdots+x^k+\cdots)\sum_{k=0}^{\infty} kS(n+k,k)x^k$$

$$= S(n+1,1)x + (2\cdot S(n+2,2)$$

$$+ 1 \cdot S(n+1,1))x^2 + (3 \cdot S(n+3,3) + 2 \cdot S(n+2,2)$$
$$+ 1 \cdot S(n+1,1))x^3 + \cdots + (k \cdot S(n+k,k)$$
$$+ (k-1) \cdot S(n+(k-1),k-1)$$
$$+ \cdots + 2 \cdot S(n+2,2) + 1 \cdot S(n+1,1))x^k + \cdots .$$

As

$$S(n+1,1) = 1 = S(n+2,1),$$
$$2 \cdot S(n+2,2) + 1 \cdot S(n+1,1) = 2S(n+2,2) + S(n+2,1)$$
$$= S(n+3,2),$$
$$3 \cdot S(n+3,3) + 2 \cdot S(n+2,2) + 1 \cdot S(n+1,1)$$
$$= 3S(n+3,3) + (2S(n+2,2) + S(n+2,1))$$
$$= 3S(n+3,3) + S(n+3,2) = S(n+4,3),$$

and in general,

$$k \cdot S(n+k,k) + (k-1) \cdot S(n+(k-1),k-1) + \cdots$$
$$+ 2 \cdot S(n+2,2) + 1 \cdot S(n+1,1) = S(n+k+1,k),$$

we get that

$$\frac{1}{1-x} \cdot \sum_{k=0}^{\infty} kS(n+k,k)x^k = \sum_{k=0}^{\infty} kS((n+1)+k,k)x^k;$$

therefore, it holds that

$$\sum_{k=1}^{\infty} S(k+(n+1),k) \cdot x^k = \frac{xU_{n+1}(x)}{(1-x)^{2(n+1)+1}}. \qquad \square$$

4.2.12. This approach allows us to get the formula

$$S(x, x-n) = \sum_{j=1}^{n-1} E^2(n,j) \binom{x+n-j-1}{2n},$$

which was proven earlier by induction using other reasons. Now, we prove this result in the following form:

• *For the Stirling numbers of the second kind, it holds that*

$$S(n+k,k) = \sum_{j=0}^{n-1} E^2(n,j)\binom{2n+k-1-j}{2n}.$$

□ In fact, it is well known (see Chapter 1) that

$$(1+x)^\alpha = 1 + \sum_{k=1}^{\infty}\binom{\alpha}{k}x^k, \quad |x|<1.$$

So, as $\binom{-t}{k} = \frac{(-t)(-t+1)\cdots(-t-k+1)}{k!} = (-1)^k\binom{t+k-1}{k}$, we have that

$$(1-x)^{-1-n} = 1+\sum_{k=1}^{\infty}\binom{-1-2n}{k}(-x)^k = \sum_{k=0}^{\infty}\binom{2n+k}{k}x^k, \quad |x|<1.$$

Then, we have that

$$\frac{xU_n(x)}{(1-x)^{2n+1}} = \left(\sum_{k=0}^{\infty}\binom{2n+k}{k}x^k\right)xU_n(x)$$

$$= \left(\sum_{k=0}^{\infty}\binom{2n+k}{2n}x^{k+1}\right)\left(\sum_{j=0}^{n-1}E^2(n,j)x^j\right)$$

$$= \left(\binom{2n}{2n}x + \binom{2n+1}{2n}x^2 + \cdots + \binom{2n+k-1}{2n}x^k + \cdots\right)$$

$$\cdot\left(E^2(n,0) + E^2(n,1)x + E^2(n,2)x^2 + \cdots + E^2(n,n-1)x^{n-1}\right)$$

$$= E^2(n,0)\binom{2n}{2n}x + \left(E^2(n,0)\binom{2n+1}{2n} + E^2(n,1)\binom{2n}{2n}\right)x^2$$

$$+ \left(E^2(n,0)\binom{2n+2}{2n} + E^2(n,1)\binom{2n+1}{2n} + E^2(n,2)\binom{2n}{2n}\right)x^3$$

$$+ \cdots + \left(E^2(n,0)\binom{2n+k-1}{2n} + E^2(n,1)\binom{2n+k-2}{2n}\right.$$

$$+\cdots+E^2(n,k-1)\binom{2n}{2n}\Bigg)\, x^k$$

$$+\cdots=\sum_{k=0}^{\infty}\left(\sum_{j=0}^{n-1}E^2(n,j)\binom{2n+k-1-j}{2n}\right)x^k.$$

Comparing the coefficients at x^k in the two decompositions of $\frac{xU_n(x)}{(1-x)^{2n+1}}$, we get that

$$S(n+k,k)=\sum_{j=0}^{n-1}E^2(n,j)\binom{2n+k-1-j}{2n}. \qquad \square$$

4.2.13. A similar result can be obtained for infinite sums, involving the Eulerian numbers of the second order and the Stirling numbers of the first kind (see [GeSt78]).

- *For the Eulerian numbers of the second order, there exists the identity*

$$\sum_{k=0}^{\infty}|s(k,k-n)|x^k$$
$$=\frac{E^2(n,n-1)x^{n+1}+E^2(n,n-2)x^{n+2}+\cdots+E^2(n,0)x^{2n}}{(1-x)^{2n+1}},$$

$n\geq 1$.

$\square$ As $|s(k,k-n)|=0$ for any positive integer n and any integer $0\leq k\leq n$, it holds that

$$\sum_{k=0}^{\infty}|s(k,k-n)|x^k$$
$$=|s(n+1,1)|x^{n+1}+|s(n+2,2)|x^{n+2}+\cdots+|s(n+k,k)|x^{n+k}$$
$$+\cdots=x^{n+1}(|s(n+1,1)|+|s(n+2,2)|x$$
$$+\cdots+|s(n+k,k)|x^{k-1}+\cdots).$$

On the other hand, it holds that

$$E^2(n,n-1)x^{n+1}+E^2(n,n-2)x^{n+2}+\cdots+E^2(n,0)x^{2n}$$
$$=x^{n+1}\left(E^2(n,n-1)+E^2(n,n-2)x^2+\cdots+E^2(n,0)x^n\right),$$

and we obtain the equality

$$|s(n+1,1)|+|s(n+2,2)|x+\cdots+|s(n+k,k)|x^{k-1}+\cdots$$
$$=\frac{E^2(n,n-1)+E^2(n,n-2)x^2+\cdots+E^2(n,0)x^n}{(1-x)^{2n+1}}.$$

The corrected version

$$\sum_{k=1}^{\infty}|s(k,k-n)|x^k$$
$$=\frac{E^2(n,n-1)x^{n+1}+E^2(n,n-2)x^{n+2}+\cdots+E^2(n,0)x^{2n}}{(1-x)^{2n+1}}$$

of the identity can also be used for $n=0$: in this case,

$$\sum_{k=1}^{\infty}|s(k,k)|x^k=|s(1,1)|x+|s(2,2)|x^2+\cdots+|s(k,k)|x^k+\cdots$$
$$=x+x^2+\cdots+x^k+\cdots=\frac{x}{1-x}=\frac{E^2(0,0)x}{1-x}.$$

It is also easy to obtain the formula for $n=1$. As $|s(k,k-1)|=\binom{n}{2}=\frac{n(n-1)}{2}$, $k\geq 2$, we get that

$$\sum_{k=0}^{\infty}|s(k,k-1)|x^k$$
$$=|s(2,1)|x^2+|s(3,2)|x^3+\cdots+|s(k,k-1)|x^k+\cdots$$
$$=\frac{x^2}{2}\left(2\cdot 1+3\cdot 2x+\cdots+k(k-1)x^{k-2}+\cdots\right)$$
$$=\frac{x^2}{2}\cdot\frac{2}{(1-x)^3}=\frac{E^2(1,0)x^2}{(1-x)^3}.$$

In general, the proof can be obtained using basic results in the theory of generating functions (see, for example, [GeSt78] and [Stan79], as well as Chapter 1). □

4.2.14. This approach allows us to get the formula

$$|s(x, x-n)| = \sum_{j=0}^{n-1} E^2(n,j)\binom{x+j}{2n},$$

which was considered earlier. Here, we prove this result in the following form:

- *For the unsigned Stirling numbers of the first kind, it holds that*

$$|s(k, k-n)| = (-1)^n s(k, k-n) = \sum_{j=0}^{n-1} E^2(n,j)\binom{k+j}{2n}$$

$$= \sum_{i=n+1}^{2n} E^2(n, 2n-i)\binom{2n+k-i}{2n}.$$

□ In fact, as $(1+x)^\alpha = 1 + \sum_{k=1}^{\infty}\binom{\alpha}{k}x^k$, and $\binom{-t}{k} = \frac{(-t)(-t+1)\cdots(-t-k+1)}{k!} = (-1)^k\binom{t+k-1}{k}$, we have that

$$(1-x)^{-1-2n} = 1 + \sum_{k=1}^{\infty}\binom{-1-2n}{k}(-x)^k = \sum_{k=0}^{\infty}\binom{2n+k}{k}x^k.$$

Then, it holds that

$$\sum_{k=0}^{\infty}|s(k,k-n)|x^k = (1-x)^{1-2n}\left(\sum_{j=n+1}^{2n} E^2(n,2n-j)x^j\right)$$

$$= \left(\sum_{k=0}^{\infty}\binom{2n+k}{k}x^k\right)\left(\sum_{j=n+1}^{2n} E^2(n,2n-j)x^j\right)$$

$$= \left(\binom{2n}{0}E^2(n,n-1)\right)x^{n+1} + \left(\binom{2n}{0}E^2(n,n-2)\right.$$
$$\left.+\binom{2n+1}{1}E^2(n,n-1)\right)x^{n+2} + \left(\binom{2n}{0}E^2(n,n-3)\right.$$
$$\left.+\binom{2n+1}{1}E^2(n,n-2) + \binom{2n+2}{2}E^2(n,n-1)\right)x^{n+3}$$
$$+\cdots+\left(\binom{2n}{0}E^2(n,2n-k) + \binom{2n+1}{1}E^2(n,n-(k-1))\right.$$
$$\left.+\cdots+\binom{2n+k-(n+1)}{k-(n+1)}E^2(n,2n-(n+1))\right)x^k+\cdots$$
$$= \sum_{k=n+1}^{\infty}\left(\sum_{j=n+1}^{k}\binom{2n+k-j}{2n}E^2(n,2n-j)\right)x^k$$
$$= \sum_{k=0}^{\infty}\left(\sum_{j=n+1}^{2n}\binom{2n+k-j}{2n}E^2(n,2n-j)\right)x^k.$$

Comparing the coefficients at x^k on the left- and right-hand side of the last expression, we get that

$$|s(k,k-n| = \sum_{j=n+1}^{2n} E^2(n,2n-j)\binom{2n+k-j}{2n}$$
$$= \sum_{i=0}^{n-1} E^2(n,i)\binom{k+i}{2n}. \qquad \square$$

Exercises

1. Represent $S(x, x-n)$, $x = 1, 2, \ldots, 8$, $n = 0, 1, \ldots, x-1$, as linear combinations of binomial coefficients.
2. Prove that $x^n = \sum_{k=0}^{n} k!S(n,k)\binom{x}{k}$; $x^n = \sum_{k=0}^{n}(-1)^{n+k}k!S(n, k)\binom{x+k-1}{k}$.

3. Prove that $|s(x, x-n)| = \sum_{k=0}^{n} E^2(n,k)\binom{x+k}{2n}$; $|s(x, x-n)| = \sum_{k=0}^{n} E^2(n, n-k-1)\binom{x+n-k-1}{2n}$; $|s(x, x-n)| = \sum_{k=0}^{n} \frac{E^2(n,k)}{(2n)!}(x+k)^{\underline{2n}}$; $|s(x, x-n)| = \sum_{k=0}^{n} \frac{E^2(n,n-k-1)}{(2n)!}(x+n-k-1)^{\underline{2n}}$.
4. Prove that $\sum_{k=0}^{\infty} |s(k,k)|x^k = \frac{U_0(x)}{1-x}$; $\sum_{k=0}^{\infty} |s(k,k-1)|x^k = \frac{x^2 U_1(x)}{(1-x)^3}$.
5. Prove that $\sum_{k=0}^{\infty} |s(k,k-2)|x^k = x^3 \cdot \frac{2+x}{(1-x)^5}$; $\sum_{k=0}^{\infty} |s(k,k-3)|x^k = x^4 \cdot \frac{6+8x+x^2}{(1-x)^5}$.

4.3. Closed Formula for Eulerian Numbers of the Second Order

4.3.1. Now, we can obtain a kind of closed formula for Eulerian numbers of the second order. It involves the Stirling numbers of the second kind.

- *For the Eulerian numbers of the second order, there exists the closed formula*

$$E^2(n,k) = \sum_{j=0}^{k+1} (-1)^j \binom{2n+1}{j} S(n+k+1-j, k+1-j), \quad n \geq 1.$$

□ It is easy to check the formula for small values of n. For $n = 1$, it holds that

$$E^2(1,0) = (-1)^0\binom{3}{0}S(2,1) + (-1)^1\binom{3}{1}S(1,0) = 1 - 0 = 1,$$

$$E^2(1,1) = (-1)^0\binom{3}{0}S(3,2) + (-1)^1\binom{3}{1}S(2,1) + (-1)^2\binom{3}{2}S(1,0) = 1\cdot 3 - 3\cdot 1 + 3\cdot 0 = 0.$$

For $n = 2$, it holds that

$$E^2(2,0) = (-1)^0\binom{5}{0}S(3,1) + (-1)^1\binom{5}{1}S(2,0) = 1 - 0 = 1,$$

$$E^2(2,1) = (-1)^0\binom{5}{0}S(4,2) + (-1)^1\binom{5}{1}S(3,1)$$
$$+ (-1)^2\binom{5}{2}S(2,0) = 1 \cdot 7 - 5 \cdot 1 + 10 \cdot 0 = 2,$$

$$E^2(2,2) = (-1)^0\binom{5}{0}S(5,3) + (-1)^1\binom{5}{1}S(4,2)$$
$$+ (-1)^2\binom{5}{2}S(3,1) + (-1)^3\binom{5}{3}S(2,0) = 1 \cdot 25 - 5 \cdot 7$$
$$+ 10 \cdot 1 - 10 \cdot 0 = 0.$$

For $n = 3$, it holds that

$$E^2(3,0) = (-1)^0\binom{7}{0}S(4,1) + (-1)^1\binom{7}{1}S(3,0) = 1 - 0 = 1,$$

$$E^2(3,1) = (-1)^0\binom{7}{0}S(5,2) + (-1)^1\binom{7}{1}S(4,1)$$
$$+ (-1)^2\binom{7}{2}S(3,0) = 1 \cdot 15 - 7 \cdot 1 + 21 \cdot 0 = 8,$$

$$E^2(3,2) = (-1)^0\binom{7}{0}S(6,3) + (-1)^1\binom{7}{1}S(5,2) + (-1)^2\binom{7}{2}S(4,1)$$
$$+ (-1)^3\binom{7}{3}S(3,0) = 1 \cdot 90 - 7 \cdot 15 + 21 \cdot 1 - 35 \cdot 0 = 6,$$

$$E^2(3,3) = (-1)^0\binom{7}{0}S(7,4) + (-1)^1\binom{7}{1}S(6,3) + (-1)^2\binom{7}{2}S(5,2)$$
$$+ (-1)^3\binom{7}{3}S(4,1) + (-1)^4\binom{7}{4}S(3,0) = 1 \cdot 350 - 7 \cdot 90$$
$$+ 21 \cdot 15 - 35 \cdot 1 + 35 \cdot 0 = 0.$$

So, we have proved the property for $n = 1, 2, 3$. However, the property fails for $n = 0$:

$$(-1)^0\binom{1}{0}S(1,1) + (-1)^1\binom{1}{1}S(0,0) = 1 - 1 = 0 \neq E^2(0,0).$$

Note also that, in general, the following holds:

$$E^2(n,0) = (-1)^0\binom{2n+1}{0}S(n+1,1) + (-1)^1\binom{2n+1}{1}S(n,0) = 1 - 0 = 1.$$

Suppose that the formula is true for a positive integer $n-1$:

$$E^2(n-1,k) = \sum_{j=0}^{k+1}(-1)^j\binom{2(n-1)+1}{j}S((n-1)+k+1-j, k+1-j),\quad k = 0, 1, \ldots, n-1.$$

Let us prove this formula for n:

$$E^2(n,k) = \sum_{j=0}^{k+1}(-1)^j\binom{2n+1}{j}S(n+k+1-j, k+1-j),\quad k = 0, 1, \ldots, n.$$

In fact, using the recurrent relations

$$E^2(n,k) = (k+1)E^2(n-1,k) + (2n-k-1)E^2(n-1,k-1)$$

and

$$S(n,k) = kS(n-1,k) + S(n-1,k-1),$$

we get that

$$
\begin{aligned}
E^2(n,k) &= (k+1)E^2(n-1,k) + (2n-k-1)E^2(n-1,k-1) \\
&= (k+1)\sum_{j=0}^{k+1}(-1)^j\binom{2n-1}{j}S(n+k-j,k+1-j) \\
&\quad + (2n-k-1)\sum_{j=0}^{k}(-1)^j\binom{2n-1}{j}S(n+k-1-j,k-j) \\
&= (k+1)\binom{2n-1}{0}S(n+k,k+1) - \left((k+1)\binom{2n-1}{1}\right. \\
&\quad \left. -(2n-k-1)\binom{2n-1}{0}\right)S(n+k-1,k) \\
&\quad + \left((k+1)\binom{2n-1}{2} - (2n-k-1)\binom{2n-1}{1}\right) \\
&\quad \times S(n+k-2,k-1) - \cdots + (-1)^k\left((k+1)\binom{2n-1}{k}\right. \\
&\quad \left. -(2n-k-1)\binom{2n-1}{k-1}\right)S(n+1,1).
\end{aligned}
$$

On the other hand, it holds that

$$
\begin{aligned}
&\sum_{j=0}^{k+1}(-1)^j\binom{2n-+1}{j}S(n+k+1-j,k+1-j) \\
&\quad = \binom{2n+1}{0}S(n+k+1,k+1) - \binom{2n+1}{1}S(n+k,k) \\
&\quad\quad + \binom{2n+1}{2}S(n+k-1,k-1) \\
&\quad\quad - \cdots + (-1)^k\binom{2n+1}{k}S(n+1,1)
\end{aligned}
$$

$$= \binom{2n+1}{0}((k+1)S(n+k,k+1) + \binom{2n+1}{0}S(n+k,k)$$
$$- \binom{2n+1}{1}S(n+k,k) + \binom{2n+1}{2}S(n+k-1,k-1)$$
$$- \cdots + (-1)^k\binom{2n+1}{k}S(n+1,1)$$
$$= \binom{2n+1}{0}((k+1)S(n+k,k+1)$$
$$+ \left(\binom{2n+1}{0} - \binom{2n+1}{1}\right) S(n+k,k)$$
$$+ \binom{2n+1}{2}S(n+k-1,k-1)$$
$$- \cdots + (-1)^k\binom{2n+1}{k}S(n+1,1)$$
$$= \binom{2n+1}{0}\Bigg((k+1)S(n+k,k+1) + \left(\binom{2n+1}{0} - \binom{2n+1}{1}\right)$$
$$\times (kS(n+k-1,k) + S(n+k-1,k-1)\Bigg)$$
$$+ \binom{2n+1}{2}S(n+k-1,k-1)$$
$$- \cdots + (-1)^k\binom{2n+1}{k}S(n+1,1)$$
$$= \binom{2n+1}{0}\Bigg((k+1)S(n+k,k+1) + \left(\binom{2n+1}{0} - \binom{2n+1}{1}\right)$$
$$\times kS(n+k-1,k) + \left(\binom{2n+1}{0} - \binom{2n+1}{1}\right)$$
$$\times \; S(n+k-1,k-1)\Bigg) + \binom{2n+1}{2}S(n+k-1,k-1)$$
$$- \cdots + (-1)^k\binom{2n+1}{k}S(n+1,1)$$

$$= \binom{2n+1}{0}\left((k+1)S(n+k,k+1) + \left(\binom{2n+1}{0} - \binom{2n+1}{1}\right)\right.$$
$$\times kS(n+k-1,k) + \left(\binom{2n+1}{0} - \binom{2n+1}{1} + \binom{2n+1}{2}\right)$$
$$\left.\times S(n+k-1,k-1) - \cdots + (-1)^k\binom{2n+1}{k}S(n+1,1)\right)$$
$$= \cdots \binom{2n+1}{0}(k+1)S(n+k,k+1) + \left(\binom{2n+1}{0} - \binom{2n+1}{1}\right)$$
$$\times kS(n+k-1,k) + \left(\binom{2n+1}{0} - \binom{2n+1}{1}\right.$$
$$\left.+\binom{2n+1}{2}\right)(k-1)S(n+k-1,k-1) - \cdots + \left(\binom{2n+1}{0}\right.$$
$$\left.-\binom{2n+1}{1} + \cdots + (-1)^k\binom{2n+1}{k}\right)S(n+1,1).$$

It is easy to show that

$$\binom{m}{0} - \binom{m}{1} + \cdots + (-1)^t\binom{m}{t} = (-1)^t\binom{m-1}{t}.$$

In fact, for $t = 0$, we have $\binom{m}{0} = 1 = (-1)^0\binom{m-1}{0}$. For $t = 1$, it holds that

$$\binom{m}{0} - \binom{m}{1} = 1 - m = (-1)^1\binom{m-1}{1}.$$

Going from $t-1$ to t, we get

$$\binom{m}{0} - \binom{m}{1} + \cdots + (-1)^{t-1}\binom{m}{t-1} + (-1)^t\binom{m}{t}$$
$$= (-1)^t\left(-\binom{m-1}{t-1} + \binom{m}{t}\right) = (-1)^t\left(-\binom{m-1}{t-1}\right.$$
$$\left.+\binom{m-1}{t-1} + \binom{m-1}{t}\right) = (-1)^t\binom{m-1}{t}.$$

So, now we can prove that

$$(k+1)\binom{2n-1}{t} - (2n-k-1)\binom{2n-1}{t-1} = (k+1-t)\binom{2n}{t},$$
$$t = 0, 1, \ldots, k.$$

For $t = 0$, we get $(k+1)\binom{2n-1}{0} = k+1 = (k=1)\binom{2n+1}{0.}$ For $t = 1$, we have

$$(k+1)\binom{2n-1}{1} - (2n-k-1)\binom{2n-1}{0}$$
$$= (k+1)(2n-1) - (2n-1-k) = 2nk = k\binom{2n}{1}.$$

In general, as

$$t\binom{m}{t} = t\frac{m!}{t!(m-t)!} = (m-t+1)\frac{m!}{(t-1)!(m-t+1)!}$$
$$= (m-t+1)\binom{m}{t-1},$$

we have that

$$(k+1-t)\binom{2n}{t} = (k+1-t)\left(\binom{2n-1}{t} + \binom{2n-1}{t-1}\right)$$
$$= (k+1)\left(\binom{2n-1}{t} - t\binom{2n-1}{t}\right) + (k+1-t)\binom{2n-1}{t-1}$$
$$= (k+1)\binom{2n-1}{t} - (2n-t)\binom{2n-1}{t-1} + (k+1-t)\binom{2n-1}{t-1}$$
$$= (k+1)\binom{2n-1}{t} - (2n-k-1)\binom{2n-1}{t-1}.$$

So, we have proven that

$$E^2(n,k) = (k+1)E^2(n-1,k) + (2n-k-1)E^2(n-1,k-1)$$
$$= (k+1)\binom{2n-1}{0}S(n+k,k+1) - \left((k+1)\binom{2n-1}{1}\right.$$

$$
\begin{aligned}
&- (2n-k-1)\binom{2n-1}{0}\Bigg) S(n+k-1,k)\\
&+ \left((k+1)\binom{2n-1}{2} - (2n-k-1)\binom{2n-1}{1}\right)\\
&\times S(n+k-2,k-1) - \cdots + (-1)^k \Bigg((k+1)\binom{2n-1}{k}\\
&-(2n-k-1)\binom{2n-1}{k-1}\Bigg) S(n+1,1)\\
= &\binom{2n+1}{0} S(n+k+1,k+1) - \binom{2n+1}{1} S(n+k,k)\\
&+ \binom{2n+1}{2} S(n+k-1,k-1)\\
&- \cdots + (-1)^k \binom{2n+1}{k} S(n+1,1). \qquad \square
\end{aligned}
$$

4.3.2. However, we can consider an alternative proof of this formula, using the decomposition

$$
\sum_{k=0}^{\infty} S(n+k,k)x^k = \frac{xU_n(x)}{(1-x)^{2n+1}}, \quad U_n(x) = \sum_{k=0}^{n} E^2(n,k)x^k.
$$

$\square$ In fact, as

$$
(1-x)^{2n+1} = \sum_{k=0}^{2n+1} \binom{n+1}{k}(-1)^k x^k = \sum_{k=0}^{\infty} \binom{2n+1}{k}(-1)^k x^k,
$$

it holds that

$$
\begin{aligned}
xU_n(x) &= (1-x)^{2n+1} \sum_{k=0}^{\infty} S(n+k,k)x^k\\
&= \left(\sum_{k=0}^{\infty} \binom{2n+1}{k}(-1)^k x^k\right)\left(\sum_{k=0}^{\infty} S(n+k,k)x^k\right),
\end{aligned}
$$

or

$$E^2(n,0)x + E^2(n,1)x^2 + \cdots + E^2(n,k)x^{k+1} + \cdots + E^2(n,n-1)x^n$$
$$= \left(\binom{2n+1}{0} \cdot S(n+1,1)\right) x + \left(\binom{2n+1}{0} \cdot S(n+2,2)\right.$$
$$\left. - \binom{2n+1}{1} \cdot S(n+1,1)\right) x^2 + \cdots + \left(\binom{2n+1}{0}\right.$$
$$\times S(n+k+1,k+1) - \binom{2n+1}{1} S(n+k,k)$$
$$+ \cdots + (-1)^k \binom{2n+1}{k}$$
$$\left. \cdot S(n+1,1)\right) x^{k+1} + \cdots .$$

Comparing the coefficients at x^{k+1}, $k = 0, 1, 2, \ldots, n-1$, we get that

$$E^2(n,k) = \binom{2n+1}{0} S(n+k+1,k+1) - \binom{2n+1}{1} S(n+k,k)$$
$$+ \cdots + (-1)^k \binom{2n+1}{k} \cdot S(n+1,1)$$
$$= \sum_{j=0}^{k} (-1)^j \binom{2n+1}{j} S(n+k+1-j, k+1-j). \qquad \square$$

4.3.3. Using this formula, we can obtain some convenient special cases.

For example, as $S(n+1,1) = 1$, we have that

$$E^2(n,0) = (-1)^0 \binom{2n+1}{0} S(n+1,1)$$
$$+ (-1)^1 \binom{2n+1}{1} S(n,0) = 1 - 0 = 1, \quad n \geq 1.$$

Similarly, as $S(n,2) = 2^{n-1} - 1$, $S(n,1) = 1$, we have that

$$E^2(n,1) = S(n+2,2) - (2n+1)S(n+1,1)$$
$$= (2^{n+1} - 1) - (2n+1) = 2^{n+1} - 2(n+1), \quad n \geq 1.$$

As $S(n,3) = \frac{1}{6}(3^n - 3 \cdot 2^n + 3)$, $S(n,2) = 2^{n-1} - 1$, $S(n,1) = 1$, we have that

$$E^2(n,2) = S(n+3,3) - (2n+1)S(n+2,2) + \binom{2n+1}{2} S(n+1,1)$$
$$= \frac{1}{2}(3^{2+n} + 1) - (2n+3)2^{n+1}(2n+1)(n+1).$$

Using the representations

$$S(n,4) = \frac{1}{24}(4^n - 4 \cdot 3^n + 6 \cdot 2^n - 4), \quad \text{and}$$
$$S(n,5) = \frac{1}{120}(5^n - 5 \cdot 4^n + 10 \cdot 3^n - 10 \cdot 2^n + 5),$$

we can get closed formulas for $E^2(n,3)$, $E^2(n,4)$, etc.

4.3.4. On the other hand, we can obtain a closed formula for the Eulerian numbers of the second order using the unsigned Stirling numbers of the first kind.

- *For the Eulerian numbers of the second order, it holds that*

$$E^2(n, 2n-k) = \sum_{t=n+1}^{k} (-1)^{k-t} |s(t, t-n)| \binom{2n+1}{k-t}$$
$$= \sum_{t=0}^{k} (-1)^{k-t} |s(t, t-n)| \binom{2n+1}{k-t}.$$

For example, it holds that

$$E^2(3,1) = E^2(3, 6-5) = \sum_{t=4}^{5} (-1)^{5-t} |s(t, t-3)| \binom{7}{5-t}$$
$$= -|s(4,1)| \binom{7}{1} + |s(5,2)| \binom{7}{0} = -6 \cdot 7 + 50 = 8;$$
$$E^2(3,2) = E^2(3, 6-4) = |s(4,1)| \binom{7}{0} = 6 \cdot 1 = 6.$$

□ We have proven that

$$|s(k,k-n)|x^k = \frac{\sum_{j=n+1}^{2n} E^2(n,2n-j)x^j}{(1-x)^{2n+1}}.$$

As $(1-x)^{2n+1} = \sum_{i=0}^{2n+1}\binom{2n+1}{i}(-1)^i x^i$, we get that

$$\sum_{j=n+1}^{2n} E^2(n,2n-j)x^j = (1-x)^{2n+1}\left(\sum_{k=0}^{\infty}|s(k,k-n)|x^k\right)$$

$$= \left(\sum_{i=0}^{2n+1}\binom{2n+1}{i}(-1)^i x^i\right)\left(\sum_{k=n+1}^{\infty}|s(k,k-n)|x^k\right)$$

$$= \left(\binom{2n+1}{0}|s(n+1,1)|\right)x^{n+1} + \left(\binom{2n+1}{0}|s(n+2,2)|\right.$$

$$\left. - \binom{2n+1}{1}|s(n+1,1)|\right)x^{n+2} + \cdots + \left(\binom{2n+1}{0}|s(k,k-n)|\right.$$

$$- \binom{2n+1}{1}|s(k-1,k-1-n)|$$

$$\left. + \cdots + (-1)^{k-(n+1)}\binom{2n+1}{h-(n+1)}\right)x^k + \cdots$$

$$= \sum_{k=n+1}^{\infty}\left(\sum_{t=n+1}^{k}|s(t,t-n)|(-1)^{k-t}\binom{2n+1}{k-t}\right)x^k.$$

By comparing the coefficients at x^k on the left- and right-hand sides of the last equation, we obtain that

$$E^2(n,2n-k) = \sum_{t=n+1}^{k}|s(t,t-n)|(-1)^{k-t}\binom{2n+1}{k-t}$$

$$= \sum_{t=0}^{k}|s(t,t-n)|(-1)^{k-t}\binom{2n+1}{k-t}. \qquad \square$$

4.3.5. Using known values of $|s(k,k-n)|$, we can obtain closed formulas for some $E^2(n,k)$.

For example, as $|s(n+1,1)| = n!$, we get that

$$E^2(n,n-1) = E^2(n,2n-(n+1)) = |s(n+1,1)|\binom{2n+1}{0} = n!.$$

As $|s(n+1,1)| = n!$ and $s(n+2,2) = (n+1)!\left(1+\frac{1}{2}+\frac{1}{3}+\cdots+\frac{1}{n+1}\right) = (n+1)!H_{n+1}$, where H_k, $k = 0,1,2,\ldots$, is the sequence of harmonic numbers, we get that

$$\begin{aligned}E^2(n,n-2) &= E^2(n,2n-(n+2)) = -|s(n+1,1)|\binom{2n+1}{1} \\ &\quad + |s(n+2,2)|\binom{2n+1}{0} \\ &= n!\left((n+1)H_{n+1} - (2n+1)\right).\end{aligned}$$

4.3.6. So, we have the following simple cases of closed formulas for the Eulerian numbers of the second order:

- $E^2(0,0) = 1$ and $E^2(n,0) = 0$, for $n \geq 1$;
- $E^2(n,n) = 0$ for $n \geq 1$, and $E^2(n,n+k) = 0$ for $n \geq 0$, $k \geq 1$;
- $E^2(n,n-1) = n!$ for $n \geq 1$;
- $E^2(n,n-2) = n!\left((n+1)H_{n+1} - (2n+1)\right)$ for $n \geq 2$;
- $E^2(n,1) = 2^{n+1} - 2(n+1)$ for $n \geq 1$;
- $E^2(n,2) = \frac{1}{2}\left(3^{2+n}+1\right) - (2n+3)2^{n+1}(2n+1)(n+1)$ for $n \geq 2$;

Exercises

1. Prove that $E^2(n,2) = \frac{9\cdot 3^n - (12+8n)2^n + (3+6n+4n^2)}{2}$.
2. Find the closed formula for the numbers $E^2(n,3)$ and for the numbers $E^2(n,4)$.
3. Calculate $E^2(n,k)$, $n = 5,6,7,8$, $k = 1,2,3,4,\ldots,n$, using the obtained closed formulas. Compare the complexity of these calculations with that of the calculations using the recurrent relation for the Eulerian numbers of the second order.
4. Formulate and check the rule for calculating the numbers $E^2(n,k)$ using Pascal's triangle and the triangle of Stirling numbers of the second kind.

5. Formulate and check the rule for calculating the numbers $E^2(n, 2n-k)$ using Pascal's triangle and the triangle of unsigned Stirling numbers of the first kind.

4.4. Generating Function of the Sequence of Eulerian Numbers of the Second Order

4.4.1. Now, we try to find the generating function of the sequence of elements of the triangle of Eulerian numbers of the second order.

It turns out that this problem is relatively simple if we consider the generating functions of the finite sequences of Eulerian numbers of the second kind $E^2(n,k)$, $k = 0, 1, 2, \ldots, n$, with a fixed n, that is, polynomials of the form

$$U_n(x) = \sum_{k=0}^{n} E^2(n,k)x^k = E^2(n,0) + E^2(n,1)x + \cdots + E^2(n,n-1)x^{n-1} + E^2(n,n)x^n.$$

These polynomials are called the *Eulerian polynomials of the second order*. Using the triangle of Eulerian numbers of the second order, we can easily get these polynomials for small n:

$$U_0(x) = E^2(0,0) \equiv 1;$$

$$U_1(x) = E^2(1,0) + E^2(1,1)x \equiv 1;$$

$$U_2(x) = E^2(2,0) + E^2(2,1)x + E^2(2,2)x^2 = 1 + 2x;$$

$$U_3(x) = E^2(3,0) + E^2(3,1)x + E^2(3,2)x^2 + E^2(3,3)x^3 = 1 + 8x + 6x^2;$$

$$U_4(x) = E^2(4,0) + E^2(4,1)x + E^2(4,2)x^2 + E^2(4,3)x^3 + E^2(4,4)x^4 = 1 + 22x + 58x^2 + 24x^3.$$

For $n \geq 1$, the values $E^2(n,n)$ are equal to zero. It means that the sum over k in the definition of $U_n(x)$, $n \geq 1$, can be written with an upper index only up to $n-1$. It also means that the polynomials $U_n(x)$ are really of degree $n-1$ for $n \geq 1$.

4.4.2. Unfortunately, a simple compact formula for the Eulerian polynomials does not exist. However, the recurrent relation for the Eulerian numbers of the second genus allows us to obtain the recurrent relation for the polynomials $U_n(x)$.

- *The Eulerian polynomials of the second order $U_n(x)$, $n \geq 1$, can be computed using the recurrent relation*

$$U_{n+1}(x) = (2nx+1)U_n(x) - x(x-1)U_n'(x), \quad U_0(x) \equiv 1.$$

□ This linear polynomial recurrence can be obtained by induction. In fact, as $U_0(x) \equiv 1$, $U_0'(x) \equiv 0$, and

$$\begin{aligned} E_1(x) \equiv 1 &= x(1-x)\cdot 0 + 1(1+0\cdot x) \\ &= x\,(1-x)\,U_0'(x) + U_0(x)\,(1+0\cdot x). \end{aligned}$$

Similarly, as $U_1(x) \equiv 1$, $U_1'(x) \equiv 0$, and

$$\begin{aligned} U_2(x) = 1+2x &= x(1-x)\cdot 0 + 1(1+2x) \\ &= x\,(1-x)\,U_1'(x) + U_1(x)\,(1+2\,x). \end{aligned}$$

As $U_2(x) = 1+2x$, $U_2'(x) \equiv 2$, and

$$\begin{aligned} U_3(x) &= 1+8x+6x^2 = x(1-x)\cdot 2 + (1+4x)(1+2x) \\ &= x\,(1-x)\,U_2'(x) + U_2(x)\,(1+2\cdot 2\,x). \end{aligned}$$

In general, as $U_n(x) = \sum_{k=0}^{n} E^2(n,k)x^k$, $U_n'(x) = \sum_{k=0}^{n} kE^2(n, k)x^{k-1}$, and the polynomial $x\,(1-x)\,U_n'(x) + U_n(x)\,(1+2n\,x)$ takes the form

$$
\begin{aligned}
x(1-x)\sum_{k=0}^{n} kE^2(n,k)x^{k-1} &+ (1+2nx)\sum_{k=0}^{n} E^2(n,k)x^k \\
&= \sum_{k=0}^{n} kE^2(n,k)x^k - \sum_{k=0}^{n} kE^2(n,k)x^{k+1} + \sum_{k=0}^{n} E^2(n,k)x^k \\
&\quad + 2n\sum_{k=0}^{n} E^2(n,k)x^{k+1} \\
&= \sum_{k=0}^{n}(k+1)E^2(n,k) \\
&\quad + \sum_{k=0}^{n}(2n-k)E^2(n,k)x^{k+1} \\
&= E^2(n,0) + \sum_{k=1}^{n}(k+1)E^2(n,k)x^k \\
&\quad + \sum_{k=1}^{n+1}(2n+1-k)E^2(n,k-1)x^k \\
&= E^2(n,0) \\
&\quad + \sum_{k=1}^{n}((k+1)E^2(n,k) + (2(n+1)-k-1)E^2(n,k-1))x^k \\
&= E^2(n+1,0) + \sum_{k=1}^{n} E^2(n+1,k)x^k \\
&= \sum_{k=0}^{n} E^2(n+1,k)x^k \\
&= \sum_{k=0}^{n+1} E^2(n+1,k)x^k \\
&= U_{n+1}(x). \qquad \square
\end{aligned}
$$

4.4.3. The last relation can be written in a more compact form:

$$\frac{U_{n+1}(x)}{(1-x)^{2n+2}} = \left(\frac{xU_n(x)}{(1-x)^{2n+1}}\right)'.$$

In other words, it holds that

$$\frac{U_{n+1}(x)}{(x-1)^{2(n+1)}} = \left(\frac{x}{1-x} \cdot \frac{U_n(x)}{(x-1)^{2n}}\right)',$$

and we can easily remember this recurrent relation by considering a rational function $u_n(x) = \frac{U_n(x)}{(x-1)^{2n}}$. This function satisfies the simple recurrent relation

$$u_{n+1}(x) = \left(\frac{x}{1-x} u_n(x)\right)', \quad u_0(x) \equiv 1.$$

Sometimes the function $u_n(x) = \frac{U_n(x)}{(x-1)^{2n}}$ is called *the Eulerian fraction of the second order*, similarly to the classic Euler fraction $\alpha_n(x) = \frac{A_n(x)}{(1-x)^{n+1}}$.

As $\frac{xU_n(x)}{(x-1)^{2n+1}} = \sum_{k=0}^{\infty} S(k+n,k)x^k, n \geq 1$, $u_n(x)$ can be represented in the following form:

$$u_n(x) = \frac{1-x}{x} \sum_{k=0}^{\infty} S(k+n,k)x^k = (1-x) \sum_{k=1}^{\infty} S(k+n,k)x^{k-1}$$

$$= \sum_{k=0}^{\infty} (S(n+k+1,k+1) - S(n+k,k))\, x^k, \quad n \geq 1.$$

4.4.4. However, the numbers $E^2(n,k)$ depend on the two parameters n and k, and it is interesting to get a two-dimensional function which generates all triangle arrays of the Eulerian numbers of the second order.

The two-dimensional exponential generating function $U(x,y)$ of the triangle array of Eulerian numbers of the second order is defined as

$$U(x,y) = \sum_{n,k=0}^{\infty} E^2(n,k)x^k \frac{y^n}{n!} = 1 + \sum_{n=1}^{\infty} \sum_{k=0}^{n-1} E^2(n,k)x^k \frac{y^n}{n!}.$$

In other words, we speak about the exponential generating function $U_x(y) = \sum_{n=0}^{\infty} U_n(x)\frac{y^n}{n!}$ of the sequence $U_n(x)$, $n = 0, 1, 2, \ldots$, of

Eulerian polynomials of the second order:

$$U_x(y) = \sum_{n=0}^{\infty} U_n(x)\frac{y^n}{n!} = U_0(x) + \sum_{n=1}^{\infty}\left(\sum_{k=0}^{n-1} E^2(n,k)x^k\right)\frac{y^n}{n!}$$

$$= \sum_{n,k=0}^{\infty} E^2(n,k)x^k\frac{y^n}{n!} = U(x,y).$$

It can be proven that the function $U(x, y)$ satisfies the partial differential equation

$$U(x,y) = (1-2xy)U_y^{'} + x(x-1)U_x^{'},$$

which, unfortunately, has no elementary solutions.

4.4.5. We have considered well-known generating functions connected with the triangle of Eulerian numbers of the second order. However, there are other generating functions representing some special sequences of Eulerian numbers of the second order.

For the zeroth column of the triangle, i.e., for the sequence $E^2(n,0)$, $n \geq 0$, consisting of only unities (sequence A000012 in the OEIS), the generating function has the form $\frac{1}{1-x}$, i.e., it holds that

$$\frac{1}{1-x} = 1 + 1\cdot x + 1\cdot x^2 + \cdots + 1\cdot x^n + \cdots, \quad |x| < 1,$$

while the exponential generation function of this sequence is e^x:

$$e^x = 1 + 1\cdot\frac{x}{1!} + 1\cdot\frac{x^2}{2} + \cdots + 1\cdot\frac{x^n}{n!} + \cdots, \quad x \in \mathbb{R}.$$

For the first column of the triangle, i.e., for the sequence $E^2(n,1)$, $n \geq 2$, starting with the elements 2, 8, 22, 52, 114, 240, 494, 1004, 2026, 4072, … (sequence A005803 in the OEIS), we have the following result:

- *The generating function of the sequence $E^2(n,1)$, $n \geq 2$, has the form $\frac{2}{(1-2x)(x-1)^2}$, i.e., it holds that*

$$\frac{2}{(1-2x)(x-1)^2}$$
$$= 2 + 8x + 22x^2 + 52x^3 + \cdots, \quad |x| < \frac{1}{2}.$$

□ The reader can obtain and study for the sequence $E^2(n,1)$, $n \geq 2$, the linear recurrent relation of the third order

$$E^2(n+3,1) - 4E^2(n+2,1) + 5E^2(n,1) - 2E^2(n,1) = 0$$

with constant coefficients $b_0 = 1$, $b_1 = -4$, $b_2 = 5$, $b_3 = -2$, and with initial conditions $E^2(2,1) = 2$, $E^2(3,1) = 8$, and $E^2(4,1) = 22$. But the simplest way is to note that $E^2(n,1) = 2E^2(n,1)$ and use a similar result for the sequence $E^2(n,1)$, $n \geq 2$ (see Chapter 3). □

Similarly, for the second column of the triangle, i.e., for the sequence $E^2(n,2)$, $n \geq 3$, starting with the elements 6, 58, 328, 1452, 5610, 19950, 67260, 218848, 695038, 2170626, ... (sequence A004301 in the OEIS), the generating function has the form $\frac{6-2x-12x^2}{(1-x)^3(1-2x)^2(1-3x)}$, i.e., it holds that

$$\frac{6-2x-12x^2}{(1-x)^3(1-2x)^2(1-3x)} = 6 + 58x + 328x^2 + 1452x^3 + \cdots, |x| < \frac{1}{3},$$

while for the third column of the triangle, i.e., for the sequence $E^2(n,3)$, $n \geq 4$, starting with the elements 24, 444, 4400, 32120, 195800, 1062500, 5326160, 25243904, 114876376, 507259276, ... (sequence A006260 in the OEIS), we obtain the generating function $\frac{24-36x-280x^2+652x^3-668x^4-288x^5}{(x-1)^4(1-2x)^3(1-3x)^2(1-4x)}$:

$$\frac{24-36x-280x^2+652x^3-668x^4-288x^5}{(x-1)^4(1-2x)^3(1-3x)^2(1-4x)} = 24 + 444x + 4400x^2 + 32120x^3 + \cdots, |x| < \frac{1}{4}.$$

Exercises

1. Find the Eulerian polynomials of the second order $U_n(x)$ for $n = 5, 6, 7, 8, 9, 10, 11$ using the triangle of Eulerian numbers of the second order.
2. Check the recurrence relation $U_{n+1}(x) = (2nx+1)U_n(x) - x(x-1)U'_n(x)$ for small values of n. Find the Eulerian polynomials of the second order $U_n(x)$ for $n = 5, 6, 7, 8, 9, 10, 11$ using the recurrent relation. Compare the complexity of two methods of calculating these polynomials.

3. Find $U_n(1)$ for $n = 1, 2, \ldots, 11$. What is the resulting number? Find $U_n(-1)$ for $n = 1, 2, \ldots, 11$. What is the resulting number?
4. For $n = 1, 2, 3, 4$, check that $\frac{U_{n+1}(x)}{(x-1)^{2n+2}} = \left(\frac{xU_n(x)}{x^{2n+1}}\right)'$.
5. Find the value of the polynomial $U_n(x)$, $n = 0, 1, 2, 3, 4{,}5$ at $x = 0$, $x = 1$, $x = -1$, $x = 0.5$, and $x = -0{,}5$.

4.5. Properties of Eulerian Numbers of the Second Order

In this section, we summarise the already known and some new properties of Eulerian numbers of the second order.

Closed formulas for Eulerian numbers of the second order

4.5.1. In Section 4.3, we have proven well-known closed formulas for $E^2(n, k)$.

The first one gives a representation of $E^2(n, k)$ using the Stirling numbers of the second kind.

- *For Eulerian numbers of the second order, there exists the closed formula*

$$E^2(n,k) = \sum_{j=0}^{k+1}(-1)^j\binom{2n+1}{j}S(n+k+1-j, k+1-j), \quad n \geq 1.$$

The second one gives a representation of $E^2(n, 2n - k)$ using the unsigned Stirling numbers of the first kind.

- *For the Eulerian numbers of the second order, it holds that*

$$E^2(n, 2n-k) = \sum_{t=n+1}^{k}(-1)^{k-t}|s(t, t-n)|\binom{2n+1}{k-t}$$
$$= \sum_{t=0}^{k}(-1)^{k-t}|s(t, t-n)|\binom{2n+1}{k-t}.$$

Using these formulas, we obtained several particular cases. So, $E^2(n,1) = 2^{n+1} - 2(n+1)$, and $E^2(n, n-1) = n!$.

A simple analysis of the last two equations allows us to obtain the following properties.

- *Every Eulerian number of the second order $E^2(n,1)$ is twice the corresponding Eulerian number of the first order $E(n,1)$:*

$$E^2(n,1) = 2E(n,1), n \geq 1.$$

□ The result is obvious, as it was proven that $E^2(n,1) = 2^n - (n+1)$, and $E^2(n,1) = 2^{n+1} - 2(n+1)$. □

- *Every Eulerian number of the second order $E^2(n, n-1)$ is the sum of the Eulerian numbers of the first order $E(n,k)$, $k = 0, 1, \ldots,$ $n-1$:*

$$E^2(n, n-1) = E(n,0) + E(n,1) + \cdots + E(n, n-1), \quad n \geq 1.$$

□ The result is also obvious, as it was proven that $E^2(n,0) + E^2(n,1) + \cdots + E^2(n, n-1) = n!$, and $E^2(n, n-1) = n!$. □

4.5.2. In [He23], a new interesting formula for calculating Eulerian numbers is given:

- *For any $n \geq 1$ and any $0 \leq k \leq n-1$, it holds that*

$$E^2(n,k) = \sum_{t_1+t_2+\cdots+t_{k+1}=n-k-1, t_i \geq 0} 1^{t_1} 2^{t_2} \cdot \ldots$$
$$\ldots \cdot (k+1)^{t_{k+1}} (2t_1+2)(2(t_1+t_2)+3)$$
$$\cdots (2(t_1+\cdots+t_k)+k+1).$$

For example,

$$E^2(5,2) = \sum_{t_1+t_2+t_3=2, t_i \geq 0} 1^{t_1} 2^{t_2} 3^{t_3} (2t_1+2)(2(t_1+t_2)+3)$$
$$= 1^0 2^0 3^2 (2 \cdot 0 + 2)(2 \cdot (0+0) + 3)$$
$$+ 1^0 2^2 3^0 (2 \cdot 0 + 2)(2 \cdot (0+2) + 3)$$

$$
\begin{aligned}
&+1^2 2^0 3^0 (2\cdot 2+2)(2\cdot(2+0)+3)\\
&+1^0 2^1 3^1 (2\cdot 0+2)(2\cdot(0+1)+3)\\
&+1^1 2^0 3^1 (2\cdot 1+2)(2\cdot(1+0)+3)\\
&+1^1 2^1 3^0 (2\cdot 1+2)(2\cdot(1+1)+3)\\
&=54+56+42+60+60+56=328.
\end{aligned}
$$

□ Let us prove this identity by induction.

It is true for small values of n and k. In fact, it holds that

$$
\begin{aligned}
E^2(1,0) &= 1 = 1^0 = \sum_{t_1=0} 1^{t_1}, \quad E^2(2,0) = 1 = 1^1 = \sum_{t_1=1} 1^{t_1},\\
E^2(3,0) &= 1 = 1^2 = \sum_{t_1=2} 1^{t_1}, \quad E^2(2,1) = 2 = 1^0\cdot 2^0(2\cdot 0+2)\\
&= \sum_{t_1+t_2=0} 1^{t_1}2^{t_2}(2\cdot t_1+2),\\
E^2(3,1) &= 8 = 1^0\cdot 2^1(2\cdot 0+2)+1^1\cdot 2^0\cdot(2\cdot 1+2)\\
&= \sum_{t_1+t_2=1} 1^{t_1}2^{t_2}(2t_1+2),\\
E^2(3,2) &= 6 = 1^0\cdot 2^0\cdot 3^0(2\cdot 0+2)(2\cdot(0+0)+3)\\
&= \sum_{t_1+t_2+t_3=0} 1^{t_1}2^{t_2}3^{t_3}(2t_1+2)(2(t_1+t_2)+3).
\end{aligned}
$$

Suppose that the identity is true for a positive integer n and all $0 \le k \le n-1$. We try to prove it for $n+1$ and all $0 \le k \le n$. As $E^2(n+1,k) = (k+1)E^2(n,k) + (2n-k+1)E^2(n,k-1)$, we have that

$$
\begin{aligned}
&E^2(n+1,k)\\
&\quad= (k+1)\sum_{t_1+\cdots+t_{k+1}=n-k-1,\, t_i\ge 0} 1^{t_1}2^{t_2}\cdots(k+1)^{t_{k+1}}(2t_1+2)\\
&\qquad\times(2(t_1+t_2)+3)\cdots(2(t_1+t_2+\cdots+t_k)+k+1)
\end{aligned}
$$

$$+ (2n-k+1) \sum_{t_1+\cdots+t_k=n-k, t_i\geq 0} 1^{t_1}2^{t_2}\cdots k^{t_k}(2t_1+2)$$

$$\times (2(t_1+t_2)+3)\cdots(2(t_1+t_2+\cdots+t_{k-1})+k)$$

$$= \sum_{t_1+\cdots+t_{k+1}=n-k-1, t_i\geq 0} 1^{t_1}2^{t_2}\cdots(k+1)^{t_{k+1}+1}(2t_1+2)$$

$$\times (2(t_1+t_2)+1)\cdots(2(t_1+t_2+\cdots+t_k)+k+1)$$

$$+ \sum_{t_1+\cdots+t_k=n-k, t_i\geq 0} 1^{t_1}2^{t_2}\cdots k^{t_k}(t_1+1)$$

$$\times (t_1+t_2+1)\cdots(t_1+t_2+\cdots+t_{k-1}+1)(2n-k+1)$$

$$= \sum_{t_1+\cdots+t_k+s_{k+1}=n-k, t_i\geq 0, s_{k+1}\geq 1} 1^{t_1}2^{t_2}\cdots(k+1)^{s_{k+1}}(2t_1+2)$$

$$\times (2(t_1+t_2)+3)\cdots(2(t_1+t_2+\cdots+t_k)+k+1)$$

$$+ \sum_{t_1+\cdots+t_k+s_{k+1}=n-k, t_i\geq 0, s_{k+1}=0} 1^{t_1}2^{t_2}\cdots k^{t_k}(k+1)^{s_{k+1}}(2l_1+2)$$

$$\times (2(t_1+t_2)+3)\cdot\cdots\cdot(2(t_1+t_2+\cdots+t_{k-1})+k)$$

$$\times (2(t_1+t_2+\cdots+t_k)+k+1)$$

$$= \sum_{t_1+\cdots+t_k+t_{k+1}=n-k, t_i\geq 0} 1^{t_1}2^{t_2}\cdots k^{t_k}(k+1)^{t_{k+1}}(2t_1+2)$$

$$\times (2(t_1+t_2)+3)$$

$$\cdots(2(t_1+t_2+\cdots+t_{k-1})+k)(2(t_1+t_2+\cdots+t_k)+k+1).$$

□

This formula is very simple for extreme values of k:

$$E^2(n+1,0) = \sum_{t_1=n} 1^{t_1} = 1^n = 1;$$

$$E^2(n+1,n) = \sum_{t_1+\cdots+t_{n+1}=0} 1^{t_1}2^{t_2}\cdots n^{t_n}(n+1)^{t_{n+1}}(2t_1+2)$$

$$\times (2(t_1+t_2)+3)\cdot\cdots\cdot(2(t_1+t_2+\cdots+t_n)+n+1)$$

$$= 1^0 2^0 \cdots (n+1)^0 (2\cdot 0+2)(2(0+0)+3)$$
$$\cdots (2(0+0\cdots+0)+n+1) = 1\cdot 2\cdot\cdots\cdot(n+1)$$
$$= (n+1)!$$

Sums with Eulerian numbers of the second order

4.5.3. The simplest sum with Eulerian numbers of the second order, e.g., the sum over all $E^2(n,k)$, $k=0,1,2,\ldots,n-1$, gives the number of all Stirling permutations of the nth order, $n\geq 1$.

- *The sum of numbers $E^2(n,0)$, $E^2(n,1),\ldots,E^2(n,n-1)$, located in the nth row of the triangle of Eulerian numbers of the second order, is equal to $(2n-1)!!$:*

$$E^2(n,0)+E^2(n,1)+E^2(n,2)+\cdots+E^2(n,n-1) = (2n-1)!, \quad n\geq 1.$$

4.5.4. Furthermore, in Section 4.2, we obtained two sums over Eulerian numbers of the second order, connecting with the Stirling numbers of the second and first kinds:

$$E^2(n,0)\binom{x+n-1}{2n} + E^2(n,1)\binom{x+n-2}{2n}$$
$$+\cdots+E^2(n,n-1)\binom{x}{2n} = S(x,x-n).$$
$$E^2(n,0)\binom{x}{2n} + E^2(n,1)\binom{x+1}{2n}$$
$$+\cdots+E^2(n,n-1)\binom{x+(n-1)}{2n} = |s(x,x-n)|.$$

4.5.5. There exist sums involving simultaneously the Eulerian numbers of the second order $E^2(n,k)$, the Stirling numbers of the second kind $S(n,k)$, and the powers of two 2^k.

- *The sum of numbers $E^2(n,0)$, $E^2(n,1),\ldots,E^2(n,k-1)$ with weights $2^{k-1},2^{k-2},\ldots,2^0$, respectively, can be represented as a sum* *over the Stirling numbers* $S(n+k,k)$, $S(n+k-1,k-1),\ldots,S(n+1,1)$ *with weights* $2^0\binom{2n+1}{0}$, $2^1\binom{2n+1}{0}-2^0\binom{2n+1}{1},\ldots,2^{k-1}\binom{2n+1}{0}-2^{k-2}\binom{2n+1}{1}+\cdots+(-1)^{k-1}2^0\binom{2n+1}{k-1}$*:*

$$\begin{aligned}&2^{k-1}E^2(n,0)+2^{k-1}E^2(n,1)+\cdots+2E^2(n,k-2)+E^2(n,k-1)\\&\quad=S(n+k,k)+S(n+k-1,k-1)\left(2\binom{2n+1}{0}-\binom{2n+1}{1}\right)\\&\qquad+\cdots+S(n+1,1)\left(2^{k-1}\binom{2n+1}{0}-2^{k-2}\binom{2n+1}{1}\right.\\&\qquad\left.+\cdots+2^0\binom{2n+1}{k-1}\right).\end{aligned}$$

For $k=1$, as $E^2(n,0)=1$ and $S(n+1,1)=1$, we get the degenerate identity

$$E^2(n,0)=S(n+1,1).$$

For $k=2$, as $E^2(n,1)=2^{n+1}-2(n+1)$ and $S(n+2,2)=2^{n+1}-1$, it holds that

$$\begin{aligned}&2E^2(n,0)+E^2(n,1)\\&\quad=S(n+2,2)+S(n+1,1)\left(2\binom{2n+1}{0}-\binom{2n+1}{1}\right).\end{aligned}$$

In general, the identity holds for any positive parameter a, taken instead of 2.

- *For any positive integer a and for any positive integers n and k, $1\le k\le n-1$, it holds that*

$$\begin{aligned}&\sum_{i=0}^{k-1}a^{k-1-i}E^2(n,i)\\&\quad=\sum_{i=0}^{k-1}S(n+k-i,k-i)\left(\sum_{j=0}^{i}(-1)^j2^{i-j}\binom{2n+1}{j}\right).\end{aligned}$$

□ We have proven that

$$\sum_{k=0}^{\infty} S(n+k,k)x^k = \frac{xU_n(x)}{(1-2x)^{2n+1}},$$

where $U_n(x) = E^2(n,0) + E^2(n,1) + \cdots + E^2(n,n-1)$. So, it holds that

$$(1-x)^{2n+1}\sum_{k=0}^{\infty} S(n+k,k)x^k = xU_n(x), \quad \text{and}$$

$$(1-ax)^{-1}(1-x)^{2n+1}\sum_{k=0}^{\infty} S(n+k,k)x^k = (1-ax)^{-1}xU_n(x).$$

As $(1-x)^{2n+1} = \sum_{i=0}^{2n+1}(-1)^i\binom{2n+1}{i}x^i$, and $(1-ax)^{-1} = 1 + ax + a^2x^2 + \cdots + a^ix^i + \cdots$, we have that

$$\begin{aligned}
&(1-ax)^{-1}(1-x)^{2n+1}\\
&\quad= (1 + ax + a^2x^2 + \cdots + a^ix^i + \cdots)\cdot\left(1 - \binom{2n+1}{1}x\right.\\
&\qquad\left.+\binom{2n+1}{2}x^2 - \cdots + (-1)^{2n+1}\binom{2n+1}{2n+1}x^{2n+1}\right)\\
&\quad= +1 + \left(a\binom{2n+1}{0} - \binom{2n+1}{1}\right)x + \left(a^2\binom{2n+1}{0}\right.\\
&\qquad\left.-a^1\binom{2n+1}{1} + \binom{2n+1}{2}\right)x^2 + \cdots\\
&\quad= \sum_{i=1}^{\infty}\left(a^i\binom{2n+1}{0} - a^{i-1}\binom{2n+1}{1} + \cdots + (-1)^i\binom{2n+1}{i}\right)x^i.
\end{aligned}$$

Therefore, it holds that

$$\begin{aligned}
&(1-ax)^{-1}(1-x)^{2n+1}\sum_{k=0}^{\infty} S(n+k,k)x^k\\
&\quad= \left(\sum_{i=0}^{\infty}\left(a^i\binom{2n+1}{0} - a^{i-1}\binom{2n+1}{1}\right.\right.
\end{aligned}$$

$$+\cdots+(-1)^i a^0\binom{2n+1}{i}\Bigg)\Bigg)x^i\Bigg)\cdot\left(\sum_{j=1}^{\infty}S(n+j,j)x^j\right)$$

$$=\sum_{k=1}^{\infty}\left(\sum_{i=0}^{k-1}S(n+k-i,k-i)\left(a^i\binom{2n+1}{0}-a^{i-1}\binom{2n+1}{1}\right.\right.$$

$$\left.\left.+\cdots+(-1)^i a^0\binom{2n+1}{i}\right)\right)x^k$$

$$=\sum_{k=1}^{\infty}\left(\sum_{i=0}^{k-1}S(n+k-i,k-i)\left(\sum_{j=0}^{i}(-1)^j 2^{i-j}\binom{2n+1}{j}\right)\right)x^k.$$

On the other hand, it holds that

$$(1-ax)^{-1}xU_n(x)=(x+ax^2+a^2x^3+\cdots)(E^2(n,0)+E^2(2(n,1)x$$

$$+\cdots+E^2(n,n-1)x^{n-1})$$

$$=\sum_{k=1}^{\infty}\left(\sum_{i=0}^{k-1}a^{k-i-1}E^2(n,i)\right)x^k.$$

Comparing the coefficients at x^k in both representations, we get that

$$\sum_{i=0}^{k-1}a^{k-i-1}E^2(n,i)$$

$$=\sum_{i=0}^{k-1}S(n+k-i,k-i)\left(\sum_{j=0}^{i}(-1)^j 2^{i-j}\binom{2n+1}{j}\right). \qquad \square$$

The polynomial $P_n^a(x)=\sum_{i=0}^{k-1}a^{k-i-1}E^2(n,i)x^i$ is connected with the Eulerian polynomial of the second order $U_n(x)$ by the equality $P_n^a(x)=a^{n-1}U_n\left(\frac{x}{a}\right)$. In this notation, we can write that

$$P_n(1)=\sum_{i=0}^{k-1}S(n+k-i,k-i)\left(\sum_{j=0}^{i}(-1)^j 2^{i-j}\binom{2n+1}{j}\right).$$

The polynomials $P_n^2(1+x)$, $n \geq 1$, are associated with fully labelled series–parallel networks (for more explanations, see [Rior76] and [Rior79]).

4.5.6. The next group of summation formulas are connected with the *generalised Bernoulli numbers* (or *Nörlund polynomials*) $B_n^{(x)}$, defined by their exponential generating function:

$$\sum_{n=0}^{\infty} B_n^{(x)} \frac{z^n}{n!} = \left(\frac{z}{e^z - 1}\right)^x.$$

For fixed n, $B_n^{(x)}$ is a polynomial in x of degree n.

It is easy to see that $B_0^{(0)} = 1$ and $B_n^{(0)} = 0$, $n \geq 1$.

On the other hand, the sequence $B_n^{(1)}$, $n = 0, 1, 2, \ldots$, gives the sequence B_n, $n = 0, 1, 2, \ldots$, of Bernoulli numbers:

$$B_n^{(1)} = B_n, n \geq 0.$$

This statement immediately follows from the representation $\sum_{n=0}^{\infty} B_n \frac{z^n}{n!} = \frac{z}{e^z-1} = \left(\frac{z}{e^z-1}\right)^1$.

If $x = n$, we get the sequence $b_n = \int_0^1 x^{\overline{n}} dx$, $n = 0, 1, 2, \ldots$, of *Cauchy numbers of the second type*:

$$B_n^{(n)} = (-1)^n b_n, \quad n \geq 0.$$

The sequence of Cauchy numbers of the second type starts with the elements $1, \frac{1}{2}, \frac{5}{6}, \frac{9}{4}, \frac{251}{30}, \frac{475}{12}, \frac{19087}{84}, \frac{36799}{24}, \frac{1070017}{90}, \frac{2082753}{20}, \ldots$ (sequences A002657 and A002790 in the OEIS). Their exponential generating function has the form

$$\sum_{n=0}^{\infty} b_n \frac{z^n}{n!} = \frac{-z}{(1-z)\log(1-z)}.$$

(See [Comt74], [Fu21], [Choi13], and [Goul60], as well as Sections 4.6 and 4.7.)

4.5.7. It is known (see the proof in Section 4.6) that, for positive integers m and n, there exists the equality

$$S(m+n, m) = \binom{n+m}{n} B_n^{(-m)}.$$

As the Stirling number $S(n+m,n)$ can be represented as a linear combination of the Eulerian numbers of the second order, we get the following decomposition:

- *For positive integers n and m, the generalised Bernoulli numbers $B_n^{(-m)}$ can be represented using the Eulerian numbers of the second order and rising powers:*

$$B_n^{(-m)} = \frac{n!}{(2n)!}\sum_{k=0}^{n-1}(-1)^{k-1}E^2(n,k)(-m)^{\overline{k+1}}(m+n+1)^{\overline{n-k-1}}.$$

□ It was proven that $S(m+n,m) = \sum_{k=0}^{n-1}\binom{2n+m-k-1}{2n}E^2(n,k)$. Therefore, we get that

$$B_n^{(-m)} = \frac{m!n!}{(m+n)!}\sum_{k=0}^{n}\binom{2n+m-k-1}{2n}E^2(n,k).$$

As $m^{\underline{k+1}} = m(m-1)\cdots(m-k) = (-1)^{k+1}(-m)(-m+1)\cdots(-m+k) = (-1)^{k+1}(-m)^{\overline{k+1}}$ and $(m+2n-k-1)(m+2n-k-2)\cdots(m+n+1) = (m+n+1)^{\overline{n-k-1}}$, we get that

$$\begin{aligned}B_n^{(-m)} &= \frac{n!}{(2n)!}\sum_{k=0}^{n-1}E^2(n,k)\frac{m!}{m-k-1}\cdot\frac{(m+2n-k-1)!}{(m+n)!}\\ &= \frac{n!}{(2n)!}\sum_{k=0}^{n-1}E^2(n,k)m(m-1)\cdots(m-k)\cdot(m+2n-k-1)\\ &\quad\times(m+2n-k-2)\cdots(m+n+1)\\ &= \frac{n!}{(2n)!}\sum_{k=0}^{n-1}(-1)^{k-1}E^2(n,k)(-m)^{\overline{k+1}}(m+n+1)^{\overline{n-k-1}}.\end{aligned}$$ □

The last equality can be considered for any integer x (as well as for any real x).

- *For any integer x, it holds that*

$$B_n^{(x)} = \frac{n!}{(2n)!}\sum_{k=0}^{n-1}(-1)^{k-1}E^2(n,k)x^{\overline{k+1}}(-x+n+1)^{\overline{n-k-1}}.$$

4.5.8. Using this result, we can obtain several sums involving the Eulerian numbers of the second order and the (classical) Bernoulli numbers.

- *The nth Bernoulli number B_n, $n \geq 2$, can be represented as a liner combination of Eulerian numbers of the second order:*

$$2B_n = \frac{E^2(n,0)}{\binom{2n-1}{0}} - \frac{E^2(n,1)}{\binom{2n-1}{1}} + \cdots + (-1)^{n-1}\frac{E^2(n,n-1)}{\binom{2n-1}{n-1}}.$$

□ We consider the equality in the form $-\frac{B_n}{n} = \frac{1}{2n}\sum_{k=0}^{n-1}(-1)^{k-1}\frac{E^2(n,k)}{\binom{2n-1}{k}}$. In order to get this equation, let us consider the well-known representation of the Bernoulli numbers:

$$\sum_{n=0}^{\infty} B_n\frac{z^n}{n!} = \frac{z}{e^z-1} = 1 + \sum_{n=1}^{\infty} B_n\frac{z^n}{n!}.$$

As

$$\left(\log\frac{e^z-1}{z}\right)' = \frac{z}{e^z-1}\cdot\frac{e^z z - e^z + 1}{z^2} = \frac{1}{z}\left(\frac{-z}{x^{-z}-1} - 1\right)$$
$$= \frac{1}{z}\left(\sum_{n=0}^{\infty} B_n\frac{(-z)^n}{n!} - 1\right) = \sum_{n=1}^{\infty}(-1)^n B_n\frac{z^{n-1}}{n!}.$$

Then, after integration, using that $B_1 = -\frac{1}{2}$ and with all other Bernoulli numbers of odd indices being equal to zero, we get that

$$\log\frac{e^z-1}{z} = \frac{z}{2} + \sum_{n=2}^{\infty}\frac{B_n}{n}\frac{z^n}{n!}.$$

As $\left(\left(\frac{z}{e^z-1}\right)^x\right)' = \left(e^{x\log\frac{z}{e^z-1}}\right)' = -e^{-x\log\frac{z}{e^z-1}}\log\frac{e^z-1}{z}$, the derivative of the polynomial $B_n^{(x)}$, taken at the point $x = 0$, is equal to $-\frac{B_n}{n}$:

$$\left(B_n^{(x)}\right)'|_{x=0} = -\frac{B_n}{n}, \quad n \geq 2.$$

On the other hand, the derivative of the function

$$f(x) = x^{\overline{k+1}}(-x+n+1)^{\overline{n-k-1}}$$
$$= x(x+1)\cdots(-x+n+1)\cdots(-x+2n-k-1),$$

taken at the point $x = 0$, is equal to $\frac{k!(2n-k-1)!}{n!}$; therefore, we get that

$$\left.\left(\frac{n!}{(2n)!}\sum_{k=0}^{n-1}(-1)^{k-1}E^2(n,k)x^{\overline{k+1}}(-z+n+1)^{\overline{n-k-1}}\right)'\right|_{x=0}$$
$$= \frac{1}{2n}\sum_{k=0}^{n-1}(-1)^{k-1}E^2(n,k)\frac{k!(2n-k-1)!}{(2n-1)!}$$
$$= \frac{1}{2n}\sum_{k=0}^{n-1}(-1)^{k-1}\frac{E^2(n,k)}{\binom{2n-1}{k}}.$$

So, we have proven that

$$-\frac{B_n}{n!} = \frac{1}{2n}\sum_{k=0}^{n-1}(-1)^{k-1}\frac{E^2(n,k)}{\binom{2n-1}{k}}, \quad \text{and} \quad 2B_n = \sum_{k=0}^{n-1}(-1)^k\frac{E^2(n,k)}{\binom{2n-1}{k}}.$$

□

4.5.9. By changing the index of the summation in the equality $2B_n = \sum_{k=0}^{n-1}(-1)^k\frac{E^2(n,k)}{\binom{2n-1}{k}}$, we get that

$$2B_{n+1} = \sum_{k=0}^{n}(-1)^k\frac{E^2(n+1,k)}{\binom{2n+1}{k}}.$$

Using the recurrent relation $E^2(n+1,k) = (k+1)E^2(n,k)+(2n-k+1)$, we obtain that

$$\sum_{k=0}^{n}(-1)^k\frac{E^2(n+1,k)}{\binom{2n+1}{k}}$$
$$= \sum_{k=0}^{n}(-1)^k\left((k+1)E^2(n,k)+(2n-k+1)E^2(n,k-1)\right)$$

$$\cdot\frac{k!(2n+1-k)!}{(2n+1)!} = \sum_{k=0}^{n}(-1)^k E^2(n,k)\frac{(k+1)!(2n-k)!(2n-k+1)}{(2n+1)!}$$

$$+\sum_{k=0}^{n}(-1)^k E^2(n,k-1)\frac{k!(2n+1-k)!(2n+1-k)}{(2n+1)!}$$

$$= \sum_{k=0}^{n-1}(-1)^k E^2(n,k)\frac{(k+1)!(2n-k)!(2n-k+1)}{(2n+1)!}$$

$$-\sum_{k=0}^{n-1}(-1)^k E^2(n,k)\frac{(k+1)!(2n-k)!(2n-k)}{(2n+1)!}$$

$$= \sum_{k=0}^{n-1}(-1)^k E^2(n,k)\frac{(k+1)!(2n-k)!}{(2n+1)!}(2n-k+1-2n+k)$$

$$= \sum_{k=0}^{n-1}(-1)^k \frac{E^2(n,k)}{\binom{2n+1}{k+1}}.$$

So, we have proven the following property:

- *The alternating sum of Eulerian numbers of the second order* $E^2(n,k)$, $k = 0,1,\ldots,n-1$, *with weights* $\frac{1}{\binom{2n+1}{k+1}}$ *is twice the Bernoulli number* B_{n+1}:

$$\frac{E^2(n,0)}{\binom{2n+1}{1}} - \frac{E^2(n,1)}{\binom{2n+1}{k+1}} + \cdots + (-1)^{n-1}\frac{E^2(n,n-1)}{\binom{2n+1}{n}} = 2B_{n+1}.$$

For small n, we have that

$$\frac{E^2(1,0)}{\binom{3}{1}} = \frac{1}{3} = 2B_2, \quad \frac{E^2(2,0)}{\binom{5}{1}} - \frac{E^2(2,1)}{\binom{5}{2}} = \frac{1}{5} - \frac{2}{10} = 0 = 2B_3,$$

$$\frac{E^2(3,0)}{\binom{7}{1}} - \frac{E^2(3,1)}{\binom{7}{2}} + \frac{E^2(3,2)}{\binom{7}{3}} = \frac{1}{7} - \frac{8}{21} + \frac{6}{35} = -\frac{1}{15} = 2B_4.$$

4.5.10. As any Bernoulli number with odd index, greater than 1, is equal to zero, we get the following:

- *The alternating sum of elements $E^2(2n, k)$, $k = 0, 1, \ldots, 2n-1$, of the $(2n)$th row of the triangle of Eulerian numbers of the second order, $n \geq 1$, with weights $\frac{1}{\binom{4n+1}{k+1}}$, is equal to zero:*

$$\frac{E^2(2n,0)}{\binom{4n+1}{1}} - \frac{E^2(2n,1)}{\binom{4n+1}{k+1}} + \cdots + - \frac{E^2(2n,2n-1)}{\binom{4n+1}{n}} = 0.$$

4.5.11. Noting that

$$\int_0^1 u^{k+1}(1-u)^{2n-k} du = \frac{(-1)^k}{2(n+1)} \frac{1}{\binom{2n+1}{k+1}},$$

we can obtain an integral representation of the Bernoulli numbers.

- *For the Bernoulli number B_{n+1}, $n \geq 0$, there exists the representation*

$$\frac{B_{n+1}}{n+1} = \int_0^1 \sum_{k=0}^{n-1} E^2(n,k) u^{k+1}(u-1)^{2n-k} du.$$

□ For $n = 1$, we have

$$\int_0^1 E^2(1,0)u(u-1)^2 du = \frac{1}{4} - \frac{2}{3} + \frac{1}{2} = \frac{1}{12} = \frac{B_2}{2}.$$

For $n = 2$, we have

$$\begin{aligned} &\int_0^1 \left(E^2(2,0)u(1-u)^5 + E^2(2,1)u^2(u-1)^3\right) du \\ &= 1 \cdot \left(\frac{1}{6} - \frac{4}{5} + \frac{6}{4} - \frac{4}{3} + \frac{1}{2}\right) - 2\left(\frac{1}{6} - \frac{3}{5} + \frac{3}{4} - \frac{1}{3}\right) \\ &= 1 \cdot \frac{1}{2 \cdot 3 \cdot 5} - 2 \cdot \frac{1}{2 \cdot 3 \cdot 10} = 0 = \frac{B_3}{2}. \end{aligned}$$

The formula also holds for $n = 0$ if we use the second Bernoulli numbers:

$$\int_0^1 E^2(0,0) u du = \frac{1}{2} = B_1^+.$$

In general, it holds that

$$\int_0^1 \sum_{k=0}^{n-1} E^2(n,k)u^{k+1}(u-1)^{2n-k}du$$

$$= \sum_{k=0}^{n-1} E^2(n,k) \int_0^1 u^{k+1}(u-1)^{2n-k}du$$

$$= \frac{1}{2(n+1)} \sum_{k=0}^{n-1} E^2(n,k) \frac{(-1)^k}{\binom{2n+1}{k+1}} = \frac{2B_{n+1}}{2(n+1)} = \frac{B_{n+1}}{n+1}. \qquad \square$$

4.5.12. Using more accurate consideration, based on the study of second derivatives of $B_n^{(x)}$ at $x = 0$, other interesting results can be obtained (see [Fu21]).

- *There exists the following formula, connecting the Bernoulli numbers, the Eulerian numbers of the second kind, and the harmonic numbers*:

$$\sum_{k=1}^{n} (-1)^k \binom{2n-1}{k-1}^{-1} (H_{2n-k} - H_{k-1})E^2(n,k-1)$$

$$= \frac{n^2}{n-1} B_{n-1} + n \sum_{k=2}^{n-2} \frac{B_k}{k} \frac{B_{n-k}}{n-k}.$$

- *For any non-negative integers N and n, with $n \geq N$, it holds that*

$$\sum_{k=1}^{n} (-1)^k \binom{2n+1}{N+k-1}^{-1} E^2(n,k-1)$$

$$= 2n \sum_{k=0}^{N-1} (-1)^{N-1-k} s(N, N-k) \frac{B_{n-k}}{n-k}.$$

In particular, for $n = N$, we get the following decomposition for the Cauchy numbers of the second type:

$$2b_n = \sum_{k=1}^{n} (-1)^{n-k} \binom{2n-1}{n+k-1}^{-1} E^2(n,k-1).$$

However, the last result can be obtained by right checking.

- *For the Cauchy numbers of the second type b_n, $n \geq 1$, there exists the following representation:*

$$2b_n = \sum_{k=0}^{n-1} (-1)^{n+k-1} \frac{E^2(n,k)}{\binom{2n-1}{n+k}}.$$

□ The proof can be obtained from the formula

$$B_n^{(x)} = \frac{n!}{(2n)!} \sum_{k=0}^{n-1} (-1)^{k-1} E^2(n,k) x^{\overline{k+1}} (-z+n+1)^{\overline{n-k-1}}.$$

Taking $x = n$, we get that

$$\begin{aligned}
B_n^{(n)} &= \frac{n!}{(2n)!} \sum_{k=0}^{n-1} (-1)^{k-1} E^2(n,k) n^{\overline{k+1}} (1)^{\overline{n-k-1}} \\
&= \sum_{k=0}^{n-1} (-1)^{k-1} E^2(n,k) \\
&\quad \times \frac{n! n(n+1) \cdots (n+k)(1 \cdot 2 \cdot \cdots \cdot (n-k-1))}{(2n)!} \\
&= \frac{1}{2} \sum_{k=0}^{n-1} (-1)^{k-1} E^2(n,k) \frac{(n+k)!(n-k-1)!}{(2n-1)!} \\
&= \frac{1}{2} \sum_{k=0}^{n-1} (-1)^{k-1} \frac{E^2(n,k)}{\binom{2n-1}{n+k}}.
\end{aligned}$$

As $B_n^{(n)} = (-1)^n b_n$, it holds that $2b_n = \sum_{k=0}^{n-1} (-1)^{n+k-1} \frac{E^2(n,k)}{\binom{2n-1}{n+k}}$. □

Exercises

1. Prove that the sum of elements of the nth row of the triangle of Eulerian numbers of the second order, $n \geq 0$, is an odd number.
2. Prove that the elements $E^2(n, n-1)$, $n \geq 1$, of the first descending diagonal of the triangle of Eulerian numbers of the second order are divisible by all prime numbers not exceeding n, as well as by the product of all prime numbers not exceeding n.

3. Give a combinatorial proof for the formula $E^2(n,1) = 2E(n,1)$, $n \geq 1$; give a combinatorial proof for the formula $E^2(n,n-1) = E(n,0) + E(n,1) + \cdots + E(n,n-1)$, $n \geq 1$.
4. Find the alternating sum of elements of the nth row of the triangle of Eulerian numbers of the second order for $n = 2,3,4,5,6,7,8,9,10$. What can we say about the divisibility by two, three, four, and other primes of the obtained number?
5. Find the connections between the weighted sums of elements of the nth row of the triangle of Eulerian numbers of the first order and the sequence of Genocci numbers $G_n = 2(1-2^n)B_n$, $n \geq 0$.
6. For $n = 3,4,5,6,7,8$, find the alternating sum of values $\frac{E^2(n,k)}{\binom{2n+1}{k+1}}$, obtained by the sequential division of elements $E^2(n,k)$, $k = 0,1,\ldots,n-1$, of the nth row of the triangle of Eulerian numbers of the second order by the elements $\binom{2n+1}{k+1}$, $k = 0,1,\ldots,n-1$, on the left-hand side of the $(2n+1)$th row of Pascal's triangle.

4.6. Polynomials Related to Eulerian Numbers of the Second Order

Eulerian polynomials of the second order

4.6.1. In this chapter, we have already considered *Eulerian polynomials of the second order* $U_n(x)$, $n = 0,1,2,\ldots$, defined as

$$U_n(x) = \sum_{k=0}^{n} E^2(n,k)x^k = E^2(n,0) + E^2(n,1)x + \cdots + E^2(n,n-1)x^{n-1} + E^2(n,n)x^n.$$

Using the triangle of Eulerian numbers of the second order, we get that

$$U_0(x) \equiv 1; \quad U_1(x) \equiv 1;$$
$$U_2(x) = 1 + 2x; \quad U_3(x) = 1 + 8x + 6x^2;$$

$$U_4(x) = 1 + 22x + 58x^2 + 24x^3;$$

$$U_5(x) = 1 + 52x + 328x^2 + 444x^3 + 120x^4, \ldots .$$

4.6.2. From the properties of Eulerian numbers of the second order follow several simple properties of the polynomials $U_n(x)$.

- *The polynomial $U_n(x)$, $n \geq 1$, is a polynomial of degree $n - 1$ with the coefficient at the highest term x^{n-1} equal to $n!$ and the coefficient at the term x^0 equal to unity; the sum of the coefficients of the Euler polynomial $U_n(x)$, $n \geq 1$, is equal to $(2n-1)!$.*

□ By definition, $U_n(x) = \sum_{k=0}^{n} E^2(n,k)x^k$, and for $n = 0$, we get the polynomial $U_0(x) \equiv 1$. For $n \geq 1$, it holds that $E^2(n,n) = 0$, and $U_n(x) = \sum_{k=0}^{n} E^2(n,k)x^k = \sum_{k=0}^{n-1} E^2(n,k)x^k$. The coefficient at the highest term x^{n-1} is $E^2(n, n-1) = n!$; the coefficient at the term x^0 is $E^2(n,0) = 1$; at last, the sum of the coefficients is $E^2(n,0) + E^2(n,1) + \cdots + E^2(n,n-1) = (2n-1)!$. □

4.6.3. We have already proved the following:

- *Eulerian polynomials of the second order can be computed using the recurrence*

$$U_{n+1}(x) = (2nx+1)U_n(x) - x(x-1)U_n'(x), \quad U_0(x) \equiv 1.$$

So, we get that

$$U_1(x) = 1 \cdot U_0(x) - 0 \equiv 1;$$

$$U_2(x) = (2x+1) \cdot 1 - 0 = 1 + 2x;$$

$$U_3(x) = (4x+1)(1+2x) - x(x-1) \cdot 2 = 1 + 8x + 6x^2;$$

$$U_4(x) = (6x+1)(1+8x+6x^2) - x(x-1) \cdot (8+12x)$$
$$= 1 + 22x + 58x^2 + 24x^3.$$

This relation can be written in a more compact form:

$$\frac{U_{n+1}(x)}{(1-x)^{2n+2}} = \left(\frac{xU_n(x)}{(1-x)^{2n+1}}\right)'.$$

In other words, we can state that the function $u_n(x) = \frac{U_n(x)}{(x-1)^{2n}}$ satisfies the simple recurrent relation

$$u_{n+1}(x) = \left(\frac{x}{1-x}u_n(x)\right)', \quad u_0(x) \equiv 1.$$

4.6.4. One of the most widely known properties of polynomials $U_n(x)$ connects them with the Stirling numbers of the second kind. In fact, in Section 4.2. we have proven that the following equality holds:

$$\sum_{k=0}^{\infty} S(n+k,k)x^k = \frac{xU_n(x)}{(1-x)^{2n+1}}, \quad n \geq 1.$$

In particular, we have that

$$S(1,0) + S(2,1)x + S(3,2)x^2 + \cdots + S(k+1,k)x^k + \cdots = \frac{x}{(1-x)^3},$$

$$S(2,0) + S(3,1)x + S(4,2)x^2 + \cdots + S(k+2,k)x^k + \cdots = \frac{x+2x^2}{(1-x)^5},$$

$$S(3,0) + S(4,1)x + S(5,2)x^2 + \cdots + S(k+3,k)x^k + \cdots$$
$$= \frac{x+8x^2+6x^3}{(1-x)^7}, \quad \text{etc.}$$

A similar decomposition of the form $\sum_{k=1}^{\infty} S(n+k,k)x^k = \frac{xU_n(x)}{(1-x)^{2n+1}}$ also holds for $n=0$:

$$S(1,1)x + S(2,2)x^2 + \cdots + S(k,k)x^k + \cdots = \frac{x}{1-x}.$$

Stirling convolution polynomials

4.6.5. We have proven that the Stirling numbers $S(n,k)$ and $s(n,k)$ can be represented as linear combinations of Eulerian numbers of the second order with binomial coefficients. In particular, it holds that

$$|s(x,x-n)| = E^2(n,0)\binom{x}{2n} + E^2(n,1)\binom{x+1}{2n} + E^2(n,2)\binom{x+2}{2n}$$
$$+ \cdots + E^2(n,n-1)\binom{x+(n-1)}{2n}.$$

This decomposition is true whenever x is an integer and n is a non-negative integer. Since the right-hand side is a polynomial in x, we can use the equality to define the unsigned Stirling numbers of the first kind for arbitrary real (or complex) values of x.

For a positive integer n, since $|s(x, x-n)| = 0$ for all non-negative integers $x \leq n$, we can state that the polynomial $|s(x, x-n)$ is zero when $x = 0, x = 1, \ldots, x = n$; therefore, it is divisible by $x, x-1, \ldots, x-n$. So, it is interesting to look at what remains after these known factors are divided out (see [GKP94]).

4.6.6. The nth *Stirling convolution polynomial* $\sigma_n(x)$ is defined as

$$\sigma_n(x) = |s(x, x-n)| \cdot \frac{1}{x(x-1)\cdots(x-n)}.$$

The following list provides several special cases of these functions for the first few $n \geq 0$:

$$\sigma_0(x) = \frac{1}{x}; \quad \sigma_1(x) = \frac{1}{2}; \quad \sigma_2(x) = \frac{1}{24}(3x-1); \quad \sigma_3(x) = \frac{1}{48}(x^2 - x);$$

$$\sigma_4(x) = \frac{1}{5760}(15x^3 - 30x^2 + 5x + 2);$$

$$\sigma_5(x) = \frac{1}{11520}(3x^4 - 10x^3 + 5x^2 + 2x);$$

$$\sigma_6(x) = \frac{1}{2903040}(63x^5 - 315x^4 + 315x^3 + 91x^2 - 42x - 16);$$

$$\sigma_7(x) = \frac{1}{5806080}(9x^6 - 63x^5 + 105x^4 + 7x^3 - 42x^2 - 16x).$$

4.6.7. Using the recurrence $|s(n,k)| = (n-1)|s(n-1,k) + |s(n-1, k-1)|$, we can obtain the following result:

- *The polynomials $\sigma_n(x)$ satisfy the recurrent relation*

$$(x+1)\sigma_n(x+1) = (x-n)\sigma_n(x) + x\sigma_{n-1}(x), \quad n \geq 1.$$

□ In fact, it holds that

$$(x+1)\sigma_n(x) = (x+1)|s(x, x-n)| \cdot \frac{1}{(x+1)x\cdots(x+1-n)}$$

$$= |s(x, x-n)| \cdot \frac{1}{x\cdots(x+(n-1))} = (x|s(x, x+1-n)|$$

$$+|s(x,x-n)|)\cdot\frac{1}{x\cdots(x+(n-1))}$$
$$=\frac{x|s(x,x-(n-1)|}{x\cdots(x+(n-1))}\frac{(x-n)|s(x,x-n)|}{x\cdots(x+(n-1))(x-n)}$$
$$=x\sigma_{n-1}(x)+(x-n)\sigma_n(x).$$ □

4.6.8. This variant of the Stirling polynomial sequence has nice ordinary generating functions of the following forms:

$$\left(\frac{ze^z}{e^z-1}\right)^x=\sum_{n=0}^{\infty}x\sigma_n(x)z^n,\quad \left(\frac{1}{z}\log\frac{1}{1-z}\right)^x=\sum_{n=0}^{\infty}x\sigma_n(x+n)z^n.$$

□ These representations can be obtained by direct checking. Since

$$(x+1)\sigma_n(x+1)=(x-n)\sigma_n(x)+x\sigma_{n-1}(x),$$

we have that

$$\sum_{n=0}^{\infty}(x+1)\sigma_n(x+1)z^n=\sum_{n=0}^{\infty}(x-n)\sigma_n(x)z^n+\sum_{n=0}^{\infty}x\sigma_{n-1}(x)z^n.$$

Introducing the function $f(x,z)=\sum_{n=0}^{\infty}x\sigma_n(x)z^n$ and noting that

$$f_z'(x,z)=\sum_{n=1}^{\infty}x\sigma_n(x)nz^{n-1},$$

we can rewrite the last equation in the form

$$f(x+1,z)=f(x,z)-\frac{z}{x}f_z'(x,z)+zf(x,z).$$

It is easy to check that the function $f(x,z)=\left(\frac{ze^z}{e^z-1}\right)^x$ satisfies these conditions. □

4.6.9. From the last result, we can derive important connections between the polynomials $\sigma_n(x)$ and the Bernoulli numbers:

- *There exist the following equalities*:

$$\sigma_n(1)=(-1)^n\frac{B_n}{n!},\quad \sigma_n(0)=-\frac{B_n}{n\cdot n!},\quad n\geq 1.$$

□ In fact, for $x = 1$, it holds that

$$\sum_{n=0}^{\infty} \sigma_n(1) z^n = \frac{ze^z}{e^z - 1} = \frac{-z}{e^{-z} - 1} = \sum_{n=0}^{\infty} \frac{B_n}{n!}(-z)^n = \sum_{n=0}^{\infty} (-1)^n \frac{B_n}{n!} z^n.$$

Comparing the coefficients of z^n on both sides of the last equation, we get the first result: $\sigma_n(1) = (-1)^n \frac{B_n}{n!}, n \geq 0$. However, as all Bernoulli numbers with odd indices greater than 1 are equal to zero, we can state that

$$\sigma_n(1) = \frac{B_n^+}{n!}, \quad n \geq 0,$$

where B_n^+ are the second Bernoulli numbers. So, $\sigma_0(1) = 1 = \frac{B_0}{0!}$; $\sigma_1(1) = \frac{1}{2} = \frac{B_1^+}{1!}$; $\sigma_2(1) = \frac{2}{24} = \frac{B_2}{2!}$; $\sigma_3(1) = 0 = \frac{B_3}{3!}$, etc.

The second result can be obtained using the study of the derivative of the function $\left(\frac{ze^z}{e^z-1}\right)^x$ at $x = 0$ (see Section 4.5). So, $\sigma_1(0) = \frac{1}{2} = -\frac{B_1}{1 \cdot 1!}$, $\sigma_2(0) = -\frac{1}{24} = -\frac{B_2}{2 \cdot 2!}$, $\sigma_3(0) = 0 = -\frac{B_3}{3 \cdot 3!}$, etc. □

For a given positive integer n, the relations between the Stirling convolution polynomials $\sigma_n(x)$, taken at m, and the *Bernoulli numbers* are given by the formula

$$\sigma_n(m) = \frac{(-1)^{m+n-1}}{m!(n-m)!} \sum_{0 \leq k < m} |s(m, m-k)| \frac{B_{n-k}}{n-k}, \quad n \geq m \geq 1.$$

4.6.10. For positive integers b and m, we also have the following:

- *The polynomials $\sigma_n(m)$ are relations to the Stirling numbers, as given by the formulas*

$$S(n+m, m) = (-1)^{n+1} \frac{(n+m)!}{(m-1)!} \sigma_n(-m),$$

$$|s(n+m, m)| = \frac{(n+m)!}{(m-1)!} \sigma_n(m).$$

□ The second equality follows immediately from the definition of $\sigma_n(x)$:

$$|s(n+m,m)| = \sigma_n(m)(n+m)(n+m-1)\cdots m$$
$$= \sigma_n(n+m)\frac{(n+m)!}{(m-1)!}.$$

The first equality uses the property $S(n,k) = |s(-k,-n)|$, $k,n \in \mathbb{Z}$:

$$S(n+m,m) = |s(-m,-m-n)|$$
$$= \sigma_n(-m)(-m)(-m-1)\cdots(-m-n)$$
$$= (-1)^{n+1}\frac{(n+m)!}{(m-1)!}\sigma_n(-m). \qquad \square$$

For example, $S(4,2) = 7 = -\frac{4!}{1!}\frac{-7}{24} = -\frac{4!}{1!}\sigma_2(-2)$, while $|s(4,2)| = 11 = \frac{4!}{1!}\frac{11}{24} = \frac{4!}{1!}\sigma_2(4)$.

4.6.11. We are interested in the connections between the polynomials $\sigma_n(x)$ and the Eulerian numbers of the second order. As, on the one hand, it holds that

$$\sigma_n(x) = |s(x,x-n)| \cdot \frac{1}{x(x-1)\cdots(x-n)},$$

and, on the other hand, it holds that

$$|s(x,x-n)| = E^2(n,0)\binom{x}{2n} + E^2(n,1)\binom{x+1}{2n} + E^2(n,2)\binom{x+2}{2n}$$
$$+ \cdots + E^2(n,n-1)\binom{x+(n-1)}{2n},$$

we get that

$$\sigma_n(x) = \frac{E^2(n,0)\binom{x}{2n} + E^2(n,1)\binom{x+1}{2n} + E^2(n,2)\binom{x+2}{2n} + \cdots + E^2(n,n-1)\binom{x+(n-1)}{2n}}{x(x-1)\cdots(x-n)}.$$

The same decomposition can also be represented as

$$\sigma_n(x) = \frac{1}{(n+1)!\binom{x}{n+1}} \sum_{k=0}^{n} E^2(n,k)\binom{x+k}{2n}, \quad n \geq 0.$$

We can rewrite this equality in the following form:

$$\begin{aligned}
\sigma_n(x) &= E^2(n,0)\frac{x!}{(2n)!(2n-x)!x(x-1)\cdots(x-n)} \\
&\quad + E^2(n,1)\frac{(x+1)!}{(2n)!(x-2n+1)!x(x-1)\cdots(x-n)} \\
&\quad + E^2(n,2)\frac{(x+2)!}{(2n)!(x-2n+2)!x(x-1)(x-n)} \\
&\quad + \cdots + E^2(n,n-1)\frac{(x+n-1)!}{(2n)!(x-n-1)!x(x-1)\cdots(x-n)} \\
&= \frac{1}{(2n)!}\left(E^2(n,0)\frac{x(x-1)(x-n)(x-n-1)\cdots(x-2n+1)}{x(x-1)\cdots(x-n)}\right. \\
&\quad +E^2(n,1)\frac{(x+1)x(x-n)(x-n-1)\cdots(x-2n)}{x(x-1)\cdots(x-n)} + E^2(n,2) \\
&\quad \times \frac{(x+2)(x+1)x\cdots(x-n)(x-n-1)\cdots(x-2n-1)}{x(x-1)\cdots(x-n)} \\
&\quad + \cdots + E^2(n,n-1) \\
&\quad \times \left.\frac{(x+n-1)(x+n-2)\cdots(x+1)x\cdots(x-n)}{x(x-1)\cdots(x-n)}\right) \\
&= \frac{1}{(2n)!}\left(E^2(n,0)(x-n-1)\cdots(x-2n+1)\right. \\
&\quad + E^2(n,1)(x+1)(x-n-1)\cdots(x-2n) \\
&\quad + E^2(n,2)(x+2)(x+1)(x-n-1)\cdots(x-2n-1) \\
&\quad \left. + \cdots + E^2(n,n-1)(x+n-1)(x+n-2)(x+1)\right).
\end{aligned}$$

So, we have proven the decomposition

$$\begin{aligned}
\sigma_n(x) &= \frac{1}{(2n)!}\left(E^2(n,0)(x-n-1)\cdots(x-2n+1)\right. \\
&\quad + E^2(n,1)(x+1)(x-n-1)\cdots(x-2n)
\end{aligned}$$

$$+ E^2(n,2)(x+2)(x+1)(x-n-1)\cdots(x-2n-1)$$
$$+\cdots+ E^2(n,n-1)(x+n-1)(x+n-2)(x+1)\big).$$

In particular, it holds that

$$\sigma_1(x) = \frac{1}{2!}E^2(1,0) = \frac{1}{2!};$$
$$\sigma_2(x) = \frac{1}{4!}\left(E^2(2,0)(x-3) + E^2(2,1)(x+1)\right)$$
$$= \frac{1}{4!}\left((x-3) + 2(x+1)\right);$$
$$\sigma_3(x) = \frac{1}{6!}\big(E^2(3,0)(x-4)(x-5) + E^2(3,1)(x+1)(x-4)$$
$$+E^2(3,2)(x+2)(x+1)\big) = \frac{1}{6!}\left(x-4)(x-5\right)$$
$$+8(x+1)(x-4) + 6(x+2)(x+1)).$$

Other polynomials related to Eulerian numbers of the second order

4.6.12. There are other classes of polynomials used in the study of Eulerian numbers of the second order, including the Bernoulli polynomials, generalised Bernoulli polynomials, Nörlund polynomials, etc.

The *Bernoulli polynomials* are defined as

$$B_n(x) = C_n^0 B_0 x^n + C_n^1 B_1 x^{n-1} + C_n^2 B_2 x^{n-2}$$
$$+\cdots+ C_n^{n-1} B_{n-1} x + C_n^n B_n,$$

where B_n, $n = 0, 1, 2, \ldots$, are Bernoulli numbers. Following the definition, we get that

$$B_0(x) \equiv 1, \quad B_1(x) = x - \frac{1}{2}, \quad B_2(x) = x^2 - x + \frac{1}{6},$$
$$B_3(x) = x^3 - \frac{3}{2}x^2 + \frac{1}{2}x, \quad B_4(x) = x^4 - 2x^3 + x^2 - \frac{1}{30}.$$

The exponential generating function of the sequence $B_0(x), B_1(x), B_2(x), \ldots, B_n(x), \ldots$ of Bernoulli polynomials has the form $\frac{te^{tx}}{e^t-1}$, i.e., it holds that

$$\frac{z}{e^z-1}e^{xz} = B_0(x) + \frac{z}{1!}B_1(x) + \frac{z^2}{2!}B_2(x) + \frac{z^3}{3!}B_3(x) + \cdots + \frac{z^n}{n!}B_n(x) + \cdots .$$

Obviously, the value of $B(x)$ at $x = 0$ is the nth Bernoulli number B_n. As $B_n(1-x) = (-1)^n B_n(x)$, we get that the value of $B_n(x)$ at $x = 1$ is the nth second Bernoulli number B_n^+:

$$B_n(0) = B_n, \quad B_n(1) = (-1)^n B_n = B_n^+.$$

4.6.13. For an arbitrary (real or complex) parameter α, the *generalised Bernoulli polynomials* $B_n^{(\alpha)}(x)$ are defined by their exponential generating function $\left(\frac{z}{e^z-1}\right)^\alpha e^{xz}$:

$$\left(\frac{z}{e^z-1}\right)^\alpha e^{xz} = B_0^{(\alpha)}(x) + \frac{z}{1!}B_1^{(\alpha)}(x) + \frac{z^2}{2!}B_2^{(\alpha)}(x) + \frac{z^3}{3!}B_3^{(\alpha)}(x) + \cdots + \frac{z^n}{n!}B_n^{(\alpha)}(x) + \cdots, \quad |z| < 2\pi.$$

It is easy to check that the function $B_n^{(\alpha)}(x)$ is a polynomial of degree n in x.

For $x = 0$, we obtain the sequence $B_0^{(\alpha)}, B_1^{(\alpha)}, \ldots, B_n^{(\alpha)}, \ldots$ of *Nörlund polynomials* (or *generalised Bernoulli numbers*) with the generating function $\left(\frac{z}{e^z-1}\right)^\alpha$:

$$\left(\frac{z}{e^z-1}\right)^\alpha = B_0^{(\alpha)} + \frac{z}{1!}B_1^{(\alpha)} + \frac{z^2}{2!}B_2^{(\alpha)} + \frac{z^3}{3!}B_3^{(\alpha)} + \cdots + \frac{z^n}{n!}B_n^{(\alpha)}(x) + \cdots .$$

So, by definition, $B_n^{(\alpha)} = B_n^{(\alpha)}(0)$. For a fixed n, $B_n^{(\alpha)}$ is a polynomial with the argument α.

It is easy to check that

$$B_n^{(\alpha)}(x) = (-1)^n B_n^{(\alpha)}(\alpha - x).$$

□ In fact, we have that

$$\sum_{n=0}^{\infty} B_n^{(\alpha)}(\alpha - x)\frac{z^n}{n!} = \left(\frac{z}{e^z - 1}\right)^{\alpha} e^{(\alpha - x)z} = \left(\frac{ze^z}{e^z - 1}\right)^{\alpha} e^{-xz}$$

$$= \left(\frac{ze^z}{e^z - 1}\right)^{\alpha} e^{-xz} = \left(\frac{-z}{1 - e^{-z}}\right)^{\alpha} e^{-xz} = \sum_{n=0}^{\infty} B_n^{(\alpha)}(x)\frac{(-z)^n}{n!}$$

$$= \sum_{n=0}^{\infty}(-1)^n B_n^{(\alpha)}(x)\frac{z^n}{n!}.$$

Comparing the coefficients at z^n in both parts of the equality, we get that

$$B_n^{(\alpha)}(\alpha - x) = (-1)^n B_n^{(\alpha)}(x), \quad \text{or } B_n^{(\alpha)}(x) = (-1)^n B_n^{(\alpha)}(\alpha - x). \ \square$$

So, we have proven that

$$B_n^{(\alpha)} = (-1)^n B_n^{(\alpha)}(\alpha).$$

4.6.14. Using the decomposition

$$\left(\frac{ze^z}{e^z - 1}\right)^x = \sum_{n=0}^{\infty} x\sigma_n(x) z^n,$$

we can prove that

$$x\sigma_n(x) = (-1)^n \frac{B_n^{(x)}}{n!}.$$

□ We can use the same idea:

$$\sum_{n=0}^{\infty} x\sigma_n(x) z^n = \left(\frac{ze^z}{e^z - 1}\right)^x = \left(\frac{-z}{e^{-z} - 1}\right)^x = \sum_{n=0}^{\infty} \frac{B^{(x)}}{n!}(-z)^n$$

$$= \sum_{n=0}^{\infty}(-1)^n \frac{B^{(x)}}{n!} z^n.$$

On the other hand, it is easy to see that

$$\sum_{n=0}^{\infty} x\sigma_n(x)z^n = \left(\frac{ze^z}{e^z-1}\right)^x = \left(\frac{z}{e^z-1}\right)^x e^{1\cdot z} = \sum_{n=0}^{\infty} \frac{B_n^{(x)}(1)}{n!} z^n,$$

i.e., $x\sigma_n(x) = \frac{B_n^{(x)}(1)}{n!}$; therefore, it holds that

$$x\sigma_n(x) = \frac{B_n^{(x)}(1)}{n!} = (-1)^n \frac{B_n^{(x)}(0)}{n!} = (-1)^n \frac{B_n^{(x)}}{n!}. \qquad \square$$

If $x = 1$, we get the known equality $\sigma_n(1) = (-1)^n \frac{B_n}{n!} = \frac{B_n^+}{n!}$.

4.6.15. Now, we can prove the following:

- *For positive integers n and m, it holds the equality*

$$S(n+m, m) = \binom{n+m}{m} B_n^{(-m)}.$$

☐ In fact, we have that

$$\begin{aligned} S(n+m, m) &= (-1)^{n+1} \frac{(n+m)!}{(m-1)!} \sigma_n(-m) \\ &= (-1)^n \frac{(n+m)!}{m!} (-m)\sigma_n(-m) \\ &= (-1)^n \frac{(n+m)!}{m!} (-1)^n \frac{B_n^{(-m)}}{n!} \\ &= \frac{(n+m)!}{m!n!} B_n^{(-m)} = \binom{n+m}{m} B_n^{(-m)}. \qquad \square \end{aligned}$$

4.6.16. One additional set of polynomials, related to the Eulerian numbers of the second order, was considered in [GeSt78].

In this paper, the functions $f_k(n) = S(n+k, n)$ $g_k(n) = |s(n, n-k)|$, considered as polynomials in argument n and referred to as the *Stirling polynomials*, were used for the construction the well-known combinatorial interpretation of the Eulerian numbers of the second order, which are defined as the coefficients of the polynomial $(1-x)^{2k+1} \sum_{n=0}^{\infty} f_k(n)$.

Exercises

1. Find the Eulerian polynomials $U_n(x)$, $n = 5, 6, 7, 8, 9, 10, 11, 12$, using the triangle of Eulerian numbers of the second order; using the recurrent relation $U_{n+1}(x) = x\,(1-x)\,U_n^{'}(x)+U_n(x)\,(1+2nx)$. Find $U_n(x)$ if $n = 0, 1, 2, 3, 4, 5$, $x = 2, 3, 4$.
2. Let $Q_n(x) = xU_n(X) = E^2(n, 0)x + E^2(n, 1)x^2 + \cdots + E^2(n, n)x^{n+1}$. Find $Q_n(x)$ if $n = 0, 1, 2, 3, 4.5$, $x = 2, 3, 4$. Check that $Q_n(1) = (2n-1)!!$. Prove that for the polynomials $Q_n(x)$, the following recurrent relation holds:

$$Q_n(x) = (2n-1)xQ_{n-1}(x) + x(1-x)Q_{n-1}^{'}(x),\ Q_0(x) \equiv 1.$$

   Prove that $Q_k(x) = (1-x)^{2k+1} \sum_{k=1}^{\infty} S(n+k, k)x^k$, $n \geq 0$.
3. Prove that $\sigma_n(x+y) = \frac{xy}{x+y} \sum_{k=0}^{n} \sigma_k(x)\sigma_{n-k}(y)$.
4. Prove that for the integers $0 \leq k \leq n$ and $r, s \in \mathbb{R}$ (in fact, for $r, s \in \mathbb{C}$), the polynomials $\sigma_n(x)$ satisfy the two *Stirling convolution formulas*, given by

$$(r+s)\sigma_n(r+s+tn) = rs \sum_{k=0}^{n} \sigma_k(r+tk)\sigma_{n-k}(s+t(n-k)),$$

$$n\sigma_n(r+s+tn) = s \sum_{k=0}^{n} k\sigma_k(r+tk)\sigma_{n-k}(s+t(n-k)).$$

5. For a fixed integer $k \geq 0$, study the functions $f_k(n) = S(n+k, n)$ $g_k(n) = |s(n, n-k)|$ as polynomials of argument n. Check that $f_0(n) = g_0(n) = 1$, $f_1(n) = \binom{n+1}{2}$, $g_1(n) = \binom{n}{2}$. Prove that both polynomials have degree $2k$ and the leading coefficient $\frac{(2k-1)!!}{2k)!}$. Prove that $\Delta g_k(n) = ng_{k-1}(n)$, $\Delta f_k(n) = (n+1)f_{k-1}(n+1)$, where Δ is the usual difference operator: $\Delta h(n) = h(n+1)-h(n)$. Check that $f_k(0) = f_l(-1) = \cdots = f_k(-k) = 0$, $k \geq 1$. Prove that $f_k(-n) = g_k(n)$, $n \in \mathbb{Z}, k \geq 1$.

4.7. Eulerian Numbers of the Second Order in the Family of Special Numbers

Eulerian numbers of the second order and elements of the Pascal's triangle

4.7.1. We considered a number of connections of the Eulerian numbers of the second order with other classes of special numbers. In particular, the construction of the triangle of Eulerian numbers of the second order is similar to the construction of Pascal's triangle, and the Eulerian numbers of the second order have many common features with binomial coefficients.

First of all, these two triangular arrays are defined by a recurrent relation giving a considering element of the nth row as a weighted sum of two elements of the previous row, placing above the element being constructed:

$$\binom{n}{k} = \binom{n-1}{k} + \binom{n-1}{k-1}, \quad \text{and}$$
$$E^2(n,k) = (k+1)E^2(n-1,k) + (2n-k-1)E^2(n-1,k-1).$$

Furthermore, both triangles have several common properties:

- the elements $\binom{n}{k}$, $k = 0, 1, \ldots, n$, of the nth row of Pascal's triangle, as well as the elements $E^2(n,k)$, $k = 0, 1, \ldots, n$, of the triangle of Eulerian numbers of the second order, form *unimodal sequences*, i.e., the sequences with the only maximum achieved at one or at two consecutive values of k;
- the elements $\binom{n}{k}$, $k = 0, 1, \ldots, n$, of the nth row of Pascal's triangle, as well as the elements $E(n,k)$, $k = 0, 1, \ldots, n$, of the triangle of Eulerian numbers of the second order, satisfy the strong

logarithmic concavity property:

$$\binom{n}{k}^2 > \binom{n-1}{k}\binom{n+1}{k}, \quad \text{and}$$
$$(E^2(n,k))^2 > E^2(n-1,k)E^2(n+1,k)$$

for all acceptable values of k;

- there exist special divisibility properties for the elements of the triangles' rows with prime indices p: all infernal elements of the pth row of Pascal's triangle are divisible by p; all internal elements of the triangle of Eulerian numbers of the second order are divisible by the prime number 2; moreover, all non-zero elements of the pth row of the triangle of Eulerian numbers of the second order form a complete residue system modulo p; in fact, $E^2(p,k)$ is congruent to $k+1$ modulo p, $k = 0, 1, \ldots, p-1$.

The list can be continued.

Besides, many properties of the Eulerian numbers of the first order are formulated using binomial coefficients, including the decompositions

$$S(x, x-n) = E^2(n,0)\binom{x+n-1}{2n} + E^2(n,1)\binom{x+n-2}{2n} + \cdots + E^2(n,n-1)\binom{x}{2n},$$

$$|s(x, x-n)| = E^2(n,0)\binom{x}{2n} + E^2(n,1)\binom{x+1}{2n} + \cdots + E^2(n,n-1)\binom{x+(n-1)}{2n},$$

and the closed formula

$$E^2(n,k) = \sum_{j=0}^{k+1}(-1)^j\binom{2n+1}{j}S(n+k+1-j, k+1-j), \quad n \geq 1.$$

Eulerian numbers of the second order and factorial numbers

4.7.2. Any binomial coefficient $\binom{n}{m}$ can be calculated through the classic formula using factorials:

$$\binom{n}{m} = \frac{n!}{m!(n-m)!}.$$

Therefore, the Eulerian numbers of the second order, connected with binomial coefficients, are also closely related to the sequence $1, 1, 2, 6, 24, 120, 720, 5040, 40320, 362880, \ldots$ (sequence A000142 in the OEIS) of *factorial numbers* $n!$, $n = 0, 1, 2, \ldots$.

Moreover, the Eulerian numbers of the second order are also closely related to the *double factorials.* The *double factorial* (or *semi-factorial*) of a positive number n, denoted by $n!!$, is the product of all the integers from 1 up to n that have the same parity (odd or even) as n (see [Call09]). For even n, the double factorial is

$$n!! = \prod_{k=1}^{\frac{n}{2}} (2k) = n(n-2)(n-4)\cdots 4\cdot 2\,,$$

while for odd n, it is

$$n!! = \prod_{k=1}^{\frac{n+1}{2}} (2k-1) = n(n-2)(n-4)\cdots 3\cdot 1\,.$$

The zero double factorial $0!! = 1$ is an empty product.

The sequence of *even double factorial numbers*, i.e., double factorials for even $n = 0, 2, 4, 6, 8, \ldots$ starts with 1, 2, 8, 48, 384, 3840, 46080, $645120, \ldots$ (sequence A000165 in the OEIS). The sequence of *odd double factorial numbers*, i.e., double factorials for odd $n = 1, 3, 5, 7, 9, \ldots$ starts with $1, 3, 15, 105, 945, 10395, 135135, \ldots$ (sequence A001147 in the OEIS).

As the number of Stirling permutations of $\{1, 1, 2, 1, \ldots, n, n\}$ is equal to $(2n-1)!!$, we can state that:

- *the sum of elements of the nth row of the triangle of Eulerian numbers of the second order, $n \geq 1$, is equal to the nth element of* *the sequence of odd double factorial numbers*:

$$E^2(n,0) + E^2(n,1) + \cdots + E^2(n,n-1) = (2n-1)!!, \quad n \geq 1.$$

In this sense, odd double factorial numbers can be considered relatives of the classic factorial numbers $n!$, which give the sums of the elements of the nth row of Euler's triangle (as well as of the triangle of unsigned Stirling numbers of the second kind), of the powers of two 2^n, which give the sums of the elements of the nth row of Pascal's triangle, and of the Bell numbers $B(n)$, which give the sums of the elements of the nth row of the triangle of Stirling numbers of the second kind:

$$E(n,0) + E(n,1) + \cdots + E(n,n-1) = n!, \quad n \geq 0;$$

$$\binom{n}{0} + \binom{n}{1} + \cdots + \binom{n}{n} = 2^n, \quad n \geq 0;$$

$$S(n,0) + S(n,1) + \cdots + S(n,n) = B(n), \quad n \geq 0;$$

$$|s(n,0)| + |s(n,1)| + \cdots + |s(n,n)| = n!, \quad n \geq 0.$$

Eulerian numbers of the second order and Stirling numbers

4.7.3. Similar connections exist between the triangle of Eulerian numbers of the second order and the triangles of Stirling numbers of the second kind and (unsigned) Stirling numbers of the first kind, including the form of the recurrent relations, the construction used for these triangles, the unimodality property, and the strong logarithmic concavity property.

Besides, these classes of numbers, as well as the set of Lah numbers $L(n,k)$, can be considered as coefficients of some transformations among the system x^k, $k = 0,1,2,\ldots$, of ordinary powers, the system $x^{\underline{k}}$, $k = 0,1,2,\ldots$, of falling powers, the system $x^{\overline{k}}$, $k = 0,1,2,\ldots$, of generalised binomial coefficients, the system

$S(x, x-k)$, $k = 0, 1, 2, \ldots$, of Stirling numbers of the second kind, and the system $s(x, x-k)$, $k = 0, 1, 2, \ldots$, of Stirling numbers of the first kind.

Moreover, the main formulas considered in this chapter include simultaneously the Eulerian numbers of the second order, the Stirling numbers of the second (or first) kind, and other numerical constructions. For example, except the formulas mentioned above, we have the equalities

$$\sum_{k=0}^{\infty} S(k+n, k) \cdot x^k = \frac{E^2(n,0)x + E^2(n,1)x^2 + \cdots + E^2(n,n-1)x^n}{(1-x)^{2n+1}},$$

$$\sum_{k=0}^{\infty} |s(k, k-n)| x^k = \frac{E^2(n,n-1)x^{n+1} + E^2(n,n-2)x^{n+2} + \cdots + E^2(n,0)x^{2n}}{(1-x)^{2n+1}},$$

$$\sum_{i=0}^{k-1} a^{k-1-i} E^2(n,i) = \sum_{i=0}^{k-1} S(n+k-i, k-i) \left(\sum_{j=0}^{i} (-1)^j 2^{i-j} \binom{2n+1}{j} \right).$$

Eulerian numbers of the second order and Bernoulli numbers

4.7.4. There exist natural connections between the Eulerian numbers of the second order and the Bernoulli numbers. They are based on the fundamental properties of the Eulerian polynomials, the Bernoulli polynomials, the Euler polynomials, and other analytic functions, including the Riemann zeta function.

For example, as $\zeta(-n) = -\frac{B_{n+1}}{n+1}$, we can state that the alternating sum of the Eulerian numbers of the second order $E^2(n,k)$, $k = 0, 1, \ldots, n-1$, with weights $\binom{2n+1}{k+1}^{-1}$, can be expressed as the value of the Riemann zeta function at the argument $-n$:

$$\frac{E^2(n,0)}{\binom{2n+1}{1}} - \frac{E^2(n,1)}{\binom{2n+1}{k+1}} + \cdots + (-1)^{n-1}\frac{E^2(n,n-1)}{\binom{2n+1}{n}}$$
$$= -2(n+1)\zeta(-n).$$

Moreover, the following equality holds:

$$-\int_0^1 \sum_{k=0}^{n-1} E^2(n,k)u^{k+1}(u-1)^{2n-k}du = \zeta(-n).$$

4.7.5. Similar deep connections exist between the Eulerian numbers of the second order and the generalised Bernoulli numbers $B_n^{(x)}$, $n = 0, 1, 2, \ldots$. We have proven that, for positive integers n and m, the generalised Bernoulli numbers $B_n^{(-m)}$ can be represented using the Eulerian numbers of the second order and the rising powers:

$$B_n^{(-m)} = \frac{n!}{(2n)!}\sum_{k=0}^{n-1}(-1)^{k-1}E^2(n,k)(-m)^{\overline{k+1}}(m+n+1)^{\overline{n-k-1}}.$$

As $B_n^{(1)} = B_n$, and $B_n^{(n)} = (-1)^n b_n$, where B_n are the Bernoulli numbers and b_n are the Cauchy numbers of the second type, the last decomposition gives several interesting equalities, which were considered in this chapter.

Eulerian numbers of the second order and Euler numbers

4.7.6. The Eulerian numbers of the first and second orders and the Euler numbers are connected not only by the name Euler.

First of all, these three classes of numbers, as well as the Euler zig-zag numbers, count some special sets of permutations.

In fact, the Euler zig-zag number A_n counts the number of all up-down permutations of the set $\{1, 2, \ldots, n\}$, while the absolute value $|E_n|$ of the Euler number E_n gives the number of proper up-down permutations of the set $\{1, 2, \ldots, n\}$, i.e., the permutations $\pi = (\pi_1, \ldots, \pi_n)$ with $\pi_1 < \pi_2 > \pi_3 < \pi_4 > \cdots < \pi_n$; therefore, $E_{2n-1} = 0$, and $|E_{2n}| = A_{2n}$.

The sequence of Euler zig-zag numbers A_n starts with the elements 1, 1, 1, 2, 5, 16, 61, 272, 1385, 7936, 50521, ... (sequence A000111 in the OEIS).

The numbers A_{2n} with even indices are called the *secant numbers* (or *zig numbers*). The first few values are 1, 1, 5, 61, 1385, 50521, 2702765, 199360981, 19391512145, 2404879675441, ... (sequence A000364 in the OEIS).

Defining $E_{2n} = (-1)^n |E_{2n}|$ and using the equations $|E_{2n}| = A_{2n}$ and $E_{2n+1} = 0$, we get the first few elements of the sequence E_n, $n = 0, 1, 2, \ldots$, of the Euler numbers; they are 1, 0, −1, 0, 5, 0, −61, 0, 1385, 0, ... (sequence A122045 in the OEIS).

The numbers A_{2n+1} with odd indices are called the *tangent numbers* (or *zag numbers*). The first few values are 1, 2, 16, 272, 7936, 353792, 22368256, 1903757312, 209865342976, 29088885112832, ... (sequence A000182 in the OEIS). The tangent numbers are closely related to the *Bernoulli numbers*. The relation is given by the formula

$$B_{2n} = (-1)^{n-1} \frac{2n}{4^{2n} - 2^{2n}} A_{2n-1}, \quad n \geq 1.$$

In the context of our consideration, it means that the non-zero alternating sums of the Eulerian numbers of the second order $E^2(n, k)$, $k = 0, 1, \ldots, n-1$, with weights $\frac{1}{\binom{2n+1}{k+1}}$, can be represented using the tangent numbers:

$$\frac{E^2(2n+1, 0)}{\binom{4n+3}{1}} - \frac{E^2(2n+1, 1)}{\binom{4n+3}{k+1}} + \cdots + - \frac{E^2(2n+1, 2n)}{\binom{4n+3}{2n+1}}$$
$$= (-1)^n 2A_{2n+1}, n \geq 0.$$

Eulerian numbers of the second order and Cauchy numbers

4.7.7. The *Cauchy numbers* are defined as the integration of rising and falling powers (factorials). These numbers can be classified into two categories: the Cauchy numbers of the first type, denoted by a_n, and the Cauchy numbers of the second type, denoted by b_n.

The nth *Cauchy number of the first type* a_n is defined as

$$a_n = \int_0^1 x^{\underline{n}} dx,$$

where $x^{\underline{n}} = x(x-1)(x-2)\cdots(x-n+1)$ represents the falling power.

The nth *Cauchy number of the second type* b_n is defined as

$$b_n = \int_0^1 x^{\overline{n}} dx,$$

where $x^{\overline{n}} = x(x+1)(x+2)(x+3)\cdots(x+n-1)$ represents the rising power.

The sequence of Cauchy numbers of the first type starts with the elements $1, \frac{1}{2}, -\frac{1}{6}, \frac{1}{4}, -\frac{19}{30}, \frac{9}{4}, -\frac{83}{84}, \frac{1375}{24}, -\frac{33953}{90}, \frac{57281}{20}, \ldots$ (sequences A006232/A006233 in the OEIS). The sequence of Cauchy numbers of the second type starts with the elements $1, \frac{1}{2}, \frac{5}{6}, \frac{9}{4}, \frac{251}{30}, \frac{475}{12}, \frac{19087}{84}, \frac{36799}{24}, \frac{1070017}{90}, \frac{2082753}{20}, \ldots$ (sequences A002657/A002790 in the OEIS).

It is well known (see the proofs in [Deza24]) that

$$\begin{aligned} x^{\underline{n}} &= s(n,0) + s(n,1)x + s(n,2)x^2 \\ &\quad + \cdots + s(n,n-1)x^{n-1} + s(n,n)x^n, \quad \text{and} \\ x^{\overline{n}} &= |s(n,0)| + |s(n,1)|x + |s(n,2)|x^2 \\ &\quad - \cdots + |s(n,n-1)|x^{n-1} + |s(n,n)|x^n. \end{aligned}$$

We can use these formulas in order to obtain the following connections between the Stirling and Cauchy numbers.

- *The Cauchy numbers of the first and second types are related to the Stirling numbers of the first kind through the formulas*

$$a_n = \sum_{k=0}^{n} \frac{s(n,k)}{k+1}, \quad b_n = \sum_{k=0}^{n} \frac{|s(n,k)|}{k+1}.$$

□ The proof is very simple. In fact,

$$\begin{aligned} a_n &= \int_0^1 x^{\underline{n}} = \int_0^1 \left(\sum_{k=0}^{n} s(n,k) x^k \right) dx \\ &= \sum_{k=0}^{n} s(n,k) \int_0^1 x^k dx = \sum_{k=0}^{n} \frac{s(n,k)}{k+1}. \end{aligned}$$ □

4.7.8. Now, we can obtain the exponential generation functions of the sequences a_n, $n = 0, 1, 2, \ldots$, and b_n, $n = 0, 1, 2, \ldots$.

- *The exponential generating function of the sequence of Cauchy numbers of the first type has the form* $\frac{z}{\ln(1+z)}$, *i.e., it holds that*

$$\frac{z}{\ln(1+z)} = \sum_{k=0}^{\infty} a_n \frac{z^n}{n!};$$

the exponential generating function of the sequence of Cauchy numbers of the second type has the form $\frac{-z}{(1-z)\log(1-z)}$, *i.e., it holds that*

$$\frac{-z}{(1-z)\log(1-z)} = \sum_{n=0}^{\infty} b_n \frac{z^n}{n!}.$$

□ It is easy to prove (see, for example, [Deza24]) that the exponential generating function of the sequence $s(n,k)$, $n = 0, 1, 2, \ldots$, of the Stirling numbers of the first kind has the form $\frac{(\log(1+z))^k}{k!}$,

$k = 0, 1, 2, \ldots$, i.e., for a fixed $k = 0, 1, 2, \ldots$, it holds that

$$\frac{(\log(1+z))^k}{k!} = s(0,k) + s(1,k)\frac{z}{1!} + s(2,k)\frac{z^2}{2!} + \cdots + s(n,k)\frac{z^n}{n!} + \cdots .$$

So, we can state that

$$\begin{aligned}\sum_{n=0}^{\infty} a_n \frac{z^n}{n!} &= \sum_{n=0}^{\infty}\sum_{k=0}^{n} \frac{s(n,k)}{k+1}\frac{z^n}{n!} = (s(0,0) + s(1,0) + \cdots) \\ &\quad + \frac{1}{2}\left(s(1,1)\frac{z}{1!} + s(2,1)\frac{z^2}{2!} + \cdots\right) \\ &\quad + \frac{1}{3}\left(s(2,2)\frac{z^2}{2!} + s(3,2)\frac{z^3}{3!} + \cdots\right) + \cdots \\ &= \sum_{k=0}^{\infty} \frac{1}{k+1} \sum_{n=l}^{\infty} s(n,k)\frac{z^k}{k!} = \sum_{k=0}^{\infty} \frac{1}{k+1} \cdot \frac{(\log(1+z))^k}{k!} \\ &= \frac{1}{\log(1+z)} \sum_{k=0}^{\infty} \frac{(\log(1+z))^{k+1}}{(k+1)!} \\ &= \frac{1}{\log(1+z)}\left(e^{\log(1+z)} - 1\right) = \frac{z}{\log(1+z)}.\end{aligned}$$

Similarly, the exponential generating function of the sequence $|s(n,k)|$, $n = 0, 1, 2, \ldots$, of the unsigned Stirling numbers of the first kind has the form $\frac{(-\log(1-z))^k}{k!}$. So, we have that

$$\begin{aligned}\sum_{n=0}^{\infty} b_n \frac{z^n}{n!} &= \sum_{n=0}^{\infty}\sum_{k=0}^{n} \frac{|s(n,k)|}{k+1}\frac{z^n}{n!} = \sum_{k=0}^{\infty} \frac{1}{k+1} \sum_{n=l}^{\infty} |s(n,k)|\frac{z^k}{k!} \\ &= \sum_{k=0}^{\infty} \frac{1}{k+1} \cdot \frac{(-\log(1-z))^k}{k!} = \frac{1}{-\log(1-z)} \\ &\quad \times \sum_{k=0}^{\infty} \frac{(-\log(1-z))^{k+1}}{(k+1)!} \\ &= \frac{1}{-\log(1-z)}\left(e^{-\log(1-z)} + 1\right) = \frac{-z}{(1-z)\log(1-z)}. \quad \square\end{aligned}$$

4.7.9. Using these results, we can state that the exponential generating function of the sequence $(-1)^n b_n$ is $\frac{z}{(1+z)\log(1+z)}$. Considering the sequence $B_0^{(0)}, B_1^{(1)}, B_2^{(2)}, B_n^{(n)}, \dots$ of generalised Bernoulli numbers, we check that its exponential generating function has the same form as the nth derivative of the function $\left(\frac{e^z}{e^z-1}\right)^n$ taken at the point $z = 0$, coinciding with the nth derivative of the function $\frac{z}{(1+z)\log(1+z)}$ taken at the point $z = 0$. It implies that

$$B_n^{(n)} = (-1)^n b_n, \quad n \geq 0.$$

In the context of our consideration, the connection of the Cauchy numbers of the second type with the generalised Bernoulli numbers is important. As $B_n^{(n)} = (-1)^n b_n$, we get the following decomposition:

$$2b_n = \sum_{k=0}^{n-1} (-1)^{n+k-1} \frac{E^2(n,k)}{\binom{2n-1}{n+k}}.$$

Eulerian numbers of the second order with negative indices

4.7.10. We built the triangle of Eulerian numbers of the second order by looking at its elements $E^2(n,k)$ for all non-negative integers n and all non-negative integers k, $k \leq n$.

The possible expansion of the set of indices is completely natural. Exactly, let us assume that, for any fixed non-negative integer n, the values $E^2(n,k)$ are defined for all integers values of k: they are equal to zero for all negative integers k and for all positive integers k, $k > n$:

$$E^2(n,k) = 0, \quad n = 0, 1, 2, 3, \dots,$$
$$k = \dots - 3, -2, -1, n+1, n+2, n+3, \dots.$$

This is fully consistent with the combinatorial interpretation of the elements of the considered triangle and with the recurrent scheme of the construction of the triangle. Placing the unity in the position $E^2(0,0)$ and using the recurrent relation $E^2(n,k) = (k+1)E^2(n-1,k)+(2n-k-1)E^2(n-1,k-1)$, we can now build the elements

$$E^2(n,k), n = 0,1,2,\ldots,k = 0,1,2,\ldots,n.$$

So, we have complete information about the behaviour of the numbers $E^2(n,k)$ for non-negative values of n and for all integer values of k.

Let us now see what happens with negative values of n if we assume that, for negative n and negative k, it holds that $E^2(n,k) = 0$:

$$E^2(n,k) = 0, \quad n = -1,-2,-3,\ldots,k = -1,-2,-3,\ldots.$$

So, for integers n and k, we know the behaviour of $E^2(n,k)$ for non-negative integers n and non-negative integers k (in fact, for $0 \leq k \leq n$), and we would like to see the values of the Eulerian numbers of the second order for negative integers n and non-negative integers k if the same recurrent relation is used.

Let us consider $k = 0$. If $n = 0$, we have $E^2(0,0) = 0$. If $n \geq 1$, we have $E^2(n,0) = 0$.

But what happens if $n \leq -1$?

- *For* $n \geq 1$, *it holds that* $E^2(-n,0) = 1$.

□ In fact, $E^2(n,k) = (k+1)E^2(n-1,k)+(2n-k-1)E^2(n-1,k-1)$. So,

$$E^2(n-1,k) = \frac{1}{k+1}\left(E^2(n,k) + (k+1-2n)E^2(n-1,k-1)\right).$$

If $k = 0$, we get, as $E^2(n,-1) = 0$ for any integer n, the recurrent relation

$$\begin{aligned} E^2(n-1,0) &= \frac{1}{0+1}(E^2(n,0) + (1-n)E^2(n-1,-1)) \\ &= E^2(n,0). \end{aligned}$$

So, staring from $n = 0$, we get $E^2(-1,0) = E^2(0,0) = 1$, $E^2(-2,0) = E^2(-1,0) = 1$, etc. □

- *For* $k \geq 0$, *it holds that* $E^2(-1,k) = 1$.

□ Using the recurrence $E^2(n-1,k) = \frac{1}{k+1}(E^2(n,k) + (k+1-2n)E^2(n-1,k-1))$, let us consider the pairs $(0,k)$ for $k \geq 0$.

For the pair $n = 0, k = 0$, we get that

$$E^2(-1,0) = E^2(0,0) = 1.$$

For the pair $n = 0, k = 1$, we have

$$E^2(-1,1) = \frac{1}{2}(E^2(0,1) + (2-0)E^2(-1,0)) = \frac{1}{2}(0 + 2\cdot 1) = 1.$$

Similarly, for $n = 0, k = 2$, we have

$$E^2(-1,2) = \frac{1}{3}(E^2(0,2) + (3-0)E^2(-1,1)) = \frac{1}{3}(0 + 3\cdot 1) = 1.$$

Suppose that $E^2(-1,k-1) = 1$. Then, for $n = 0, k$, we have

$$E^2(-1,k) = \frac{1}{k+1}(E^2(0,k) + (k+1-0)E^2(-1,k-1))$$
$$= \frac{1}{k+1}(0+)k+1)\cdot 1) = 1.$$ □

- *For $k \geq 0$, it holds that*

$$E^2(-2,k) = \frac{2+3+\cdots+(k+2)}{2} = \frac{(k+2)(k+3)-2}{4}.$$

□ Using the recurrence $E^2(n-1,k) = \frac{1}{k+1}(E^2(n,k) + (k+1-2n)E^2(n-1,k-1))$, let us consider the pairs $(-1,k)$ for $k \geq 0$.

For the pair $n = -1, k = 0$, we get that

$$E^2(-2,0) = E^2(0,0) = 1 = \frac{2}{2}.$$

For the pair $n = -1, k = 1$, we have

$$E^2(-2,1) = \frac{1}{2}(E^2(-1,1) + (1+3)E^2(-2,0)) = \frac{1}{2}(1 + 4\cdot 1)$$
$$= \frac{5}{2} = \frac{2+3}{2}.$$

Similarly, for $n = -1, k = 2$, we have

$$E^2(-2,2) = \frac{1}{3}(E^2(-1,2) + (2+3)E^2(-2,1))$$
$$= \frac{1}{3}(1 + 5 \cdot \frac{5}{2}) = \frac{9}{2} = \frac{2+3+4}{2}.$$

Suppose that $E^2(-2, k-1) = \frac{2+3+\cdots+(k+1)}{2}$. Then, for $n = -1, k$, we have

$$E^2(-2,k) = \frac{1}{k+1}(E^2(-1,k) + (k+1)E^2(-2,k-1))$$
$$= \frac{1}{k+1}(1 + (k+3) \cdot \frac{2+3+\cdots+(k+1)}{2})$$
$$= \frac{1}{2(k+1)} \cdot (2 + (k+3)\frac{(k+1)(k+2)}{2} - (k+3))$$
$$= \frac{1}{2(k+1)} \cdot ((k+3)\frac{(k+1)(k+2)}{2} - (k+1))$$
$$= \frac{1}{2} \cdot (\frac{(k+2)(k+3)}{2} - 1)$$
$$= \frac{2+3+\cdots+(k+2)}{2}.$$

□

Eulerian numbers of the *m*th order

4.7.11. The Eulerian numbers of the mth order $E^m(n,k)$, $m = 1,2,3,\ldots$, give a natural generalisation of both classes of Eulerian numbers, considered in detail in this book.

In order to define these numbers (see, for example, [He23]), let us consider the set of all permutations of the multiset $Q_n^m = \{1,\ldots,1,2,\ldots,2,\ldots,n,\ldots,n\}$, with a multiplicity of m for each integer. The number of such permutations is equal to

$$P_{(m,m,\ldots,m)} = \frac{(mn)!}{(m!)^n}.$$

We are interested in the so-called *m-Stirling permutations*, i.e., permutations of the set Q_n^m such that for each value i appearing in the permutation, any values between the two copies of i are larger than i.

- *The number of permutations of a multiset* $\{1, 1, 2, 2, \ldots, n, n\}$ *such that for each* m, $m = 1, 2, \ldots, n$, *all numbers located between two values of* m *will be greater than* m *is equal to*

$$1 \cdot (m + 1)(2m + 1) \cdots (m(n - 1) + 1).$$

□ This fact can be easy proven by induction. The statement is true for $n = 1$: $(1, 1, \ldots, 1)$ is the only permutation of the multiset $\{1, , \ldots, 1\}$ that has the considered property. The statement also holds for $n = 2$: there are exactly $m + 1$ permutations of the multiset $\{1, \ldots, 1, 2, \ldots, 2\}$ with the considered property, as we should put the string $2, 2, \ldots, 2$ as the fixed block, and there are exactly $m - 1 + 2 = m + 1$ positions for this block between, before, or after m unities.

Let the statement hold for a positive integer $n - 1$, i.e., there are exactly

$$1 \cdot 3 \cdot \cdots \cdot (m(n - 2) + 1)$$

permutations of the multiset $\{1, \ldots, 1, 2, \ldots, 2, 3, \ldots, 3, \ldots, n - 1, \ldots, n - 1\}$ with the considered property. Any m-Stirling permutation of the multiset $\{1, \ldots, 1, 2, \ldots, 2, \ldots, n - 1, \ldots, n - 1, n, \ldots, n\}$ can be obtained from a m-Stirling permutation of the multiset $\{1, \ldots, 1, 2, \ldots, 2, \ldots, n - 1, \ldots, n - 1\}$ by properly placing m elements n. As by condition there can be no elements between two elements n, $n \ldots n$ should be placed together at any possible place in the fixed m-Stirling permutation of $\{1, \ldots, 1, 2, \ldots., 2, \ldots, n - 1, \ldots, n - 1\}$: before the first element, between the first and the second elements, ..., between the $(m(n-1)-1)$th and the $(m(n-1))$th (the last) elements, after the last element; altogether, there are $m(n - 1) + 1$ possible places for $nn \ldots n$. So, from any m-Stirling permutation of the $(n - 1)$th order, we get exactly $m(n - 1) + 1$ m-Stirling permutations of the nth order. The result follows obviously. □

The products of the form $1 \cdot (m+1)(2m+1) \cdots (km+1)$ are called *multifactorial numbers* (or *multifactorials*). For $m = 1$, we get classical factorials $1 \cdot 2 \cdot \cdots \cdot (k + 1) = (k + 1)!$. For $m = 2$, we get double factorials $1 \cdot 3 \cdot \cdots \cdot (2k + 1) = (2k + 1)!!$. For $m = 3$, the sequence $1 \cdot 4 \cdot 7 \cdot \dots \cdot (3k+1) = (3k+1)!!!$ of triple factorials starts with the elements 1, 1, 4, 28, 280, 3640, 58240, 1106560, 24344320, $608608000, \dots$ (sequence A007559 in the OEIS). For $m = 4$, the sequence $1 \cdot 5 \cdot 9 \cdot \dots \cdot (4k+1) = (4k+1)!!!!$ of quartic factorial numbers starts with the elements 1, 1, 5, 45, 585, 9945, 208845, 5221125, 151412625, $4996616625, \dots$ (sequence A007696 in the OEIS). So, we have proven the following property.

- *The m-Stirling permutations of the nth order are counted by the multifactorial* $(m(n-1)+1)!! \cdots !$.

4.7.12. For a given permutation $\pi = (\pi_1, \pi_2, \dots, \pi_{mn})$ of the multiset $\{1, \dots, 1, 2, \dots, 2, \dots, n, \dots, n\}$, we define an ascent as the position i, $1 \leq i \leq mn-1$, such that $\pi_i < \pi_{i+1}$. Similarly, we define a descent as the position i, $1 \leq i \leq mn-1$, such that $\pi_i > \pi_{i+1}$.

Now, we can define the *Eulerian number of the mth order* $E^m(n,k)$ as the number of m-Stirling permutations with exactly k ascents.

So, by definition, $E^1(n,k) = E(n,k)$ are the Eulerian numbers of the first order, considered in Chapter 3, while the numbers $E^2(n,k)$ are the Eulerian numbers of the second order; we study them in this chapter. It turns out that the numbers $E^m(n,k)$ represent a common generalisation of the two number sets $E(n,k)$ and $E^2(n,k)$.

4.7.13. It is easy to prove the following fact.

- *For the Eulerian numbers of the mth order, there exists the recurrent relation*

$$E^m(n,k) = (k+1)E^m(n-1,k) + (mn-k-m+1)E^m(n-1,k-1), \quad E^m(1,0) = 1.$$

In particular, it holds that

$$E^1(n,k) = (k+1)E^m(n-1,k) + (n-k)E^1(n-1,k-1),$$

$$E^2(n,k) = (k+1)E^2(n-1,k) + (2n-k-1)E^2(n-1,k-1),$$

$$E^3(n,k) = (k+1)E^3(n-1,k) + (3n-k-2)E^3(n-1,k-1),$$

$$E^4(n,k) = (k+1)E^4(n-1,k) + (4n-k-3)E^4(n-1,k-1).$$

☐ The proof for a general m is similar to the proofs for $m = 1, 2$. In fact, each m-Stirling permutation $\rho = (\rho_1, \rho_2, \ldots, \rho_{m(n-1)})$ of the $(n-1)$th order leads to $m(n-1)+1$ m-Stirling permutations of the nth order if we insert the block of new elements $nn\ldots n$ in all possible ways: between two elements ρ_i, ρ_{i+1}, $i = 1, 2, 3, \ldots, m(n-1)-1$; before the first element ρ_1; or after the last element $\rho_{m(n-1)}$.

Upon inserting the block $nn\ldots n$ in some position, we get three possibilities:

- a permutation $\pi = (n, \ldots, n, \rho_1, \rho_2, \ldots, \rho_{m(n-1)})$;
- a permutation $\pi = (\rho_1, \rho_2, \ldots, \rho_i, n, \ldots, n, \rho_{i+1}, \ldots, \rho_{m(n-1)})$, where $i = 1, 2, 3, \ldots, m(n-1)-1$;
- a permutation $\pi = (\rho_1, \rho_2, \ldots, \rho_{m(n-1)}, n, \ldots, n)$.

The number of ascents in the permutation π is equal to the number of ascents in the permutation ρ for the first possibility, or for the second possibility, if $\rho_i < \rho_{i+1}$. The number of ascents in the permutation π is greater than the number of ascents in the permutation ρ for the third possibility, or for the second possibility, if $\rho_i \geq \rho_{i+1}$. Therefore, in order to obtain a Stirling permutation π of $\{1, \ldots, 1, 2, \ldots, 2, \ldots, n, \ldots, n\}$ with k ascents, we can:

- use any permutation ρ of $\{1, \ldots, 1, 2, \ldots, 2, \ldots, n-1, \ldots, n-1\}$ with k ascents, putting the elements $nn\ldots n$ in the first position, or between two elements, forming an ascent, altogether giving $k+1$ possibilities;
- use any permutation ρ of $\{1, \ldots, 1, 2, \ldots, 2, \ldots, n-1, \ldots, n-1\}$ with $k-1$ ascents (and, hence, with $(m(n-1)-(k-1)-1 = mn-m-k$ non-ascents), putting the elements $nn\ldots n$ in the last position, or between two elements, forming an non-ascent, altogether giving $mn-k-m+1$ possibilities.

In other words, there are $(k+1)E^m(n-1, k)$ ways to construct a permutation π with k ascents from the permutations ρ of $\{1, \ldots, 1, 2, \ldots, 2, \ldots, n-1, \ldots, n-1\}$ with k ascents and $(mn-k-m+1)E^2(n-1, k-1)$ ways to construct a permutation π from the permutations ρ with $k-1$ ascents. The sum of all possibilities gives the recurrent formula

$$E^m(n,k) = (k+1)E^m(n-1,k) + (mn-k-m+1)E^m(n-1,k-1). \qquad \square$$

The values of $E^m(n,k)$ for small n and k ($0 \le n \le 5$, $0 \le k \le 5$) are given in the following table:

$n\backslash k$	0	1	2	3	4	5
0	1					
1	1	0				
2	1	m	0			
3	1	$4m$	$m(2m-1)$	0		
4	1	$11m$	$m(18m-7)$	$m(6m^2-7m+2)$	0	
5	1	$26m$	$2m(49m-16)$	$2m(48m^2-46m+11)$	$m(4m-3)$ $(6m^2-7m+2)$	0

In particular, the values of $E^3(n,k)$, $n \ge 1$, $1 \le k \le n-1$, start with the numbers $1, 1, 3, 1, 12, 15, 1, 33, 141, 105, \ldots$ (sequence A219512 in the OEIS):

$$\begin{array}{ccccccccccccc}
 & & & & & & 1 & & & & & & \\
 & & & & & 1 & & 3 & & & & & \\
 & & & & 1 & & 12 & & 15 & & & & \\
 & & & 1 & & 233 & & 141 & & 105 & & & \\
 & & 1 & & 78 & & 786 & & 1830 & & 945 & & \\
 & 1 & & 171 & & 3450 & & 17538 & & 26685 & & 10395 & \\
1 & & 360 & & 13257 & & 125352 & & 396495 & & 435960 & & 135135 \\
\vdots & & \vdots & & \vdots & & \vdots & & \vdots & & \vdots & & \vdots
\end{array}$$

The small values of $E^4(n,k)$, $1 \le n \le 4$, $0 \le k \le n-1$, are given in the following triangle ([He23]):

$$\begin{array}{ccccccccccc}
 & & & & & 1 & & & & & \\
 & & & & 1 & & 4 & & & & \\
 & & & 1 & & 16 & & 28 & & & \\
 & & 1 & & 44 & & 260 & & 280 & & \\
 & 1 & & 104 & & 1440 & & 4760 & & 3640 & \\
\vdots & \vdots & & \vdots & & \vdots & & \vdots & & \vdots & \vdots
\end{array}$$

4.7.14. From the construction of the considered triangles of Eulerian numbers of the mth order, it follows that:

- *the sum of the elements of the* $(n+1)$*th row of the triangle is equal to the multifactorial*

$$1 \cdot (m+1) \cdot \dots \cdot (m(n-1)+1) = (m(n-1)+1)!! \cdots ! \, .$$

□ The proof is obvious: the sum gives all m-Stirling permutations of the nth order. □

From this property, we immediately obtain the following simple fact:

- *The sum of the elements of the* $(n+1)$*th row of the triangle is equal to the sum of the elements of the* n*th row multiplied by* $(mn+1)$:

$$\sum_{k=0}^{n} E^m(n+1,k) = (1+mn) \sum_{k=0}^{n-1} E^m(n,k).$$

In particular, it holds that

$$\sum_{k=0}^{n} E(n+1,k) = (n+1)! = (n+1) \cdot n! = (n+1) \sum_{k=0}^{n-1} E(n,k);$$

$$\sum_{k=0}^{n} E^2(n+1,k) = (2n+1)!! = (2n+1)(2n-1)!!$$

$$= (2n+1) \sum_{k=0}^{n-1} E^2(n,k);$$

$$\sum_{k=0}^{n} E^3(n+1,k) = (3n+1)!!! = (3n+1) \cdot (3n-2)!!!$$

$$= (3n+1) \sum_{k=0}^{n-1} E^3(n,k);$$

$$\sum_{k=0}^{n} E^4(n+1,k) = (4n+1)!!!! = (4n+1)(4n-3)!!!!$$
$$= (4n+1)\sum_{k=0}^{n-1} E^4(n,k).$$

4.7.15. For the Eulerian numbers of the mth order, a closed formula can be obtained (see [He23]):

- *For the Eulerian number $E^m(n,k)$, $0 \leq k \leq n-1$, the following equality holds:*

$$E^m(n,k) = \sum_{t_1+t_2+\cdots+t_{k+1}=n-k-1, t_i\geq 0} 1^{t_1}2^{t_2}$$
$$\cdots (k+1)^{t_{k+1}}(mt_1+m)(m(y_1+t_2)$$
$$+2(m-1)+1)\cdots(m(t_1+\cdots+t_k)+k)m-1)+1).$$

□ The proof for the general case coincides with the proofs for $m=1$, and $m=2$, considered in Sections 3.5 and 4.5. □

This formula allows us to obtain two important special cases:

- *The zeroth column $E^m(n,0)$, $n \geq 0$, of the triangle of Eulerian numbers of the mth order consists of unities; the first column $E^m(n,1)$, $n \geq 1$, of the triangle consists of the numbers $m(2^n-n-1)$; the first descending diagonal $E^m(n,n-1)$, $n \geq 1$, of the triangle consists of multifactorial numbers $((m-1)n+1)!!\ldots!$:*

$$E^m(n,0) = 1; \quad E^m(n,1) = m(2^n-n-1);$$
$$E^m(n,n-1) = 1\cdot m\cdot(2(m-1)+1)\cdots((n-1)(m-1)+1).$$

□ There are many ways to prove these facts. Using the above formula, we can state that, for $k=0$, the sum $t_1+t_2+\cdots+t_{k+1} = n-k-1$ is the degenerate sum $t_1 = n-2$, and

$$E^m(n,0) = 1^{n-2} = 1.$$

Similarly, for $k = n-1$, the sum $t_1 + t_2 + \cdots + t_{k+1} = n-k-1$ is the degenerate sum $t_1 + t_2 + \cdots + t_{k+1} = 0$, and

$$E^m(n, n-1) \\ = 1^0 2^0 \cdots (k+1)^0 m(2(m-1)+1) \cdots ((n-1)(m-1)+1).$$

As for $k = 1$, we have that

$$E^m(n,1) = \sum_{t_1+t_2=n-2, t_i \geq 0} 1^{t_1} 2^{t_2}(mt_1 + m)$$

$$= m \sum_{k=0}^{n-2} 2^k (n-1-k) = m(2^n - n - 1).$$

The equality $\sum_{k=0}^{n-2} 2^k(n-1-k) = 2^n - n - 1$ can be proven by induction. □

4.7.16. So, we can state that there exist the following simple properties:

- *The elements $E^m(n,1)$, $n \geq 1$, of the first column of the triangle of Eulerian numbers of the mth order are connected to the elements $E(n,1)$, $n \geq 1$, of the first column of Euler's triangle*:

$$E^m(n,k) = mE(n,k).$$

As the first elements of the first column of Euler's triangle are 1, 4, 11, 26, 57, 120, 247, 502, 1013, 2036, ... (sequence A000295 in the OEIS), the first elements of the first column of the triangle of Eulerian numbers of the second order are 2, 8, 22, 52, 114, 240, 494, 1004, 2026, 4072, ... (sequence A005803 in the OEIS); the first elements of the first column of the triangle of Eulerian numbers of the third order are 3, 12, 33, 78, 171, 360, 741, 1506, 3039, 6108, ...; and the first elements of the first column of the triangle of Eulerian numbers of the fourth order are 4, 16, 44, 104, 228, 480, 988, 2008, 4052, 6144, ...

- *The sum of the elements of the nth row of the triangle of Eulerian numbers of the mth order is equal to the last non-zero element of the triangle of Eulerian numbers of the $(m+1)$st order*:

$$E^m(n,0) + E^m(n,1) + \cdots + E^m(n,n-1) = E^{m+1}(n,n-1).$$

In particular, it holds that

$E(n,0) + E(n,1) + \cdots + E(n,n-1) = n! = E^2(n,n-1)$,

$E^2(n,0) + E^2(n,1) + \cdots + E^2(n,n-1) = (2n-1)!! = E^3(n,n-1)$,

$E^3(n,0) + E^3(n,1) + \cdots + E^3(n,n-1) = (3n-2)!!! = E^4(n,n-1)$.

4.7.17. Using the numbers $E^m(n,k)$, we can construct the *Eulerian polynomials of the mth order*:

$$A_n^m(x) = \sum_{k=0}^{n-1} E^m(n,k)x^k.$$

This standard algorithm allows us to obtain from the recurrent relation for the Eulerian numbers $E^m(n,k)$ the following recurrence for the Eulerian polynomials $A_n^m(x)$:

$$A_{n+1}^m(x) = (1+mnx)A_n^m(x) + x(1-x)(A_n^m)'(x).$$

Using this approach, it is convenient to introduce two classes of functions:

$$\alpha_n^m(x) = \frac{A_n^m(x)}{(1-x)^{mn-m+1}}, \quad \text{and} \quad \beta_n^m(x) = \frac{xA_n^m(x)}{(1-x)^{mn+1}}.$$

In other words, it holds that

$$\beta_n^m(x) = \frac{x}{(1-x)^{1-m}} \cdot \alpha_n^m(x).$$

The recurrent relation for the polynomials $A_n^m(x)$ allows us to prove that

$$(\beta_n^m(x))' = \left(\frac{x}{(1-x)^{1-m}} \cdot \alpha_n^m(x)\right)' = \alpha_{n+1}^m(x).$$

For $m = 1$, we get the classic Euler fractions $\alpha_n(x) = \frac{A_n(x)}{(1-x)^{n+1}}$; in this case, $\beta_n(x) = x\alpha_n(x)$, and $(\beta_n(x))' = (x\alpha_n(x))' = \alpha_{n+1}(x)$.

For $m = 2$, we get the *Euler fractions of the second order* $\alpha_n^2(x) = u_n(x) = \frac{A_n(x)}{(1-x)^{2n}}$; in this case, $\beta_n^2(x) = \frac{x}{1-x} \cdot \alpha_n^2(x)$, and $(\beta_n^2(x))' = (\frac{x}{1-x} \cdot \alpha_n^2(x))' = \alpha_{n+1}^2(x)$.

For a given $m \geq 2$, it is reasonable to consider the function $\alpha_n^m(x)$ as the *Eulerian fraction of the mth order*.

Exercises

1. Construct the nth row, $n = 5, 6, 7, 8$, of the Eulerian triangle of the third order and of the Eulerian triangle of the fourth order. Construct the first eight rows of the Eulerian triangle of the fifth order. Check the simplest properties of these triangles.
2. Construct the Eulerian triangles of the mth order, $m = 1, 2, 3, 4, 5$, modulo 2; modulo 3; and modulo 5. What can be said about these triangles?
3. Find the first 10 elements of the sequence of Cauchy numbers of the first type using the definition; using the decomposition involving the Stirling numbers of the first kind; and using the exponential generating function of these sequence. Find the first 10 elements of the sequence of Cauchy numbers of the second type using the same methods.
4. Prove that $a_n = \int_0^1 x^{\underline{n}} dx = \sum_{k=1}^{n} n^{\underline{k}} \frac{a_{n-k}}{k+1}$, and $b_n = \int_0^1 x^{\overline{n}} dx = \sum_{k=1}^{n} n^{\underline{k}} \frac{b_{n-k}}{k+1}$.
5. Find the first 10 elements of the sequence $E^2(-3, k)$, $k = 0, 1, 2, \ldots$. Which formula could be used for calculating these numbers?

References

[AbSt72], [Abra74], [Adam97], [Apos86], [BeSl95], [Bron01], [BuGr94], [BHM23], [Carl65], [Choi13], [ChMo75], [CoGu96], [Comt74], [CYY20], [Dede63], [DeZa80], [DeDe12], [Deza17], [Deza18], [Deza21], [Deza24], [Dick05], [Eule36], [Eule48], [Eule49], [Eule55], [Fu21], [Gelf59], [Gess92], [GeSt78], [Goul60], [GKP94], [He23], [HCD19], [Jord65], [Kord95], [Kurt72], [Lah54], [Pasc54], [Qi11], [QiBa17], [Sloa25], [SlPl95], [Stan97], [Stan86], [Stei94], [Uspe76], [Weis25], [Wiki25].

Chapter 5

Other Numbers Related to Leonard Euler

5.1. Construction of Euler Numbers

Combinatorial definition of Euler numbers

5.1.1. Define an *alternating permutation* (or *zigzag permutation*) of the set $\{1, 2, 3, \ldots, n\}$ as an arrangement of those numbers so that each entry is alternately greater or less than the previous entry.

If $n = 2$, there are two permutations, $(1, 2)$ and $(2, 1)$, which, formally, are alternating: $1 < 2$ and $2 > 1$. For $n = 3$, there are six permutations, four of which are alternating: $(1, 3, 2)$ and $(2, 3, 1)$ have the form $\pi_1 < \pi_2 > \pi_3$, i.e., they are *up-down permutations*, while $(2, 1, 3)$ and $(3, 1, 2)$ have the form $\pi_1 > \pi_2 < \pi_3$, i.e., they are *down-up permutations*. For $n = 4$, there are 24 permutations, 10 of which are alternating: $(1, 3, 2, 4)$, $(1, 4, 2, 3)$, $(2, 3, 1, 4)$, $(2, 4, 1, 3)$, and $(3, 4, 1, 2)$ have the form $\pi_1 < \pi_2 > \pi_3 < \pi_4$, i.e., they are *up-down permutations*, while $(3, 2, 3, 1)$, $(3, 2, 4, 1)$, $(4, 1, 3, 2)$, $(3, 1, 4, 2)$ and $(2, 1, 4, 3)$ have the form $\pi_1 > \pi_2 < \pi_3 > \pi_4$, i.e., they are *down-up permutations*.

Authors have used the term *alternating permutation* slightly differently: some require that the second entry in an alternating permutation be larger than the first, others require that the alternation be reversed (so that the second entry is smaller than the first, then the third larger than the second, and so on), while others call both types by the name "alternating permutation".

We use the term *alternating permutation* for a permutation of the form $\pi_1 < \pi_2 > \pi_3 < \cdots$, i.e., for up-down permutations. The down-up permutations, i.e., permutations of the form $\pi_1 > \pi_2 < \pi_3 > \cdots$, we call *reverse alternating permutations.* Since a permutation $\pi = (\pi_1, \pi_2, \ldots, \pi_n)$ is alternating if and only if the permutation $\rho = (n+1-\pi_1, n+1-\pi_2, \ldots, n+1-\pi_n)$ is reverse alternating, it follows that the numbers of alternating (up-down) and reverse alternating (down-up) permutations coincide.

We call an alternating permutation a *proper alternating permutation* if it has the form $\pi_1 < \pi_2 > \pi_3 < \pi_4 > \cdots < \pi_n$. Similarly, a reverse alternating permutation is called *proper* if it has the form $\pi_1 > \pi_2 < \pi_3 > \pi_4 < \cdots > \pi_n$. It is obvious that proper alternating permutations (as well as proper reverse alternating permutations) exist only for even values of n.

With this approach, we can state that, for $n = 2, 3, 4$, the number of alternating (up-down) permutations is 1, 2, and 5, the number of proper alternating (up-down) permutations is 1, 0, and 5, and the number of both (up-down and down-up) permutations is twice the number of alternating permutations, i.e., it is equal to 2, 4, and 10, correspondingly. For $n = 1$, the only permutation (1) is considered simultaneously an up-down and down-up permutation. It is not proper. As always, the number of permutations of the empty set is supposed to be 1. The empty permutation is also considered simultaneously an up-down and as down-up proper permutation.

5.1.2. Denote by $\tilde{E}_n$ the number of all proper alternating (i.e., proper up-down) permutations of the set $\{1, 2, \ldots, n\}$. As noted earlier, for an odd number $2n+1$, the value $\tilde{E}_{2n+1} = 0$, as, starting with $\pi_1 < \pi_2$, we should end with $\pi_{2n} > \pi_{2n+1}$. On the other hand, for even number $2n$, any alternating (up-down) permutation should be proper: starting with $\pi_1 < \pi_2$, we should finish with $\pi_{2n-1} < \pi_{2n}$.

Moreover, $\tilde{E}_0 = 1$ and $\tilde{E}_1 = 0$. The sequence $\tilde{E}_n$, $n = 0, 1, 2, \ldots$, starts with the elements 1, 0, 1, 0, 5, 0, 61, 0, 1385, 0, ...

Define the *Euler numbers* as $E_{2n} = (-1)^n \tilde{E}_{2n}$, while $E_{2n+1} = \tilde{E}_{2n+1} = 0$, $n \geq 0$. In other words, the absolute value of the nth Euler number E_n is equal to the number of proper alternating (up-down) permutations on the set $\{1, 2, \ldots, n\}$; moreover, the number E_n with even index n is positive for $n \equiv 0(mod\ 4)$ and is negative for $n \equiv 2(mod\ 4)$.

So, the sequence E_n, $n = 0, 1, 2, \ldots$, starts with the elements $1, 0, -1, 0, 5, 0, -61, 0, 1385, 0, \ldots$ (sequence A122045 in the OEIS).

5.1.3. The numbers A_n of all alternating (up-down) permutations (not necessary proper) on the set $\{1, 2, \ldots, n\}$ are known as the *Euler zigzag numbers* (or *up-down numbers*).

The first few values of A_n, $n = 0, 1, 2, \ldots$, are 1, 1, 1, 2, 5, 16, 61, 272, 1385, 7936, 50521, ... (sequence A000111 in the OEIS). The determination of the number A_n of alternating permutations of the set $\{1, \ldots, n\}$ is called *André's problem.*

The numbers Z_n count permutations of the set $\{1, \ldots, n\}$ that are either up-down or down-up (or both, for $n = 0, 1$). So, $Z_n = 2A_n$ for $n \geq 2$; the first few values of Z_n, $n = 0, 1, 2, \ldots$, are $1, 1, 2, 4, 10, 32, 122, 544, 2770, 15872, 101042, \ldots$ (sequence A001250 in the OEIS).

As for even n any up-down permutation is proper, we have that $\tilde{E}_{2n} = A_{2n}$ and, hence,

$$E_{2n} = (-1)^n A_{2n},$$

where A_{2n} is the number of alternating (up-down and, moreover, proper) permutations of the set $\{1, 2, \ldots, 2n\}$.

The numbers A_n satisfy a simple recurrent relation, similar to that of Catalan numbers: by splitting the set of alternating permutations (both down-up and up-down) of the set $\{1, 2, \ldots, n, n+1\}$ according to the position k of the largest entry $n+1$, one can show that

$$Z_n = 2A_{n+1} = \sum_{k=0}^{n} \binom{n}{k} A_k A_{n-k}, \quad n \geq 1.$$

This recurrent relation allows us to obtain the exponential generating function $A(x) = \sum_{n=0}^{\infty} A_n \frac{x^n}{n!}$ for the sequence A_n, $n = 0, 1, 2, \ldots$. In fact, it holds that

$$A(x) = \tan\left(\frac{\pi}{4} + \frac{x}{2}\right) = \sec x + \tan x,$$

i.e., $A(x)$ is the sum of the secant and tangent functions. This result is known as *André's theorem* (see [Stan10]).

5.1.4. Since the secant function is even and the tangent function is odd, we obtain two additional classes of numbers.

The numbers A_{2n} with even indices are called the *secant numbers* (or *zig numbers*); they are the numerators in the Maclaurin series of $\sec x$:

$$\sec x = \frac{1}{\cos x} = \sum_{n=0}^{\infty} \frac{A_{2n}x^{2n}}{(2n)!} = 1 + \frac{x^2}{2!} + \frac{5x^4}{4!} + \cdots + \frac{A_{2n}x^{2n}}{(2n)!} + \cdots,$$
$$|x| < \frac{\pi}{2}.$$

The first few values are 1, 1, 5, 61, 1385, 50521, 2702765, 199360981, 19391512145, 2404879675441, ... (sequence A000364 in the OEIS).

As we have noted before, the secant numbers are related to the Euler numbers E_n by the formula

$$E_{2n} = (-1)^n A_{2n}, \quad \text{or} \quad A_{2n} = |E_{2n}|.$$

The numbers A_{2n+1} with odd indices are called the *tangent numbers* (or *zag numbers*); they are the numerators in the Maclaurin series of $\tan x$:

$$\tan x = \frac{\sin x}{\cos x} = \sum_{n=0}^{\infty} \frac{A_{2n+1}x^{2n+1}}{(2n+1)!}$$
$$= x + \frac{2x^3}{3!} + \frac{16x^5}{5!}$$
$$+ \cdots + \frac{A_{2n+1}x^{2n+1}}{(2n+1)!} + \cdots, \quad |x| < \frac{\pi}{2}.$$

The first few values are 1, 2, 16, 272, 7936, 353792, 22368256, 1903757312, 209865342976, 29088885112832, ... (sequence A000182 in the OEIS). The tangent numbers are closely related to the *Bernoulli numbers.* The relation is given by the formula

$$B_{2n} = (-1)^{n-1} \frac{2n}{4^{2n} - 2^{2n}} A_{2n-1}, \quad n \geq 1.$$

5.1.5. Alternating permutations were first considered by the French mathematician Désiré André [Andr79], who obtained the basic and elegant decomposition $\sec x + \tan x = \sum_{n=0}^{\infty} A_n \frac{x^n}{n!}$. However, it is possible (see [Stan97] and [Gins65]) that Jacques Philippe Marie Binet was aware before André that the coefficients of $\sec x$ count alternating permutations.

Recurrent relation for Euler numbers

5.1.6. In this section, we derive the recurrent relation for the Euler numbers.

- *The Euler numbers E_n satisfy the recurrent relation*

$$E_n + \binom{n}{n-2} E_{n-2} + \binom{n}{n-4} E_{n-4} + \binom{n}{n-6} E_{n-6} + \cdots = 0,$$
$$E_0 = 1, \quad E_1 = 0.$$

This relation has an extremely simple symbolic form: if we consider the symbol E^n as a formal record of the number E_n, then we can write this recurrent relation in the form

$$(E+1)^n + (E-1)^n = 0.$$

For odd indices, we get the equality $E_{2n+1} + \binom{2n+1}{2n-1} E_{2n-1} + \binom{2n+1}{2n-3} E_{2n-3} + \cdots \binom{2n+1}{1} E_1 = 0.$

This equality is trivial, as all numbers $E_1, \ldots, E_{2n+1}$ are equal to zero. For even indices, we get the equality

$$E_{2n} + \binom{2n}{2n-2} E_{2n-2} + \binom{2n}{2n-4} E_{2n-4} + \cdots + \binom{2n}{0} E_0 = 0.$$

As $E_{2n} = (-1)^n A_{2n}$, where A_{2n} is the number of (proper) up-down permutations of $\{1, 2, \ldots, 2n\}$, we can simply prove the following recurrent relation for the numbers A_{2n} (see [BLM20]):

- *For the secant numbers A_{2n}, $n = 0, 1, 2, \ldots$, there exists the recurrent relation*

$$A_{2n} - \binom{2n}{2n-2} A_{2n-2} + \binom{2n}{2n-4} A_{2n-4} + \cdots + (-1)^n \binom{2n}{0} A_0 = 0, \quad A_0 = 1.$$

□ In fact, for $n = 1$, we have $A_2 - \binom{2}{0} A_0 = 1 - 1 = 0$.

For $n = 2$, we have $A_4 - \binom{4}{2} A_2 + \binom{4}{0} A_0 = 5 - 6 \cdot 1 + 1 \cdot 1 = 0$.

In general, for a given positive integer $n \geq 2$, consider the sum

$$A_{2n} - \binom{2n}{2n-2} A_{2n-2} + \binom{2n}{2n-4} A_{2n-4} + \cdots + (-1)^n \binom{2n}{0} A_0.$$

By the definition of the numbers $A_{2n}, \ldots, A_0$, this sum can be considered as the construction, counting all up-down permutations of all even subsets $\{i_1, \ldots, i_{2k}\}$ of the set $\{1, 2, \ldots, n\}$. In fact, the summand $\binom{2n}{2k} A_{2k}$, $k = 0, 1, \ldots, n$, can be interpreted as the number of possibilities to choose $2k$ elements $\{i_1, \ldots, i_{2k}\}$ from the set $\{1, 2, \ldots, n\}$ ($\binom{2n}{2k}$ possibilities) and then construct from the elements $i_1, \ldots, i_{2k}$ all possible up-down permutations (A_{2k} possibilities). Moreover, the contribution of a permutation to the set $\{i_1, \ldots, i_{2k}\}$ is positive if k is even and is negative if k is odd.

So, consider the set U of all up-down permutations of the sets $\{i_1, i_2, \ldots, i_{2k}\} \subset \{1, 2, \ldots, n\}$, $k = 0, 1, 2, \ldots, n$, and construct a bijection $f : U \to U$ between such permutations, which is an inversion, i.e., for any permutation $\pi \in U$, $f(f(\pi)) = w$. The main property of the proposed bijection is as follows: if a permutation π is a permutation on the set $\{i_1, \ldots, i_{2k}\}$, then $f(\pi)$ is a permutation on the set $\{t_1, t_2, \ldots, t_{2k+2}\}$ or a permutation on the set $\{t_1, t_2, t_{2k-2}\}$, i.e., the signs of the unities, corresponding to these two permutations in the sum

$$A_{2n} - \binom{2n}{2n-2} A_{2n-2} + \binom{2n}{2n-4} A_{2n-4} + \cdots + (-1)^n \binom{2n}{0} A_0,$$

are opposite, giving zero as the resulting value.

In order to construct this bijection f, fix a set $X = \{i_1, \ldots, i_{2k}\}$, $X \subset V_n$, $0 \leq j \leq n$, and consider any up-down permutation $\pi = (\pi_1, \pi_2, \ldots, \pi_{2k})$, $\pi_1 < \pi_2 > \cdots < \pi_{2k}$, on $X = \{i_1, i_2, \ldots, i_{2k}\}$. Let $S = V_n \backslash X$, i.e., S contains all the elements of V_n which do not belong to X.

A. Let $X \neq \emptyset$, and $X \neq V_n$. Then, $S \neq \emptyset$, and $S \neq V_n$' moreover, by the construction, S has at least two elements. Let x and y be the two greatest elements of S, $x < y$:

1. If $y < \pi_1$, delete in the permutation $\pi = (\pi_1, \pi_2, \ldots, \pi_{2k})$ the first two elements, obtaining an up-down permutation $\pi^* = (\pi_3, \ldots, \pi_{2k})$, $\pi_3 < \pi_4 > \cdots < \pi_{2k}$, on the set $X^* = \{\pi_3, \ldots, \pi_{2k}\}$ of size $2k - 2$ (X^* can be empty set). In this case, let $S^* = S \cup \{\pi_1, \pi_2\}$. Define $f(\pi) = \pi^*$.
2. If $z > \pi_1$, add to the permutation $w = (\pi_1, \pi_2, \ldots, \pi_{2k})$ elements $x < y$, obtaining an up-down permutation $\pi^* = (x, y, \pi_1, \ldots, \pi_{2k})$, $x < y > \pi_3 < \pi_4 > \cdots < \pi_{2k}$, on the set $X^* = \{x, y, \pi_1, \ldots, \pi_{2k}\}$ of size $2k + 2$ (X^* can be V_n). In this case, let $S^* = S \backslash \{x, y\}$. Define $f(\pi) = \pi^*$.

B. If $X = V_n$ (and, hence, $S = \emptyset$) delete in the permutation $w = (\pi_1, \pi_2, \ldots, \pi_{2n})$ the first two elements, obtaining an up-down permutation $\pi^* = (\pi_3, \ldots, \pi_{2n})$, $\pi_3 < \pi_4 > \cdots < \pi_{2n}$, on the set $X^* = \{\pi_3, \ldots, \pi_{2n}\}$ of size $2n - 2$. In this case, let $S^* = \{\pi_1, \pi_2\}$. Define $f(\pi) = \pi^*$.

C. If $X = \emptyset$ (and, hence, $S = V_n$), add to the empty permutation π the largest elements $n - 1 < n$, obtaining an up-down permutation $\pi^* = (n-1, n)$, $n - 1 < n$, on the set $X^* = \{n-1, n\}$ of size 2. In this case, let $S^* = V_n \backslash \{n-1, n\}$. Define $f(\pi) = \pi^*$.

It is easy to check that $(X^*)^* = X$, $(S^*)^* = S$, and $f(f(\pi)) = \pi$. Moreover, in any case, the length of the permutation $\pi^* = f(\pi)$, is $2(k \pm 1)$, where $2k$ is the length of the permutation π, so the contributions of the permutations π and π^* have different signs, resulting zero as the sum.

In particular, for $n = 2$, we have the set $\{1, 2\}$. So, we have one empty permutation $\emptyset$ and one up-down permutation of V_2: $(1, 2)$. In this degenerative case, $f((1, 2)) = \emptyset$, and vice versa.

For $n = 4$, we have one empty permutation $\emptyset$, $\binom{4}{2} = 6$ up-down permutations (1,2), (1,3), (1,4), (2,3), (2,4), (3,4) on the two-element subsets of V_4, and five up-down permutations on V_4: $(1, 3, 2, 4)$, $(1, 4, 2, 3)$, $(2, 3, 1, 4)$, $(2, 4, 1, 3)$, $(3, 4, 1, 2)$.

If, for example, $X = \{1, 2\}$, $\pi = (1, 2)$, then $S = \{3, 4\}$, $3 < 4$. As $4 > 2$, we construct $\pi^* = (3, 4, 1, 2)$, $f(\pi) = \pi^*$, and $X^* = \{1, 2, 3, 4\}$, $S^* = \emptyset$. Conversely, if $X^* = \{1, 2, 3, 4\}$ and $\pi^* = (3, 4, 1, 2)$, then, for $S^* = \emptyset$, we get $(\pi^*)^* = (1, 2) = \pi$, and $(X^*)^* = \{1, 2\} = X$, $(S^*)^* = \{3, 4\} = S$.

So, we can distribute all the considered permutations into pairs, giving opposite contributions to the sum $A_4 - \binom{4}{2}A_2 + \binom{4}{0}A_0$: $(1, 3, 2, 4) \leftrightarrow (2, 4)$, $(, 14, 2, 3) \leftrightarrow (2, 3)$; $(2, 3, 1, 4) \leftrightarrow (1, 4)$, $(2, 4, 1, 3) \leftrightarrow (1, 3)$, $(3, 4, 1, 2) \leftrightarrow (12,)$, $\emptyset \leftrightarrow (3, 4)$. It means that the sum is equal to zero. □

So, we can state that, starting from $E_0 = 0$, we can obtain all Euler numbers E_{2n} with even index using the formula

$$E_{2n} = -\left(\binom{2n}{2n-2}E_{2n-2} + \binom{2n}{2n-4}E_{2n-4} + \cdots + \binom{2n}{0}E_0\right).$$

In particular, $E_2 = -\binom{2}{0}E_0 = -1$, $E_4 = -\binom{4}{2}E_1 - \binom{4}{0}E_0 = 6 \cdot 1 - 1 \cdot 1 = 5$, $E_6 = -\binom{6}{4}E_4 - \binom{6}{2}E_2 - \binom{6}{0}E_0 = -15 \cdot 5 + 15 \cdot 1 - 1 \cdot 1 = 5 - 61$, etc.

5.1.7. However, for the numbers A_n of all up-down (not necessary proper) permutations on the set V_n, $n = 1, 2, \ldots$, we can obtain the following recurrent relation:

- *For the numbers* A_n, $n = 0, 1, 2, \ldots$, *there exists the recurrent relation*

$$Z_{n+1} = 2A_{n+1} = \sum_{k=0}^{n} \binom{n}{k} A_k A_{n-k}, \quad n \geq 1, \quad \text{with}$$

$$A_0 = 1, A_1 = 1.$$

□ In fact, consider the set $V_n = \{1, 2, \ldots, n\}$. Choose from the set V_n k the elements $i_1, \ldots, i_k$. Let $X = \{i_1, \ldots, i_k\}$ and $S = V_n \backslash X = \{m_1, \ldots, m_{n-k}\}$. Consider any construction of the form $\pi = (a_1, \ldots, a_k, n+1, b_1, \ldots, b_{n-k})$, where $a = (a_1, \ldots, a_k)$ is some permutation on X and $b = (b_1, \ldots, b_{n-k})$ is some permutation on S. It is easy to see that any permutation π on $\{1, 2, \ldots, n, n+1\}$ can de obtained in this way. Counting by this algorithm all permutations of the set $\{1, 2, \ldots, n+1\}$, we get the simple identity

$$(n+1)! = \sum_{k=0}^{n} \binom{n}{k} n!(n-k)! = \sum_{k=0}^{n} n! = n! \sum_{n=0}^{n} 1 = n!(n+1).$$

If we wish to obtain in this way all alternating permutations on the set $\{1, 2, \ldots, n+1\}$, we should be more accurate. In fact, as $n+1 > b_1$ for any possible b_1, we should use in our construction only up-down permutations $b = (b_1, \ldots, b_{n-k})$ in order to get the up-down (or down-up) permutation π. On the other hand, as $a_k < n+1$ for all possible a_k, we should use for even k some down-up permutation $a = (a_1, \ldots, a_k)$, $a_1 > a_2 < \cdots > a_k$, and we should use for odd k an up-down permutation $a_1 < a_2 > a_3 < \cdots > a_k$. In the first case, π is a down-up permutation, while in the second case, π is an up-down permutation. As, for $n \geq 2$, the number of up-down permutation of V_n is equal to the number of down-up permutations of V_n, we get all $2A_n$ up-down and down-up permutations of V_{n+1}. Since there are A_k "good" permutations of the set X and A_{n-k} "good" permutations of the set S, we get that, for a fixed $X\{i_1, \ldots, i_k\}$, there are $A_k A_{n-k}$ possibilities to construct a "good" permutation of the set $\{1, 2, \ldots, n+1\}$. Distinct choices of X ($\binom{n}{k}$ possibilities) and summation over all such constructions give the value $\sum_{k=0}^{n} \binom{n}{k} A_k A_{n-k}$. So, we have proved that

$$Z_{n+1} = 2A_{n+1} = \sum_{k=0}^{n} \binom{n}{k} A_k A_{n-k}, \quad n \geq 1.$$

The initial conditions $A_0 = A_1 = 1$ follow from the definition. □

In particular, starting with $A_0 = A_1 = 1$, we get that $2A_2 = \binom{1}{0} A_0 A_1 + \binom{1}{1} A_1 A_0 = 2$, i.e., $A_2 = 1$; $2A_3 = \binom{2}{0} A_0 A_2 + \binom{2}{1} A_1 A_1 + \binom{2}{1} A_2 A_0 = 1 + 2 + 1 = 4$, i.e., $A_3 = 2$, $2A_4 = \binom{3}{0} A_0 A_3 + \binom{3}{1} A_1 A_2 + \binom{3}{2} A_2 A_1 + \binom{3}{3} A_3 A_0 = 2 + 3 + 3 + 2 = 10$, i.e., $A_4 = 5$, etc.

Triangle of Euler numbers

5.1.8. The Euler numbers E_n, $n = 0, 1, 2, \ldots$, or, more exactly, their absolute values $|E_n|$, $n = 0, 1, 2, \ldots$, can be obtained using a simple arithmetic triangle $E_{n,k}$, $n \geq 0$, $0 \leq k \leq n$, constructed using the following (see [Stan97]):

$$\begin{array}{ccccccccccccc}
 & & & & & & E_{0,0} & & & & & & \\
 & & & & & E_{1,0} & \rightarrow & E_{1,1} & & & & & \\
 & & & & E_{2,2} & \leftarrow & E_{2,1} & \leftarrow & E_{2,0} & & & & \\
 & & & E_{3,0} & \rightarrow & E_{3,1} & \rightarrow & E_{3,2} & \rightarrow & E_{3,3} & & & \\
 & & E_{4,4} & \leftarrow & E_{4,3} & \leftarrow & E_{4,2} & \leftarrow & E_{4,1} & \leftarrow & E_{4,0} & & \\
 & E_{5,0} & \rightarrow & E_{5,1} & \rightarrow & E_{5,2} & \rightarrow & E_{5,3} & \rightarrow & E_{5,4} & \rightarrow & E_{5,5} & \\
E_{6,6} & \leftarrow & E_{6,5} & \leftarrow & E_{6,4} & \leftarrow & E_{6,3} & \leftarrow & E_{6,2} & \leftarrow & E_{6,1} & \leftarrow & E_{6,0} \\
 & & \vdots & & \vdots & & \vdots & & \vdots & & \vdots & &
\end{array}$$

In order to construct the triangle, we should start from zero ($E_{0,0} = 0$) and read the entries in the direction of the arrows from the zeroth row, using zero as the first number read in each row ($E_{n,0} = 0$, $n \geq 1$) and constructing each subsequent element of the row as the sum of the previous element in the same row and the element above in the previous row.

In other words, we construct the triangular array $E_{n,k}$ such that

$$E_{0,0} = 1, \quad E_{n,0} = 0 \quad \text{for} \quad n \geq 1, \quad \text{and}$$

$$E_{n+1,k+1} = E_{n+1,k} + E_{n,n-k}.$$

Several first rows of the triangle are represented in the following depiction. It is easy to find there the numbers $|E_n|$: $1, 0, 1, 0, 2, 0, 5, 0, 61, \ldots$. They form the left-hand side of the triangle.

$$\begin{array}{ccccccccccccc}
 & & & & & & 0 & & & & & & \\
 & & & & & 0 & \rightarrow & 1 & & & & & \\
 & & & & 1 & \leftarrow & 1 & \leftarrow & 0 & & & & \\
 & & & 0 & \rightarrow & 1 & \rightarrow & 2 & \rightarrow & 2 & & & \\
 & & 5 & \leftarrow & 5 & \leftarrow & 4 & \leftarrow & 2 & \leftarrow & 0 & & \\
 & 0 & \rightarrow & 5 & \rightarrow & 10 & \rightarrow & 14 & \rightarrow & 16 & \rightarrow & 16 & \\
61 & \leftarrow & 61 & \leftarrow & 56 & \leftarrow & 46 & \leftarrow & 32 & \leftarrow & 16 & \leftarrow & 0 \\
 & & \vdots & & \vdots & & \vdots & & \vdots & & \vdots & &
\end{array}$$

In fact, the triangle contains all the Euler zig-zag numbers A_n: $E_{n,n} = A_n$. As $A_{2n} = |E_{2n}|$, we obtain that $E_{2n,2n} = |E_{2n}|$. By the construction of the triangle, these numbers are the left-most entries of each even row. On the other hand, the left-most entries $E_{2n+1,0}$ of each odd row are equal to zero, so they coincide with the numbers $|E_{2n+1}|$.

The numbers $E_{n,k}$, forming the constructed arithmetic triangle, are called the *Entringer numbers*. It is easy to prove that:

- *the number $E_{n,k}$ counts the number of reverse alternating (down-up) permutations of the set $\{1, 2, \ldots, n, n+1\}$, with the first term being $k+1$.*

□ By considering the reverse alternating permutations $(k + 2, \pi_2, \ldots, \pi_{n+2})$, corresponding to the number $E_{n+1,k+1}$, we can obtain two cases. In the first case, $\pi_2 \neq k+1$. The number of such $(n+2)$-permutations $(k+2, \pi_2, \ldots, \pi_{n+2}) = (k+2, \pi_2, \ldots, k+1, \ldots, \pi_{n+2})$, $k+2 > \pi_2 < \cdots$, is equal to the number $E_{n+1,k}$ of $(n+2)$-permutations $(k+1, \pi_2, \ldots, k+2, \ldots, \pi_{n+2})$, $k+1 > \pi_2 < \cdots$. In the second case, $\pi_2 = k+1$. The number of such $(n+2)$-permutations $(k+2, k+1, \pi_3, \ldots, \pi_{n+2})$, $k+2 > k+1 < \pi_3 > \cdots$, is equal to the number $E_{n,n-k}$ of $(n+1)$-permutations $(n+2-(k+1), n+2-\pi_3, \ldots, n+2-\pi_{n+2}) = (n-k+1, n+2-\pi_3, \ldots, n+2-\pi_{n+2})$, $n-k+1 > n+2-\pi_3 < \cdots$. □

Now, it is obvious that the number $E_{n,n}$ of the reverse alternating $(n+1)$-permutations $(n+1, \pi_2, \pi_3, \ldots, \pi_{n+1})$, $n+1 > \pi_2 < \pi_3 > \cdots$, coincides with the number A_n of alternating n-permutations $(\pi_2, \pi_3, \ldots, \pi_{n+1})$, $\pi_2 < \pi_3 > \cdots$.

Exercises

1. Find all alternating and reverse alternating permutations of the set $\{1, 2, \ldots, n\}$, $n = 5, 6$. Calculate the numbers E_n, $n = 5, 6$.
2. Using the recurrent relation $E_n + \binom{n}{n-2} E_{n-2} + \cdots + \binom{n}{n-4} E_{n-4} + \cdots = 0$ and the initial conditions $E_0 = 1$ and $E_1 = 0$, find the numbers E_n, $n = 2, 3, \ldots, 8$.

3. Using the recurrent relation $A_n - \binom{n}{n-2}A_{n-2} + \cdots + \binom{n}{n-4}A_{n-4} - \cdots = 0$ and the initial conditions $A_0 = 1$ and $A_1 = 1$, find the numbers A_n, $n = 2, 3, \ldots, 8$. Check that $A_{2n} = |E_{2n}|$, $n = 0, 2, 3, 6, 8$.
4. Find the *Entringer numbers* $E_{n,k}$, $n = 0, 1, 2, 3, 4, 5, 6$, $0 \leq k \leq n$, as the numbers of the reverse alternating permutations of the set $\{1, 2, \ldots, n+1\}$, with the first element being $k+1$; find the *Entringer numbers* $E_{n,k}$, $n = 0, 1, 2, 3, 4, 5, 6$, $0 \leq k \leq n$, using the recurrent relation $E_{n,k} = E_{n,k-1} + E_{n-1,n-k}$, with the initial conditions $E_{0-0} = 1$ and $E_{n,0} = 0$, $n \geq 1$.
5. Find the first 10 rows of the triangle of Entringer numbers. Prove that $|E_{2n}| = E_{n,n-1}$. Formulate and prove several other properties of these triangle.
6. Prove that $E_{n,n-1} = E_{n,n} = |E_{2n}|$, $n \geq 2$, using alternating permutations and the recurrent formula for $E_{n,k}$. Prove that $E_{n,1} = 1 = E_0$ using alternating permutations and the recurrent formula for $E_{n,k}$. Formulate and prove several other properties of the triangle of Entringer numbers.

5.2. Generating Function of the Sequence of Euler Numbers

5.2.1. The recurrent relation

$$E_{2n} + \binom{2n}{2n-2}E_{2n-2} + \binom{2n}{2n-4}E_{2n-4} + \cdots + \binom{2n}{0}E_0 = 0$$

allows us to obtain the exponential generating function for the sequence E_n, $n = 0, 1, 2, \ldots$, of Euler numbers.

- *The exponential generating function of the sequence* E_n, $n = 0, 1, 2, \ldots$, *of Euler numbers has the form* $\frac{2}{e^x+e^{-x}}$, *i.e., it holds that*

$$\frac{2}{e^x + e^{-x}} = \sum_{n=0}^{\infty} E_n \frac{x^n}{n!}.$$

□ Consider for the function $\frac{2}{e^x+e^{-x}}$ the decomposition

$$\frac{2}{e^x+e^{-x}} = b_0\frac{x^0}{0!} + b_1\frac{x^1}{1!} + b_2\frac{x^2}{2!} + \cdots + b_n\frac{x^n}{n!} + \cdots .$$

As the function $f(x) = \frac{2}{e^x+e^{-x}}$ is even $(f(-x) = f(x))$,

$$b_1 = b_3 = \cdots = b_{2n+1} = \cdots = 0.$$

So, it holds that

$$\frac{2}{e^x+e^{-x}} = b_0\frac{x^0}{0!} + b_2\frac{x^2}{2!} + \cdots + b_{2n}\frac{x^{2n}}{(2n)!} + \cdots .$$

In other words, we have that

$$2 = \left(e^x + e^{-x}\right)\left(b_0\frac{x^0}{0!} + b_2\frac{x^2}{2!} + \cdots + b_{2n}\frac{x^{2n}}{(2n)!} + \cdots\right).$$

As

$$e^x = \frac{x^0}{0!} + \frac{x^1}{1!} + \cdots + \frac{x^n}{n!} + \cdots \quad \text{and}$$

$$e^{-x} = \frac{x^0}{0!} - \frac{x^1}{1!} + \cdots + (-1)^n\frac{x^n}{n!} + \cdots,$$

$$e^x + e^{-x} = 2\left(\frac{x^0}{0!} + \frac{x^2}{2!} + \cdots + \frac{x^{2n}}{(2n)!} + \cdots\right),$$

and

$$\begin{aligned}&\left(e^x + e^{-x}\right)\left(b_0\frac{x^0}{0!} + b_2\frac{x^2}{2!} + \cdots + b_{2n}\frac{x^{2n}}{(2n)!} + \cdots\right)\\ &= 2\left(\frac{x^0}{0!} + \frac{x^2}{2!} + \cdots + \frac{x^{2n}}{(2n)!} + \cdots\right)\\ &\quad\times\left(b_0\frac{x^0}{0!} + b_2\frac{x^2}{2!} + \cdots + b_{2n}\frac{x^{2n}}{(2n)!} + \cdots\right).\end{aligned}$$

So, it holds that

$$\begin{aligned}&\left(\frac{x^0}{0!} + \frac{x^2}{2!} + \cdots + \frac{x^{2n}}{(2n)!} + \cdots\right)\\ &\quad\times\left(b_0\frac{x^0}{0!} + b_2\frac{x^2}{2!} + \cdots + b_{2n}\frac{x^{2n}}{(2n)!} + \cdots\right) = 1.\end{aligned}$$

It is easy to see that the left-hand side of the last equality has the form

$$\left(\frac{b_0 0!}{0!0!}\right)\frac{x^0}{0!}+\left(\frac{b_0 2!}{0!2!}+\frac{b_2 2!}{2!0!}\right)\frac{x^2}{2!}+\left(\frac{b_0 4!}{0!4!}+\frac{b_2 4!}{2!2!}+\frac{b_4 4!}{4!0!}\right)\frac{x^4}{4!}$$
$$+\cdots+\left(\frac{b_0(2n)!}{0!(2n)!}+\frac{b_2(2n)!}{2!(2n-2)!}+\frac{b_4(2n)!}{4!(2n-4)!}+\cdots+\frac{b_{2n}(2n)!}{(2n)!0!}\right)$$
$$\times\frac{x^{2n}}{(2n)!}=b_0+\left(b_0\binom{2}{0}+b_2\binom{2}{2}\right)\frac{x^2}{2!}+\left(b_0\binom{4}{0}+b_2\binom{4}{2}\right.$$
$$\left.+b_4\binom{4}{4}\right)\frac{x^4}{4!}+\cdots+\left(b_0\binom{2n}{0}+b_2\binom{2n}{2}\right.$$
$$\left.+\cdots b_{2n}\binom{2n}{2n}\right)\frac{x^{2n}}{(2n)!}+\cdots.$$

As this decomposition should be equal to unity, we get that

$$b_0=1,\quad\text{and}\quad b_0\binom{2n}{0}+b_2\binom{2n}{2}+\cdots+b_{2n}\binom{2n}{2n}=0\quad\text{for}\quad n\geq 1.$$

Recalling that $b_{2n+1}=0$ for all odd values of the indices, we can state that the coefficients b_n, $n\geq 0$, of the decomposition

$$\frac{2}{e^x+e^{-x}}=b_0\frac{x^0}{0!}+b_1\frac{x^1}{1!}+b_2\frac{x^2}{2!}+\cdots+n_n\frac{x^n}{n!}+\cdots$$

satisfy the same recurrent relation with the same initial conditions as the numbers E_n, $n\geq 0$. It means that $b_n=E_n$ for all $n\geq 0$ and, therefore, $\frac{2}{e^x+e^{-x}}=\sum_{n=0}^{\infty}E_n\frac{x^n}{n!}$. □

Since $\frac{e^x+e^{-x}}{2}=\cosh x$, where $\cosh x$ is the hyperbolical cosine, the exponential generating function for the sequence of Euler numbers E_n has the form $\frac{1}{\cosh x}$.

5.2.2. Using the recurrent relation

$$2A_{n+1}=\sum_{k=0}^{n}\binom{n}{k}A_kA_{n-k},\quad A_0=1,$$

we can obtain the exponential generating function for the sequence A_n, $n=0,1,2,\ldots$, of Euler zig-zag numbers.

- *The exponential generating function for the sequence A_n, $n = 0, 1, 2, \ldots$, of Euler zig-zag numbers has the form* $\sec x + \tan x$, *i.e., it holds that*

$$\sec x + \tan x = \sum_{n=0}^{\infty} A_n \frac{x^n}{n!}, \quad |x| < \frac{\pi}{2}.$$

□ Consider the function $A(x) = \sum_{n=0}^{\infty} A_n \frac{x^n}{n!}$. Its derivative has the form

$$A'(x) = \left(\sum_{n=0}^{\infty} A_n \frac{x^n}{n!}\right)' = \sum_{n=1}^{\infty} A_n n \frac{x^{n-1}}{n!} = \sum_{n=0}^{\infty} A_{n+1} \frac{x^n}{n!}$$
$$= A_1 + A_2 \frac{x^1}{1!} + \cdots + A_{n+1} \frac{x^n}{n!} + \cdots .$$

On the other hand, using the recurrent relation for the numbers A_n, $n \geq 1$, and the values $A_0 = A_1 = 1$, we can check that

$$(A(x))^2 = \left(\sum_{n=0}^{\infty} A_n \frac{x^n}{n!}\right)^2 = A_0 + \left(\frac{A_0 A_1 1!}{0!1!} + \frac{A_1 A_0 1!}{1!0!}\right) \frac{x^1}{1!}$$
$$+ \left(\frac{A_0 A_2 2!!}{0!2!} + \frac{A_1 A_1 2!}{1!1!} + \frac{A_2 A_0 2!}{2!0!}\right) \frac{x^2}{2!}$$
$$+ \cdots + \left(\frac{A_0 A_n n!}{0!n!} + \frac{A_1 A_{n-1} n!}{1!(n-1)!} + \cdots \frac{A_n A_0 n!}{n!0!}\right) \frac{x^n}{n!} + \cdots$$
$$= A_0 + \left(A_0 A_1 \binom{1}{0} + A_1 A_0 \binom{1}{1}\right) \frac{x^1}{1!} + \left(A_0 A_2 \binom{2}{0}\right.$$
$$\left. + A_1 A_1 \binom{2}{1} + A_2 A_0 \binom{2}{2}\right) \frac{x^2}{2!} + \cdots + \left(A_0 A_n \binom{n}{0}\right.$$
$$\left. + A_1 A_{n-1} \binom{n}{1} + \cdots A_n A_0 \binom{n}{n}\right) \frac{x^n}{n!} + \cdots$$
$$= A_1 + 2A_2 \frac{x^1}{1!} + 2A_3 \frac{x^2}{2!} + \cdots + 2A_n \frac{x^n}{n!} + \cdots$$
$$= 2\left(A_1 + 2A_2 \frac{x^1}{1!} + 2A_3 \frac{x^2}{2!}\right.$$
$$\left. + \cdots + 2A_n \frac{x^n}{n!} + \cdots\right) - 1$$
$$= 2A'(x) - 1.$$

So, we get that $(A(x))^2 = 2A^{'}(x) - 1$, i.e., the function $A(x) = \sum_{n=0}^{\infty} A_n \frac{x^n}{n!}$ satisfies the differential equation $2y^{'} - 1 = y^2$ with the initial condition $y(0) = 1$.

But the only solution for this differential equation is the function $f(x) = \sec x + \tan x$. In fact, $f(0) = 1$, and as

$$f^{'}(x) = (\sec x + \tan x)^{'} = \left(\frac{1 + \sin x}{\cos x}\right)^{'}$$
$$= \frac{\cos^2 x + (1 + \sin x)\sin x}{\cos^2 x} = \frac{1 + \sin x}{\cos^2 x},$$

we get that

$$(f(x))^2 = (\sec x + \tan x)^2 = \left(\frac{1 + \sin x}{\cos x}\right)^2 = \frac{1 + 2\sin x + \sin^2 x}{\cos^2 x}$$
$$= \frac{2 + 2\sin x - \cos^2}{\cos^2 x} = 2 \cdot \frac{1 + \sin x}{\cos^2 x} - 1 = 2f^{'}(x) - 1. \quad \square$$

5.2.3. Now, it is easy to prove that

$$A_{2n-1} = (-1)^{n-1} \frac{4^{2n} - 2^{2n}}{2n} B_{2n}, \ n \geq 1,$$

where B_n is the nth Bernoulli number.

$\square$ As $\frac{e^x}{e^x - 1} = \sum_{n=0}^{\infty} B_n \frac{x^n}{n!}$, we have that

$$\tanh x = \frac{e^x - e^{-x}}{e^x + e^{-x}} = \frac{e^{2x} - 1}{e^{2x} + 1} = 1 - \frac{2}{e^{2x} - 1} + \frac{4}{e^{4x - 1}},$$

and using the values $B_0 = 1$, $B_1 = -\frac{1}{2}$, and $B_{2n+1} = 0$, $n \geq 1$, the following decomposition can be obtained:

$$\tanh x = \frac{e^x - e^{-x}}{e^x + e^{-x}} = 1 - \frac{1}{x} \cdot \frac{2x}{e^{2x} - 1} + \frac{1}{x} \cdot \frac{4x}{e^{4x} - 1}$$
$$= 1 - \frac{1}{x} B_0 - 2B_1 - \sum_{n=2}^{\infty} B_n 2^n \frac{x^{n-1}}{n!} + \frac{1}{x} B_0 + 4B_1$$
$$+ \sum_{n=2}^{\infty} B_n 4^n \frac{x^{n-1}}{n!}$$

$$= \sum_{k=1}^{\infty} B_{2k}(4^{2k} - 2^{2k})\frac{x^{2k-1}}{(2k)!}$$

$$= \sum_{n=0}^{\infty} \frac{1}{2n+2} B_{2n+2}(4^{2n+2} - 2^{2n+2})\frac{x^{2n+1}}{(2n+1)!}.$$

As $\tanh x = i \tan ix$ and therefore $\tan x = -i \tanh ix$, we get that

$$\begin{aligned}\tan x &= -i \tanh ix \\ &= -i\left(\sum_{n=0}^{\infty} \frac{1}{2n+2} B_{2n+2}(4^{2n+2} - 2^{2n+2})\frac{(ix)^{2n+1}}{(2n+1)!}\right) \\ &= \sum_{n=0}^{\infty} \frac{1}{2n+2}(-1)^n B_{2n+2}(4^{2n+2} - 2^{2n+2})\frac{x^{2n+1}}{(2n+1)!}.\end{aligned}$$

Comparing the coefficients at the degree x^{2n+1} in the two decompositions of $\tan x$, we obtain that

$$A_{2n+1} = (-1)^n \frac{4^{2n+2} - 2^{2n+2}}{2n+2} B_{2n+2}, \quad n \geq 0, \quad \text{and}$$

$$A_{2n-1} = (-1)^{n-1} \frac{4^{2n} - 2^{2n}}{2n} B_{2n}, \quad n \geq 1. \qquad \square$$

5.2.4. The Euler numbers E_n (as well as the Euler zig-zag numbers A_n, the Bernoulli numbers B_n, and many other classes of numbers and polynomials) are most often determined by their exponential generating function $\frac{2}{e^x+e^{-x}}$, that is, as coefficients of the decomposition

$$\frac{2}{e^x + e^{-x}} = \sum_{n=0}^{\infty} \frac{E_n}{n!} \cdot x^n.$$

Looking at the resulting numerical sequence, we can immediately make several conjectures on its structure. Namely, we can prove the:

- *all Euler numbers with odd indices are equal to zero*: $E_{2k+1} = 0$, $k = 0, 1, 2, 3, \ldots$;
- *all Euler numbers are integers*: $E_n \in \mathbb{Z}, n = 0, 1, 2, 3, \ldots$;
- *Euler numbers with even indices have alternating signs*: $E_{4k} > 0$, $E_{4k+2} < 0$, $k = 0, 1, 2, 3, \ldots$.

Exercises

1. Using the function $\frac{2}{e^x+e^{-x}}$, find the first 10 Euler numbers, defined as the coefficients of the decomposition $\frac{2}{e^x+e^{-x}}$ into the Maclaurin series. Using the properties of the function $\frac{2}{e^x+e^{-x}}$, prove that all Eulerian numbers with odd indices are equal to zero.
2. Using the properties of the function $\frac{2}{e^x+e^{-x}}$, prove the recurrent relation for the Euler numbers, defined as the coefficients of the decomposition $\frac{2}{e^x+e^{-x}}$ into the Maclaurin series. Check that it has the symbolic form $(E+1)^n + (E-1)^n = 0$. Calculate the first 10 Euler numbers using the obtained recurrent relation. Using the obtained recurrent relation for the Euler numbers, prove that all Eulerian numbers, defined as the coefficients of the decomposition $\frac{2}{e^x+e^{-x}}$ into the Maclaurin series, are integers.
3. Prove that the exponential generating function of the sequence A_n, $n = 0, 1, 2, \ldots$, has the form $\tan\left(\frac{\pi}{4} + \frac{x}{2}\right)$, i.e., it holds that $\tan\left(\frac{\pi}{4} + \frac{x}{2}\right) = \sum_{n=0}^{\infty} \frac{A_n}{(2n)!} x^{2n}$, $|x| < \frac{\pi}{2}$.
4. Using the decomposition $\sec x = \sum_{n=0}^{\infty} \frac{A_{2n}}{(2n)!} x^{2n}$, compute the first 10 secant numbers $A_{2n} = |E_{2n}|$. Using the similar decomposition $\tan x$, compute the first 10 tangent numbers A_{2n+1}.
5. Prove that, for even n, there is a relationship $A_n = (-1)^{\frac{n}{2}} E_n$, and, for odd n, there is a relationship $A_n = (-1)^{\frac{n-1}{2}} \frac{2^{n+1}(2^{n+1}-1)B_{n+1}}{n+1}$, where E_n are the Euler numbers, and B_n are the Bernoulli numbers.

5.3. Closed Formula for Euler Numbers

5.3.1. Unfortunately, there is no simple closed formula for the Euler numbers. The most well-known closed formula for calculating the Euler numbers with even indices E_{2n} includes the imaginary unity i, $i^2 = -1$, and has the following form ([Sloa25]):

- *The Euler number E_{2n}, $n \geq 0$, can be calculated using the formula*

$$E_{2n} = i \sum_{k=1}^{2n+1} \sum_{j=0}^{k} \binom{k}{j} \frac{(-1)^j (k-2j)^{2n+1}}{2^k i^k k}.$$

In fact, we have a more general result, which is true for all Euler zig-zag numbers A_n (see https://oeis.org/A000111/a000111.pdf):

- *There exists a closed formula for the Euler zig-zag numbers* A_n:

$$A_n = i^{n+1} \sum_{k=1}^{n+1} \sum_{j=0}^{k} \binom{k}{j} \frac{(-1)^j (k-2j)^{n+1}}{2^k i^k k}.$$

□ As $\sum_{n=0}^{\infty} A_n \frac{x^n}{n!} = \sec x + \tan x$, we get that

$$\sum_{n=0}^{\infty} \frac{A_n}{n!} \int_0^y x^n dx = \sum_{n=0}^{\infty} A_n \frac{y^{n+1}}{(n+1)!} = \sum_{j=1}^{\infty} A_{j-1} \frac{y^j}{j!}.$$

On the other hand, we have that

$$\sum_{n=0}^{\infty} \frac{A_n}{n!} \int_0^y x^n dx = \int_0^y (\sec x + \tan x) dx = -\log(1 - \sin y).$$

It is easy to see that

$$\begin{aligned} &-\log(1-\sin y) \\ &= \sum_{n=1}^{\infty} \frac{(\sin y)^n}{n} = \sum_{n=1}^{\infty} \frac{(e^{iy} - e^{-iy})^n}{2^n i^n n} \\ &= \sum_{n=1}^{\infty} \sum_{k=0}^{n} \frac{\binom{n}{k} e^{(n-k)iy} e^{-kiy} (-1)^k}{2^n i^n n} = \sum_{n=1}^{\infty} \sum_{k=0}^{n} \frac{\binom{n}{k} e^{(n-2k)iy} (-1)^k}{2^n i^n n} \\ &= \sum_{n=1}^{\infty} \sum_{k=0}^{n} \frac{\binom{n}{k}(-1)^k}{2^n i^n n} \sum_{j=0}^{\infty} \frac{(n-2k)^j i^j y^j}{j!} \\ &= \sum_{j=0}^{\infty} \sum_{n=1}^{\infty} \sum_{k=0}^{n} \frac{\binom{n}{k}(-1)^k (n-2k)^j i^j}{2^n i^n n} \frac{y^j}{j!}. \end{aligned}$$

Comparing the coefficients in the two representations of the sum $\sum_{n=0}^{\infty} \frac{A_n}{n!} \int_0^y x^n dx$, we get that

$$A_{j-1} = \sum_{n=1}^{\infty} \sum_{k=0}^{n} \frac{\binom{n}{k}(-1)^k (n-2k)^j i^j}{2^n i^n n}.$$

As $\sum_{k=0}^{n}\binom{n}{k}(n-2k)^j(-1)^k = 0$ for $n > j$, we get that

$$A_{j-1} = \sum_{n=0}^{j}\sum_{k=0}^{n}\frac{\binom{n}{k}(-1)^k(n-2k)^j i^j}{2^n i^n n}, \quad \text{or}$$

$$A_n = i^{n+1}\sum_{k=1}^{n+1}\sum_{j=0}^{k}\frac{\binom{k}{j}(-1)^j(k-2j)^{n+1}}{2^k i^k k}.$$

For even indices, noting that $i^{2n+1} = (-1)^n i$ and that $A_{2n} = (-1)^n E_{2n}$, we obtain now the above-mentioned result for the Euler number E_{2n}. □

For example, we have that

$$E_2 = i\left(\sum_{j=0}^{1}\binom{1}{j}\frac{(-1)^j(1-2j)^3}{2i} + \sum_{j=0}^{2}\binom{2}{j}\frac{(-1)^j(2-2j)^3}{2^2\cdot i^2\cdot 2}\right.$$
$$\left. + \sum_{j=0}^{3}\binom{3}{j}\frac{(-1)^j(3-2j)^3}{2^3\cdot i^3\cdot 3}\right).$$

Direct checking shows that

$$i\sum_{j=0}^{1}\binom{1}{j}\frac{(-1)^j(1-2j)^3}{2i} = 1, \quad i\sum_{j=0}^{2}\binom{2}{j}\frac{(-1)^j(2-2j)^3}{2^2\cdot i^2\cdot 2} = 0,$$
$$i\sum_{j=0}^{3}\binom{3}{j}\frac{(-1)^j(3-2j)^3}{2^3\cdot i^3\cdot 3} = -2,$$

which gives the expected result: $E_2 = 1 + 0 - 2 = -1$.

5.3.2. On the other hand, there exist formulas expressing the Euler numbers as double sums without imaginary elements (see [Comt74]):

- *For the Euler numbers, the following formula holds*:

$$E_{2n} = \sum_{k=1}^{n}\frac{1}{2^{k-1}}\sum_{j=1}^{k}(-1)^j\binom{2k}{k-j}j^{2n}.$$

For example, $E_2 = -\frac{1}{2^0}\binom{2}{0}1^2 = -1$, while

$$E_4 = -\frac{1}{2^0}\binom{2}{0}1^2 + \frac{1}{2}\left(-\binom{4}{1}\cdot 1^4 + \binom{4}{0}\cdot 2^4\right) = -1 + \frac{12}{2} = 5.$$

There exist several variants of such formulas ([WeQi15]). In particular, it holds that

$$E_{2n} = \sum_{k=1}^{2n}(-1)^k \frac{1}{2^k}\sum_{j=0}^{2k}(-1)^j \binom{2k}{k-j}(k-j)^{2n}.$$

In the simplest case of $n = 1$, we have that

$$\begin{aligned} E_2 = &-\frac{1}{2}\left(\binom{2}{0}\cdot 1^2 - \binom{2}{1}0^2 + \binom{2}{2}(-1)^2\right) \\ &+\frac{1}{4}\left(\binom{4}{0}\cdot 2^2 - \binom{4}{1}1^2 + \binom{4}{2}0^2 - \binom{4}{2}\cdot(-1)^2 + \binom{4}{4}(-2)^2\right) \\ = &-1 + 0 = -1. \end{aligned}$$

On the other hand, we can calculate E_{2n} using the formula

$$E_{2n} = (2n+1)\sum_{k=1}^{2n}(-1)^k \frac{1}{2^k(k+1)}\binom{2n}{k}\sum_{j=0}^{k}\binom{k}{j}(2j-k)^{2n}.$$

For example, we have that

$$\begin{aligned} E_2 = 3&\left(-\frac{1}{2\cdot 2}\binom{2}{1}\left(\binom{1}{0}\cdot(-1)^2 + \binom{1}{1}1^2\right)\right. \\ &\left.+\frac{1}{4\cdot 3}\binom{2}{2}\left(\binom{2}{0}(-2)^2 + \binom{2}{1}\cdot 0^2 + \binom{2}{2}2^2\right)\right) \\ = 3&\left(-1 + \frac{8}{12}\right)\binom{4}{2} = -1. \end{aligned}$$

5.3.3. Using the above formulas and the equality $A_{2n} = (-1)^n E_{2n}$, we can say that there are several formulas for calculating the secant numbers A_{2n}. Similar representations exist also for the tangent numbers A_{2n+1}. For example, in [Comt74], the following formulas are represented:

- *For the secant numbers A_{2n} and the tangent numbers A_{2n+1}, the following formulas hold:*

$$A_{2n} = \sum_{k=1}^{n} \frac{1}{2^{k-1}} \sum_{j=1}^{k} (-1)^{n-j} \binom{2k}{k-j} j^{2n};$$

$$A_{2n+1} = \sum_{k=1}^{n} \frac{1}{2^{k-1}} \sum_{j=1}^{k} (-1)^{n-j} \binom{2k}{k-j} (k+1) j^{2n}.$$

In particular, $A_2 = \frac{1}{2^0}\binom{2}{0}1^2 = 1$, $A_3 = \frac{1}{2^0}\binom{2}{0}2 \cdot 1^2 = 2$, while

$$A_4 = -\frac{1}{2^0}\binom{2}{0}1^2 + \frac{1}{2}\left(-\binom{4}{1} \cdot 1^4 + \binom{4}{0} \cdot 2^4\right) = -1 + \frac{12}{2} = 5,$$

$$A_5 = -\frac{1}{2^0}\binom{2}{0}2 \cdot 1^2 + \frac{1}{2}\left(-\binom{4}{1} \cdot 3 \cdot 1^4 + \binom{4}{0}3 \cdot 2^4\right)$$
$$= -2 + \frac{36}{2} = 16.$$

5.3.4. In terms of the Stirling numbers of the second kind and rising powers, we can construct the following formulas (see [WeQi15]):

- *For the Euler numbers E_{2n}, it holds that*

$$E_{2n} = -4^{2n} \sum_{k=1}^{2n} (-1)^k \cdot \frac{S(2n,k)}{k+1} \cdot \left(\frac{3}{4}\right)^{\overline{k}}.$$

However, the analogue of this formula holds for any positive integer n:

$$E_n = 2^{2n-1} \sum_{k=1}^{n} \frac{(-1)^k S(n,k)}{k+1} \left(3\left(\frac{1}{4}\right)^{\overline{k}} - \left(\frac{3}{4}\right)^{\overline{k}}\right).$$

Closed formula in terms of the Stirling numbers of the second kind can also be obtained for the Euler zig-zag numbers:

$$A_n = -\frac{4^n}{a_n}\sum_{k=1}^{n}\frac{(-1)^k S(n,k)}{k+1}\left(\frac{3}{4}\right)^{\overline{k}}, \quad \text{where}$$

$$a_n = \begin{cases} (-1)^{\frac{n-1}{2}}(1+2^{-n}), & \text{if } n \text{ is odd,} \\ (-1)^{\frac{n}{2}}, & \text{if } n \text{ is even.} \end{cases}$$

5.3.5. The Euler numbers E_{2n} can also be expressed as a sum over even partitions of a $(2n)$-set:

$$E_{2n} = (2n)! \sum_{(k_1,k_2,\ldots,k_n)} \frac{(-1)^{k_1+k_2+\cdots+k_n}(k_1+k_2+\cdots+k_n)!}{k_1!k_2!\cdots k_n!(2!)^{k_1}(4!)^{k_2}\cdots((2n)!)^{k_n}},$$

where $k_1, k_2, \ldots, k_n$ are non-negative integers, such that $2k_1 + 4k_2 + \cdots + 2nk_n = 2n$.

A similar formula holds for odd partitions of a $(2n-1)$-set:

$$E_{2n} = (-1)^{n-1}(2n-1)! \times \sum_{(k_1,k_2,\ldots,k_n)} \frac{(-1)^{k_1+2k_2+\cdots+nk_n}(k_1+k_2+\cdots+k_n)!}{k_1!k_2!\cdots k_n!(1!)^{k_1}(3!)^{k_2}\cdots((2n-1)!)^{k_n}},$$

where $k_1, k_2, \ldots, k_n$ are non-negative integers, such that $k_1 + 3k_2 + \cdots + (2n-1)k_n = 2n-1$. For example,

$$E_{10} = 10!\left(-\frac{1}{10!} + \frac{2}{2!8!} + \frac{2}{4!6!} - \frac{3}{2!^2 6!} - \frac{3}{2!4!^2} + \frac{4}{2!^3 4!} - \frac{1}{2!^5}\right)$$

$$= 9!\left(-\frac{1}{9!} + \frac{3}{1!^2 7!} + \frac{6}{1!3!5!} + \frac{1}{3!^3} - \frac{5}{1!^4 5!} - \frac{10}{1!^3 3!^2} + \frac{7}{1!^6 3!} - \frac{1}{1!^9}\right)$$

$$= -50521.$$

Exercises

1. Using the formula $E_{2n} = i\sum_{k=1}^{2n+1}\sum_{j=0}^{k}\binom{k}{j}\frac{(-1)^j(k-2j)^{2n+1}}{2^k i^k k}$, find the numbers E_4, E_6, E_8, and E_{10}.

2. Using the formula $E_{2n} = \sum_{k=1}^{n} \frac{1}{2^{k-1}} \sum_{j=1}^{k} (-1)^j \binom{2k}{k-j} j^{2n}$, find the numbers E_6, E_8, and E_{10} .
3. Using the formula $E_{2n} = (2n+1) \sum_{k=1}^{2n} (-1)^k \frac{1}{2^k(k+1)} \binom{2n}{k} \sum_{j=0}^{\ell} \binom{k}{j} (2j-k)^{2n}$, find the numbers E_4, E_6, E_8, and E_{10}.
4. Find the Euler zig-zag numbers A_n, $n = 6, 7, 8, 9, 10$, using the formulas

$$A_{2n} = \sum_{k=1}^{n} \frac{1}{2^{k-1}} \sum_{j=1}^{k} (-1)^{n-j} \binom{2k}{k-j} j^{2n},$$

$$A_{2n+1} = \sum_{k=1}^{n} \frac{1}{2^{k-1}} \sum_{j=1}^{k} (-1)^{n-j} \binom{2k}{k-j} (k+1) j^{2n}.$$

5. Using the representations of E_{2n} as sums with the Stirling numbers of the second kind, find the Euler numbers E_{2n}, $n = 1, 2, 3, 4$. Using the representations of A_n as sums with the Stirling numbers of the second kind, find the Euler zig-zag numbers A_n, $n = 1, 2, 3, \ldots, 10$.
6. Write down the numbers E_{2n}, $n = 2, 3, 4$, as a sum over even partitions of a $(2n)$-set. Write down the numbers E_{2n}, $n = 2, 3, 4$, as a sum over odd partitions of a $(2n-1)$-set.

5.4. Properties of Euler Numbers

5.4.1. Euler numbers E_n have many interesting properties and are connected with different classes of mathematical objects, including numbers (secant numbers, tangent numbers, Bernoulli numbers, Genocci numbers, Entringer numbers, etc.), functions (e^x, $\tan x$, $\sec x$, $\log x$, etc.), special polynomials, permutations, and graphs.

Looking at the considered numerical sequence, we can make several simple statements on its structure. Namely, defining this number set as a numerical construction, closely connected with the counting of some special permutations, we can state that, by definition, *all Euler numbers are integers*: $E_n \in \mathbb{Z}, n = 0, 1, 2, 3, \ldots$; *all Euler numbers with odd indices are equal to zero*: $E_{2k+1} = 0,\ k = 0, 1, 2, 3, \ldots$;

the Euler numbers with even indices have alternating signs: $E_{4k} > 0$, $E_{4k+2} < 0$, $k = 0, 1, 2, 3, \dots$.

5.4.2. The Euler numbers (which are integers) are closely connected with the Euler zig-zag numbers A_n (which are positive integers), especially with the tangent numbers A_{2n}:

$$E_{2n} = (-1)^n A_{2n}.$$

Moreover, we have proven that:

- *any tangent number A_{2n} can be represented as a linear combination of the Euler numbers E_{2k}, $k = 0, 1, \dots, 2n-2$*:

$$A_{2n} = (-1)^{n+1} \left(\binom{2n}{2n-2} E_{2n-2} + \binom{2n}{2n-4} E_{2n-4} + \cdots + \binom{2n}{0} E_0 \right).$$

Using similar considerations (see [BLM20]), we can obtain an identity connecting the Euler numbers E_{2n} and the tangent numbers A_{2n+1}.

- *Any tangent number A_{2n+1} can be represented as a linear combination of the Euler numbers E_{2k}, $k = 0, 1, \dots, 2n$*:

$$A_{2n+1} = (-1)^n \left(\binom{2n+1}{2n} E_{2n} + \binom{2n}{2n-2} E_{2n-2} + \cdots \binom{2n}{0} E_0 \right).$$

□ Consider the set U of all up-down permutations of all even subsets $\{i_1, \dots, i_{2j}\}$ of the set $\{1, 2, \dots, 2n+1\}$. In this case, the sum

$$A_{2n} - \binom{2n+1}{2n-2} A_{2n-2} + \binom{2n+1}{2n-4} A_{2n-4} + \cdots + (-1)^n \binom{2n+1}{0} A_0$$

can be considered as the construction counting all such up-down permutations in some alternating mode: the summand $\binom{2n+1}{2j} A_{2j}$ can be interpreted as the number of possibilities to choose $2k$ elements $\{i_1, \dots, i_{2j}\}$ from the set $\{1, 2, \dots, 2n+1\}$ ($\binom{2n+1}{2j}$ possibilities) and then construct from the elements $i_1, \ldots, i_{2j}$ all possible up-down permutations (A_{2j} possibilities); moreover, the contribution of a permutation to $\{i_1, \ldots, i_{2j}\}$ is $+1$ if j is even and is -1 if j is odd.

By constructing a bijection $f : U \to U$ between such permutations, as was represented in Section 5.1, we obtain the same situation with only one exception. In fact, if the set X has $2n$ elements, $X = \{i_1, \ldots, i_{2n}\}$, then S has the single element $S = \{x\}$. If $x < \pi_1$, where $\pi = (\pi_1, \pi_2, \ldots, \pi_{2n})$, $\pi_1 < \pi_2 > \pi_3 \cdots < \pi_{2n}$, is some up-down permutation on X, the algorithm works. But if $x > \pi_1$, the algorithm gives a permutation $(x, \pi_1, \ldots, \pi_{2n})$, with $x > \pi_1 < \pi_2 > \cdots < \pi_{2n}$. It is a down-up permutation of odd length $2n+1$, containing all the considered elements $\{1, 2, \ldots, 2n+1\}$. As the number of up-down permutations on a set is equal to the numbers of down-up permutations on the same set, we can state that the number of such exceptions is A_{2n+1}, and the contribution of each exceptional permutation is $(-1)^n$. So, we have proven that

$$(-1)^n A_{2n+1} = \binom{2n+1}{2n}(-1)^n A_{2n} + \binom{2n+1}{2n-2}(-1)^{n-1} A_{2n-2} + \cdots + \binom{2n+1}{0} A_0, \quad \text{or}$$

$$A_{2n+1} = (-1)^n \left(\binom{2n+1}{2n} E_{2n} + \binom{2n+1}{2n-2} E_{2n-2} + \binom{2n+1}{2n-4} E_{2n-4} + \cdots + \binom{2n+1}{0} E_0 \right). \qquad \square$$

For example, $A_3 = -\left(\binom{3}{2}E_2 + \binom{3}{0}E_0\right) = -(3(-1)+1) = 2$, while

$$A_5 = \binom{5}{4}E_4 + \binom{5}{2}E_2 + \binom{5}{0}E_0 = 5 \cdot 5 + 10 \cdot (-1) + 1 \cdot 1 = 16.$$

5.4.3. The last formula allows us to connect the Euler numbers and the Bernoulli numbers.

We have proven that

$$B_{2n} = (-1)^{n-1} \frac{2n}{4^{2n} - 2^{2n}} A_{2n-1}, \quad n \geq 1.$$

As $(-1)^{n-1}A_{2n-1} = \sum_{k=0}^{n-1}\binom{2n-1}{2k}E_{2k}$, we obtain the identity

$$B_{2n} = \frac{2n}{4^{2n} - 2^{2n}} \sum_{k=0}^{n-1} \binom{2n-1}{2k} E_{2k}, \quad n \geq 1.$$

However, the Euler and Bernoulli numbers have many other natural connections.

First of all, both sequences give zeros for odd indices greater the unity:

$$B_{2n+1} = E_{2n+1} = 0, \quad n \geq 1.$$

Both sequences have alternating signs for the entries with even indices:

$$B_{4n} < 0 < E_{4n}, \quad B_{4n+2} > 0 > E_{4n}.$$

These properties are closely connected with the form of exponential functions of the considered sequences; both functions are even and depend on e^x:

$$\frac{2}{e^x + e^{-x}} = \frac{2e^x}{e^{2x}+1} = \sum_{n=0}^{\infty} E_n \frac{x^n}{n!}, \quad \frac{x}{e^x - 1} = \sum_{n=0}^{\infty} B_n \frac{x^n}{n!}.$$

The Bernoulli and Euler numbers, B_n and E_n, are the particular values $B_n(0)$ and $2^n E_n\left(\frac{1}{2}\right)$ of the Bernoulli and Euler polynomials, $B_n(x)$ and $E_n(x)$, which are usually defined via their exponential generating functions:

$$\frac{ze^{xz}}{e^z - 1} = \sum_{n=0}^{\infty} B_n(x)\frac{z^n}{n!}, \quad |z| < 2\pi, \quad \frac{2e^{xz}}{e^z+1} = \sum_{n=0}^{\infty} E_n(x)\frac{z^n}{n!}, \quad |z| < \pi.$$

The analysis of these polynomials leads to important identities, connecting these two classes of numbers with analytic number theory.

- *For the Riemann zeta function $\zeta(s)$ and the Dirichlet beta function $\beta(x)$,*

$$\zeta(x) = \sum_{n=0}^{\infty} \frac{1}{n^s}, \quad Re\, s \geq 1, \ \beta(s) = \sum_{n=0}^{\infty} \frac{(-1)^n}{(2n+1)^s}, \quad Re\, s > 0,$$

we have the equalities

$$\zeta(2m) = (-1)^{m+1}(2m)^{2m}\frac{B_{2m}}{2(2m)!},$$

$$\beta(2m+1) = (-1)^m\left(\frac{\pi}{2}\right)^{2m+1}\frac{E_{2m}}{2(2m)!}.$$

In particular, it follows that

$$\zeta(2) = \frac{\pi^2}{6}; \quad \zeta(4) = \frac{4}{3\cdot 5!}\pi^4; \ \zeta(6) = \frac{16}{3\cdot 7!}\pi^6;$$

$$\zeta(8) = \frac{192}{5\cdot 9!}\pi^8; \quad \zeta(10) = \frac{5\cdot 2^8}{3\cdot 11!}\pi^{10};$$

$$\beta(1) = \frac{\pi}{4}; \quad \beta(3) = \frac{1}{2^4\cdot 2!}\pi^3; \quad \beta(5) = \frac{5}{2^6\cdot 4!}\pi^5;$$

$$\beta(7) = \frac{61}{2^8\cdot 6!}\pi^7; \quad \beta(9) = \frac{1385}{2^{10}\cdot 8!}\pi^9.$$

5.4.4. The Euler numbers closely connected with counting problems in combinatorics.

By definition, the absolute value $|E_n|$ of the nth Euler number E_n gives the number of proper up-down permutations of the set $\{1, 2, \ldots, n\}$.

Furthermore, the numbers $|E_{2n}|$ are a special case of the Entringer numbers $E_{n,k}$, giving the number of alternating permutations of $\{1, 2, \ldots, n+1\}$, with the first term being $k+1$. Obviously,

$$A_n = E_{n,n}, \quad \text{and} \quad |E_{2n}| = E_{2n,2n}.$$

We define a *simsun permutation* as a permutation π of the set $\{1, 2, \ldots, n\}$ such that π has no *proper double descents* $\pi_i > \pi_{i+1} > \pi_{i+2}$, $i = 1, 2, \ldots, n-2$, and such that, for all $0 \le k \le n-1$, if we remove $n, n-1, \ldots, n-k$ from π (written as a word), then the resulting permutation also has no proper double descents.

For instance, the permutations $\pi = (1, 2, 3, 4)$ and $w = (1, 3, 2, 4)$ are simsun permutations, while $\rho = (3, 2, 41,)$ is not a simsun permutation since if we remove 4 from ρ, we obtain the permutation $(3, 2, 1)$, which has a proper double descent.

It can be shown (see [Stan97]), that:

- *the number of all simsun permutations of the set* $\{1, 2, \ldots, n\}$ *is equal to* A_{n+1}*; therefore, the number of all simsun permutations of the set* $\{1, 2, \ldots, 2n-1\}$ *is equal to* $|E_{2n}|$.

□ In fact, let $f_k(n)$ be the number of simsun permutations of $\{1, 2, \ldots, n\}$, with k descents. Then, inserting $n+1$ into a simsun permutation of $\{1, 2, \ldots, n\}$ gives the recurrence $f_k(n+1) = (n-2k+2)f_{k-1}(n)+(k+1)f_k(n)$ with the initial conditions $f_0(1) = 1$, and $f_k(n) = 0$ for $k > \lfloor \frac{n}{2} \rfloor$ (see [Stan97] for a combinatorial proof). □

5.4.5. In Chapter 2, we considered several ways to represent a permutation π of the set $\{1, 2, \ldots, n\}$ as a rooted tree.

The first method gives a bijection between the set S_n of all permutations of $\{1, 2, \ldots, n\}$ and the set of all *increasing binary trees* on n vertices, i.e., rooted binary trees with n vertices, labelled $1, 2, \ldots, n$ such that the labels along any path from the root are increasing.

The increasing binary tree $T(\pi)$, corresponding to a permutation π, can be constructed inductively. In fact, we can: represent π as $\pi = \rho 1 \delta$, where ρ and δ are permutations of some sets A_1 and A_2, forming a partition (A_1, A_2) of the set $\{2, 3, \ldots, n\}$; choose 1 as the root of $T(\pi)$; use the trees $T(\rho)$ and $T(\delta)$ as the left and right subtrees of (π). Here, the tree $T(\rho)$ (as well as $T(\delta)$) is constructed using the same algorithm if we choose as the root of $T(\rho)$ the least element of ρ; $T(\emptyset) = \emptyset$. For more details, see Chapter 2.

By the construction, the tree $T(\pi)$, corresponding to a permutation $\pi = (\pi_1, \pi_2, \ldots, \pi_n)$, has the following properties (for convenience, consider the permutation π as the string $(0, \pi_1, \pi_2, \ldots, \pi_n, 0)$, i.e., set $\pi_0 = \pi_{n+1} = 0$):

- a vertex i of $T(\pi)$ has precisely the successor below right if $\pi_{i-1} < \pi_i < \pi_{i+1}$ (the element i is a *double ascent*);
- a vertex i of $T(\pi)$ has precisely the successor below left if $\pi_{i-1} > \pi_i > \pi_{i+1}$ (the element i is a *double descent*);
- a vertex i of $T(\pi)$ has precisely the successors below left and below right if $\pi_{i-1} > \pi_i < \pi_{i+1}$ (the element i is a *valley*);

- a vertex i of $T(\pi)$ has no successors if $\pi_{i-1} < \pi_i > \pi_{i+1}$ (the element i is a *peak*).

So, we can state that:

- *there are exactly $n!$ rooted binary trees on n vertices,* and
- *the number of complete (i.e., every vertex is either an endpoint or has two successors) increasing binary trees with $2n+1$ vertices is equal to the number A_{2n+1}.*

5.4.6. An *even tree* is a rooted tree such that every vertex has an even number of children. Obviously, every such tree must have an odd number of vertices.

It can be proven (see [Stan97] and [KPP94]) that:

- *the number of even trees on $2n+1$ vertices is equal to $A_{2n} = |E_{2n}|$.*

Moreover, since there are no even trees on $2n$ vertices, we can state that:

- *for any positive integer n, the number of even trees on n vertices is equal to $|E_n|$.*

5.4.7. An *Euler number prime* is an Euler number E_n such that the absolute value $|E_n|$ is prime. These numbers are distinct from a different type of primes known as *Euler primes* (see *lucky numbers of Euler* in Section 5.7).

The first few Euler number primes E_n occur for $n = 4, 6, 38, 454, 510, \dots$ (sequence A103234 in the OEIS), corresponding to $5, -61, -2348958052704310825201782857619894774 1, \dots$ (sequence A092823 in the OEIS).

As of April 2025, the largest known Euler number prime is $|E_{510}|$, which has 10^{62} decimal digits; it was proven to be prime by David Broadhurst in 2002 (see [Weis25]). There are no larger Euler number primes up to a limit of at least $n = 100000$.

5.4.8. Euler numbers have an interesting divisibility property. It is proven (see [Zhan98]) that:

- *for any prime p, it holds that*

$$(-1)^{\frac{p-1}{2}}|E_{p-1}| \equiv \begin{cases} 0 \ (mod\ p), & \text{if } p \equiv 1(mod\ 4); \\ -2 \ (mod\ p), & \text{if } p \equiv 3(mod\ 4). \end{cases}$$

As a simple corollary of this theorem, we get that

$$|E_{2n+2}| + |E_{2n}| \equiv 0(mod\ 6),$$
$$|E_{2n+4}| + 10|E_{2n+2}| + 9|E_{2n}| \equiv 0(mod\ 24),$$
$$|E_{2n+6}| + |E_{2n}| \equiv 0(mod\ 42).$$

The proof can be obtained by studying the behaviour of the kth derivatives of the function $\sec x = \sum_{n=0}^{\infty} |E_{2n}| \frac{(2x)^n}{(2n)!}$ and their connections with the so-called central factorial numbers (see [Rior79] and [Zhan98]).

5.4.9. Euler numbers grow very quickly. It can be proved that they have the lower bound

$$|E_{2n}| > 8\sqrt{\frac{n}{\pi}}\left(\frac{4n}{\pi e}\right)^{2n}.$$

Moreover, for large n, this lower bound gives an asymptotic of the numbers $|E_{2n}|$:

$$|E_{2n}| \sim 8\sqrt{\frac{n}{\pi}}\left(\frac{4n}{\pi e}\right)^{2n}.$$

Exercises

1. Check the identity

$$E_{2n} = (-1)^n (2n)! \begin{vmatrix} \frac{1}{2!} & 1 & & & \\ \frac{1}{4!} & \frac{1}{2!} & 1 & & \\ \vdots & & \ddots & \ddots & \\ \frac{1}{(2n-2)!} & \frac{1}{(2n-4)!} & & \frac{1}{2!} & 1 \\ \frac{1}{(2n)!} & \frac{1}{(2n-2)!} & \cdots & \frac{1}{4!} & \frac{1}{2!} \end{vmatrix}$$

for small values of n. Prove this identity.

2. Prove the symbolic formulas $E_{2n} = \frac{(4B-1)^{2n+1}}{2n+1}$ and $B_{2n} = \frac{2n(E+1)^{2n-1}}{4^n(4^n-1)}$. Here, E^k and B^k should be read as the numbers E_k and B_k, respectively.
3. Let $F(2k+1) = 1 - \frac{1}{3^{2k+1}} + \cdots + (-1)^n \frac{1}{(2n+1)^{2k+1}} + \cdots$. Prove that $F(2k+1) = \frac{(-1)^k \pi^{2k+1}}{4^{k+1}(2k)!} E_{2k}$. Show that the series $1 - \frac{1}{3} + \cdots + (-1)^n \frac{1}{2n+1} + \cdots$ converges to $\frac{\pi}{4}$. Obtain the same result using the decomposition of $\arctan x$ in the Taylor series.
4. Using the fact that all partial sums of the series $F(2k+1)$ are positive, prove that the sequence E_{2n}, $n = 0, 1, 2, \ldots$, has alternating signs.
5. Using the asymptotic property $F(2k+1) \to 1$ for $k \to \infty$, prove that for large n the following asymptotic formula holds: $|E_{2n}| = 4\sqrt{e}\left(\frac{4n}{\pi e}\right)^{\frac{2n+1}{2}}(1 + o(1))$.

5.5. Polynomials Related to Euler Numbers

5.5.1. The Euler numbers are closely connected with the *Euler polynomials* $E_n(x)$, $n = 0, 1, 2, \ldots$, which are usually determined using their the exponential generating function $\frac{2e^{xz}}{e^z+1}$:

$$\frac{2e^{xz}}{e^z+1} = E_0(x) + \frac{z}{1!}E_1(x) + \frac{z^2}{2!}E_2(x) + \frac{z^3}{3!}E_3(x) + \cdots + \frac{z^n}{n!}E_n(x) + \cdots .$$

The first Euler polynomials have the form

$$E_0(x) \equiv 1, \quad E_1(x) = x - \frac{1}{2}, \quad E_2(x) = x^2 - x,$$

$$E_3(x) = x^3 - \frac{3}{2}x^2 + \frac{1}{4}, \quad E_4(x) = x^4 - 2x^3 + x,$$

$$E_5(x) = x^5 - \frac{5}{2}x^4 + \frac{5}{2}x^2 - \frac{1}{2}, \quad E_6(x) = x^6 - 3x^5 + 5x^3 - 3x.$$

Taking $x = \frac{1}{2}$, we get on the left-hand side of the equation the function $\frac{2e^{\frac{z}{2}}}{e^z+1} = \frac{2}{e^{\frac{z}{2}}+e^{-\frac{z}{2}}}$:

$$\frac{2}{e^{\frac{z}{2}}+e^{-\frac{z}{2}}} = \sum_{n=0}^{\infty} \frac{E_n\left(\frac{1}{2}\right)}{n!} z^n.$$

On the other hand, it holds that

$$\frac{2}{e^{\frac{z}{2}}+e^{-\frac{z}{2}}} = \sum_{n=0}^{\infty} \frac{E_n}{n!} \cdot \left(\frac{z}{2}\right)^n = \sum_{n=0}^{\infty} \frac{E_n}{2^n n!} z^n.$$

Comparing the coefficients at z^n in the two obtained decompositions, we get that:

- *the number E_n is proportional to the value of the polynomial $E_n(x)$ at $x = \frac{1}{2}$:*

$$E_n = 2^n E_n\left(\frac{1}{2}\right).$$

5.5.2. The Euler polynomials are very close to *Bernoulli polynomials* $B_n(x)$, $n = 0, 1, 2, \ldots$, which are also determined using their exponential generating function $\frac{ze^{xz}}{e^z-1}$:

$$\frac{ze^{xz}}{e^z-1} = B_0(x) + \frac{z}{1!}B_1(x) + \frac{z^2}{2!}B_2(x) + \frac{z^3}{3!}B_3(x) + \cdots + \frac{z^n}{n!}B_n(x) + \cdots.$$

The first Bernoulli polynomials have the form

$$B_0(x) \equiv 1, \quad B_1(x) = x - \frac{1}{2}, \quad B_2(x) = x^2 - x + \frac{1}{6},$$

$$B_3(x) = x^3 - \frac{3}{2}x^2 + \frac{1}{2}x, \quad B_4(x) = x^4 - 2x^3 + x^2 - \frac{1}{30},$$

$$B_5(x) = x^5 - \frac{5}{2}x^4 + \frac{5}{3}x^3 - \frac{1}{6}x,$$

$$B_6(x) = x^6 - 3x^5 + \frac{5}{2}x^4 - \frac{1}{2}x^2 + \frac{1}{42}.$$

For $x=0$, we get that $\frac{z}{e^z-1}=\sum_{n=0}^{\infty}B_n(0)\frac{z^n}{n!}$. Using the decomposition $\frac{z}{e^z-1}=\sum_{n=0}^{\infty}B_n\frac{z^n}{n!}$ for the Bernoulli numbers and comparing the coefficients at z^n, we obtain that:

- *the number B_n is equal to the value of the polynomial $B_n(x)$ at $x=0$:*

$$B_n = B_n(0).$$

5.5.3. However, the Euler and Bernoulli numbers can be obtained as the values of corresponding polynomials at $x=1$:

- *For the sequences $E_n(x)$, $n=0,1,2,\ldots$, and $B_n(x)$, $n=0,1,2,\ldots$, of the Euler and Bernoulli polynomials, we have the identities*

$$B_n(1-x)=(-1)^nB_n(x), \quad E_n(1-x)=(-1)^nE_n(x).$$

□ In fact, it holds that $\frac{ze^{(1-x)z}}{e^z-1}=\frac{ze^ze^{-xz}}{e^z-1}=\frac{ze^{-xz}}{1-e^{-z}}=\frac{(-z)e^{-xz}}{e^{-z}-1}$ and, therefore,

$$B_0(1-x)+\frac{z}{1!}B_1(1-x)+\frac{z^2}{2!}B_2(1-x)+\cdots+\frac{z^n}{n!}B_n(1-x)$$
$$+\cdots=B_0(x)+\frac{(-z)}{1!}B_1(x)+\frac{(-z)^2}{2!}B_2(x)+\frac{(-z)^3}{3!}B_3(x)$$
$$+\cdots+\frac{(-z)^n}{n!}B_n(x)+\cdots.$$

Comparing the elements at z^n in the two representations, we get $B_n(1-x)=(-1)^nB_n(x)$. Similarly, we have that $\frac{2e^{(1-x)z}}{e^z+1}=\frac{2e^ze^{-xz}}{e^z+1}=\frac{2e^{-xz}}{e^{-z}+1}$; therefore, it holds that

$$E_0(1-x)+\frac{z}{1!}E_1(1-x)+\frac{z^2}{2!}E_2(1-x)+\cdots+\frac{z^n}{n!}E_n(1-x)$$
$$+\cdots=E_0(x)+\frac{(-z)}{1!}E_1(x)+\frac{(-z)^2}{2!}E_2(x)+\frac{(-z)^3}{3!}E_3(x)$$
$$+\cdots+\frac{(-z)^n}{n!}E_n(x)+\cdots.$$

Comparing the elements at z^n in the two representations, we get $E_n(1-x)=(-1)^nE_n(x)$. □

In the context of our consideration, the last property means that $B_n(1) = (-1)^n B_n(0) = (-1)^n B_n$. As all Bernoulli numbers with odd indices greater than 1 are equal to zero, we can state that

$$B_n(1) = B_n^+.$$

On the other hand, for $x = \frac{1}{2}$, we have that $1 - x = \frac{1}{2}$, i.e., $A_n\left(\frac{1}{2}\right) = \frac{E_n}{2^n} = (-1)^n A_n\left(\frac{1}{2}\right)$. This fact completely agrees with the properties of Euler numbers: $E_{2k+1} = 0$.

5.5.4. By expanding the left-hand sides of the decompositions for $E_n(x)$ and $B_n(x)$, we can obtain recurrent relations for the Euler and Bernoulli polynomials.

- *For the sequences* $E_n(x)$, $n = 0, 1, 2, \ldots$, *and* $B_n(x)$, $n = 0, 1, 2, \ldots$, *of the Euler and Bernoulli polynomials, we have the following recurrent relations:*

$$B_n(x) = x^n - \frac{1}{n+1}\left(\binom{n+1}{0}B_0(x) + \binom{n+1}{1}B_1(x) + \cdots + \binom{n+1}{n-1}B_{n-1}(x)\right);$$

$$E_n(x) = x^n - \frac{1}{2}\left(\binom{n}{0}E_0(x) + \binom{n}{1}E_1(x) + \cdots + \binom{n}{n-1}E_{n-1}(x)\right).$$

□ Using the decomposition $e^z = 1 + \frac{z}{1!} + \frac{z^2}{2!} + \frac{z^3}{3!} + \cdots + \frac{z^n}{n!} + \cdots$, we get that $e^z - 1 = \frac{z}{1!} + \frac{z^2}{2!} + \frac{z^3}{3!} + \cdots + \frac{z^n}{n!} + \cdots$, and $ze^{xz} = z + \frac{x}{1!}z^2 + \frac{x^2}{2!}z^3 + \cdots + \frac{x^n}{n!}z^{n+1} + \cdots$. Now, we can state that

$$ze^{xz} = (e^z - 1)\left(B_0(x) + \frac{z}{1!}B_1(x) + \frac{z^2}{2!}B_2(x) + \frac{z^3}{3!}B_3(x) + \cdots + \frac{z^n}{n!}B_n(x) + \cdots\right);$$

therefore, it holds that

$$z+\frac{x}{1!}z^2+\frac{x^2}{2!}z^3+\frac{x^3}{3!}z^4+\cdots+\frac{x^n}{n!}z^{n+1}+\cdots$$
$$=\left(\frac{z}{1!}+\frac{z^2}{2!}+\cdots+\frac{z^n}{n!}+\cdots\right)\left(B_0(x)+\frac{z}{1!}B_1(x)+\frac{z^2}{2!}B_2(x)\right.$$
$$\left.+\cdots+\frac{z^n}{n!}B_n(x)+\cdots\right)$$
$$=z\left(\frac{B_0(x)}{1!0!}\right)+z^2\left(\frac{B_0(x)}{2!0!}+\frac{B_1(x)}{1!1!}\right)$$
$$+z^3\left(\frac{B_0(x)}{3!0!}+\frac{B_1(x)}{2!1!}+\frac{B_2(x)}{1!2!}\right)+\cdots+z^{n+1}\left(\frac{B_0(x)}{(n+1)!0!}\right.$$
$$\left.+\frac{B_1(x)}{n!1!}+\frac{B_2(x)}{(n-1)!2!}+\cdots+\frac{B_n(x)}{0!n!}\right)+\cdots.$$

Comparing the coefficients at z^n on both sides of the equality, we get that

$$\frac{x^n}{n!}=\frac{B_0(x)}{(n+1)!0!}+\frac{B_1(x)}{n!1!}+\frac{B_2(x)}{(n-1)!2!}+\cdots+\frac{B_n(x)}{1!(n+1)!},\quad \text{or}$$

$$(n+1)x^n=\binom{n+1}{0}B_0(x)+\binom{n+1}{1}B_1(x)$$
$$+\binom{n+1}{2}B_2(x)+\cdots+\binom{n+1}{n}B_n(x).$$

Similarly, on the one hand, it holds that

$$2e^{xz}=2+\frac{2x}{1!}z+\frac{2x^2}{2!}z^2+\frac{2x^3}{3!}z^3+\cdots+\frac{2x^n}{n!}z^n+\cdots,$$

and on the other hand, it holds that

$$2e^{xz}=(e^z+1)\left(E_0(x)+\frac{z}{1!}E_1(x)+\frac{z^2}{2!}E_2(x)+\frac{z^3}{3!}E_3(x)\right.$$
$$\left.+\cdots+\frac{z^n}{n!}E_n(x)+\cdots\right)$$
$$=2E_0(x)+z\left(\frac{E_0(x)}{0!1!}+\frac{2E_1(x)}{1!0!}\right)$$

$$+ z^2 \left(\frac{E_0(x)}{0!2!} + \frac{E_1(x)}{1!1!} + \frac{2E_2(x)}{2!0!} \right) + \cdots + z^n \left(\frac{E_0(x)}{0!n!} \right.$$

$$\left. + \frac{E_1(x)}{1!(n-1)!} + \frac{E_2(x)}{2!(n-2)!} + \cdots + \frac{2E_n(x)}{n!0!} \right) + \cdots .$$

Comparing the coefficients at z^n on both sides of the equality, we get that

$$\frac{2x^n}{n!} = \frac{E_0(x)}{0!n!} + \frac{E_1(x)}{1!(n-1)!} + \frac{E_2(x)}{2!(n-2)!} + \cdots + \frac{2B_n(x)}{n!0!}, \quad \text{or}$$

$$2x^n = \binom{n}{0} E_0(x) + \binom{n}{1} E_1(x) + \binom{n}{2} E_2(x) + \cdots + 2\binom{n}{n} E_n(x).$$

So, it holds that

$$E_n(x) = x^n - \frac{1}{2} \left(\binom{n}{0} E_0(x) + \binom{n}{1} E_1(x) + \binom{n}{2} E_2(x) \right.$$

$$\left. + \cdots \binom{n}{n-1} E_{n-1}(x) \right). \qquad \square$$

5.5.5. Using similar arguments, we can obtain the representations of the Euler and Bernoulli polynomials using the Euler and Bernoulli numbers, respectively.

- *For the sequences* $E_n(x)$, $n = 0, 1, 2, \ldots$, *and* $B_n(x)$, $n = 0, 1, 2, \ldots$, *of Euler and Bernoulli polynomials, there are the following representations:*

$$B_n(x) = \sum_{k=0}^{n} \binom{n}{k} B_{n-k} x^k, \quad B_0(x) \equiv 1;$$

$$E_n(x) = \sum_{k=0}^{n} \binom{n}{k} E_{n-k} 2^{k-n} \left(x - \frac{1}{2} \right)^k, \quad E_0(x) \equiv 1.$$

□ In fact, as $\frac{t}{e^z-1} = B_0 + \frac{z}{1!}B_1 + \frac{z^2}{2!}B_2 + \frac{z^3}{3!}B_3 + \cdots + \frac{z^n}{n!}B_n + \cdots$ and $e^{xz} = 1 + \frac{z}{1!}x + \frac{z^2}{2!}x^2 + \frac{z^3}{3!}x^3 + \cdots + \frac{z^n}{n!}x^n + \cdots$, we get that

$$\frac{ze^{xz}}{e^z-1} = \left(B_0 + \frac{z}{1!}B_1 + \frac{z^2}{2!}B_2 + \cdots + \frac{z^n}{n!}B_n + \cdots\right)$$
$$\cdot \left(1 + \frac{z}{1!}x + \frac{z^2}{2}x^2 + \cdots + \frac{z^n}{n!}x^n + \cdots\right)$$
$$= B_0 + t\left(\frac{B_0}{0!1!}x + \frac{B_1}{1!0!}\right) + \cdots + t^n\left(\frac{B_0}{0!n!}x^n + \frac{B_1}{1!(n-1)!}x\right.$$
$$\left. + \cdots + \frac{B_n}{n!0!}\right) + \cdots .$$

Comparing the coefficients at z^n in the two decompositions of $\frac{ze^{xz}}{e^z-1}$, we get that

$$\frac{B_n(x)}{n!} = \frac{B_0}{0!n!}x^n + \frac{B_1}{1!(n-1)!}x + \cdots + \frac{B_n}{n!0!},$$

or, in other words, that

$$B_n(x) = \frac{n!}{0!n!}B_0x^n + \frac{n!}{1!(n-1)!}B_1x + \cdots + \frac{n!}{n!0!}$$
$$B_n = \binom{n}{0}B_0x^n + \binom{n}{1}B_1x^{n-1} + \cdots + \binom{b}{n}B_n.$$

Similarly, on the one hand, it holds that

$$\frac{2e^{(x+\frac{1}{2})z}}{e^z+1} = E_0\left(x+\frac{1}{2}\right) + \frac{z}{1!}E_1\left(x+\frac{1}{2}\right) + \frac{z^2}{2!}E_2\left(x+\frac{1}{2}\right)$$
$$+ \cdots + \frac{z^n}{n!}E_n\left(x+\frac{1}{2}\right) + \cdots ;$$

on the other hand, we have that $\frac{2e^{(x+\frac{1}{2})z}}{e^z+1} = \frac{2}{e^{\frac{z}{2}}+e^{-\frac{z}{2}}}$, and

$$\frac{2}{e^{\frac{z}{2}}+e^{-\frac{z}{2}}}e^{xz} = \left(E_0 + \frac{\left(\frac{z}{2}\right)}{1!}E_1 + \frac{\left(\frac{z}{2}\right)^2}{2!}E_2 + \cdots + \frac{\left(\frac{z}{2}\right)^n}{n!}E_n + \cdots\right)$$
$$\cdot\left(1 + \frac{z}{1!}x + \frac{z^2}{2}x^2 + \cdots + \frac{z^n}{n!}x^n + \cdots\right)$$
$$= E_0 + z\left(\frac{E_0}{0!1!}x + \frac{\frac{E_1}{2}}{1!0!}\right)$$
$$+ \cdots + z^n\left(\frac{E_0}{0!n!}x^n + \frac{\frac{E_1}{2}}{1!(n-1)!}x + \cdots + \frac{\frac{E_n}{2^2}}{n!0!}\right) + \cdots.$$

Comparing the coefficients at z^n in the two decompositions of $\frac{2e^{(x+\frac{1}{2})z}}{e^z+1}$, we get that

$$\frac{E_n\left(x+\frac{1}{2}\right)}{n!} = \frac{E_0}{0!n!}x^n + \frac{\frac{E_1}{2}}{1!(n-1)!}x + \cdots + \frac{\frac{E_n}{2^n}}{n!0!},$$

or, in other words, that

$$E_n\left(x+\frac{1}{2}\right) = \frac{n!}{0!n!}\frac{E_0}{2^0}x^n + \frac{n!}{1!(n-1)!}\frac{E_1}{2^1}x + \cdots + \frac{n!}{n!0!}\frac{E_n}{2^n}$$
$$= \binom{n}{0}\frac{E_0}{2^0}x^n + \binom{n}{1}\frac{E_1}{2^1}x^{n-1} + \cdots + \binom{n}{n}\frac{E_n}{2^n}.$$

Going from $x+\frac{1}{2}$ to x, we obtain that

$$E_n(x) = \binom{n}{0}\frac{E_0}{2^0}\left(x-\frac{1}{2}\right)^n + \binom{n}{1}\frac{E_1}{2^1}\left(x-\frac{1}{2}\right)^{n-1} + \cdots + \binom{n}{n}\frac{E_n}{2^n}.$$

□

5.5.6. The last decompositions allows us to obtain simple properties of the derivatives of the considered polynomials:

- *For the Euler and Bernoulli polynomials, it holds that*

$$B_n'(x) = nB_{n-1}(x), \; E_n'(x) = nE_{n-1}(x), \quad n \geq 1.$$

□ These properties are a direct consequence of the property of binomial coefficients:

$$k\binom{n}{k} = \frac{n!k}{k!(n-k)!} = \frac{n(n-1)!}{(k-1)!(n-k)!} = n\binom{n-1}{k-1}, \quad 1 \le k \le n.$$

In fact, we get that

$$\begin{aligned} B_n'(x) &= \sum_{k=1}^{n} k\binom{n}{k} B_{n-k}x^{k-1} = n\sum_{k=1}^{n}\binom{n-1}{k-1}B_{n-k}x^{k-1} \\ &= n\sum_{j=0}^{n-1}\binom{n-1}{j}B_{n-1-j}x^j = nB_{n-1}(x). \end{aligned}$$

Similarly, it holds that

$$\begin{aligned} E_n'(x) &= \sum_{k=1}^{n} k\binom{n}{k} E_{n-k}2^{k-n}\left(x-\frac{1}{2}\right)^{k-1} \\ &= n\sum_{k=1}^{n}\binom{n-1}{k-1}E_{n-k}2^{k-n}\left(x-\frac{1}{2}\right)^{k-1} \\ &= n\sum_{j=0}^{n-1}\binom{n-1}{j}E_{n-1-j}2^{j-(n-1)}\left(x-\frac{1}{2}\right)^{j} = nE_{n-1}(x). \quad \square \end{aligned}$$

5.5.7. There exist closed formulas for the Euler and Bernoulli polynomials which use only binomial coefficients (see, for example, [Deza18]).

- *For the Euler and Bernoulli polynomials, it holds that*

$$B_n(x) = \sum_{k=0}^{n}\frac{1}{k+1}\sum_{j=0}^{k}(-1)^j\binom{k}{j}(x+j)^n,$$

$$E_n(x) = \sum_{k=0}^{n}\frac{1}{2^n}\sum_{j=0}^{k}(-1)^j\binom{k}{j}(x+j)^n.$$

5.5.8. The Bernoulli and Euler polynomials may be inverted to express the monomial in terms of the polynomials. Specifically, it follows that

$$x^n = \frac{1}{n+1}\sum_{k=0}^{n}\binom{n+1}{k}B_k(x), \quad \text{and}$$

$$x^n = E_n(x) + \frac{1}{2}\sum_{k=0}^{n-1}\binom{n}{k}E_k(x).$$

5.5.9. The *Fourier series* of the Euler and Bernoulli polynomials are given by the expansions

$$B_n(x) = -2n!\sum_{k=1}^{\infty}\frac{\cos\left(2k\pi x - \frac{n\pi}{2}\right)}{(2k\pi)^n};$$

$$\frac{(-1)^n}{4(2n)!}\pi^{2n+1}E_{2n}(x) = \sum_{k=0}^{\infty}\frac{\sin((2k+1)\pi x)}{(2k+1)^{2n+1}},$$

$$\frac{(-1)^n}{4(2n-1)!}\pi^{2n}E_{2n-1}(x) = \sum_{k=0}^{\infty}\frac{\cos((2k+1)\pi x)}{(2k+1)^{2n}}.$$

We use these decompositions in the following section.

Exercises

1. For small values of n, check and then prove that

$$(-1)^n B_n(-x) = B_n(x) + nx^{n-1},$$
$$(-1)^n E_n(-x) = -E_n(x) + 2x^n.$$

2. For small values of n, check and then prove that $B_n\left(\frac{1}{2}\right) = \left(\frac{1}{2^{n-1}} - 1\right)B_n$.
3. Prove that $B_n(x+y) = \sum_{k=0}^{n}\binom{n}{k}B_k(x)y^{n-k}$, $E_n(x+y) = \sum_{k=0}^{n}\binom{n}{k}E_k(x)y^{n-k}$. These identities are also equivalent to stating that these polynomial sequences are Appell sequences; Hermite polynomials are another example.

4. For small positive integers n and m, check and then prove the *multiplication theorems*:

$$B_n(mx) = m^{n-1} \sum_{k=0}^{m-1} B_n\left(x + \frac{k}{m}\right);$$

$$E_n(mx) = m^n \sum_{k=0}^{m-1} (-1)^k E_n\left(x + \frac{k}{m}\right) \quad \text{for odd} \quad m;$$

$$E_n(mx) = \frac{-2}{n+1} m^n \sum_{k=0}^{m-1} (-1)^k B_{n+1}\left(x + \frac{k}{m}\right) \quad \text{for even} \quad m.$$

5. For small positive integers n, check and then prove the formulas

$$\int_a^x E_n(t)\, dt = \frac{E_{n+1}(x) - E_{n+1}(a)}{n+1},$$

$$\int_a^x B_n(t)\, dt = \frac{B_{n+1}(x) - B_{n+1}(a)}{n+1}.$$

6. For small positive integers n and m, check and then prove that

$$\int_0^1 B_n(t) B_m(t)\, dt = (-1)^{n-1} \frac{m!\, n!}{(m+n)!} B_{n+m},$$

$$\int_0^1 E_n(t) E_m(t)\, dt = (-1)^n 4(2^{m+n+2} - 1) \frac{m!\, n!}{(m+n+2)!} B_{n+m+2}.$$

5.6. Problems Related to Euler Numbers

5.6.1. A famous application of the Euler numbers is the explicit evaluation of the *Dirichlet beta function* $\beta(s)$ at the odd integer values of the argument s. This result is closely connected with a similar result concerning the Riemann zeta function $\zeta(s)$ and the Bernoulli numbers: using the Bernoulli numbers, the explicit evaluation of $\zeta(s)$ at the even integer values of the argument s can be obtained.

5.6.2. The *Riemann zeta function* (or the *Euler–Riemann zeta function*) *zeta*(s) is a function of a complex variable $s = \sigma + it$, defined as

$$\zeta(s) = \sum_{n=1}^{\infty} \frac{1}{n^s}, \quad \text{for} \quad Re \quad s = \sigma > 1.$$

The Riemann zeta function plays an important role in analytic number theory, and has applications in physics, probability theory, and applied statistics.

Euler first introduced and studied this function over the real numbers in the first half of the 17th century. In 1737, he discovered the connection between the zeta function and the prime numbers and proved the identity (now called *Euler's product formula*)

$$\sum_{n=1}^{\infty} \frac{1}{n^s} = \prod_{p \text{ prime}} \frac{1}{1 - p^{-s}},$$

where the infinite product on the right-hand side extends over all prime numbers.

Bernhard Riemann ([Riem59]) extended the Euler definition of a complex variable and proved its meromorphic continuation and functional equation

$$\zeta(s) = 2^s \pi^{s-1} \sin\left(\frac{\pi s}{2}\right) \Gamma(1-s)\, \zeta(1-s),$$

where $\Gamma(s)$ is the *gamma function*. Moreover, he established a relationship between the zeros of the zeta function and the distribution of the prime numbers.

The series $\sum_{n=1}^{\infty} \frac{1}{n^s}$ is a *Dirichlet series* that converges absolutely to an analytic function for s, with $Re\ s = \sigma > 1$, and diverges for all other values of s. The functional equation shows that the function, defined by the series on the half-plane of convergence, can be continued analytically to all complex values $s \neq 1$. For $s = 1$, the series is the harmonic series which diverges to $+\infty$, and $\lim_{s \to 1}(s-1)\zeta(s) = 1$. Thus, the Riemann zeta function is a meromorphic function on the whole complex plane, which is holomorphic everywhere except for a simple pole at $s = 1$ with residue 1.

The values of the Riemann zeta function at even positive integers were computed by Euler. They are connected with Bernoulli numbers. The values at negative integer points, also found by Euler, are rational numbers and connected with Bernoulli sequence, too. They play an important role in the theory of modular forms.

5.6.3. The *Dirichlet beta function* (or *Catalan beta function*) $\beta(s)$ is a number-theoretic function of a complex variable $s = \sigma + it$, closely related to the Riemann zeta function. The Dirichlet beta function is defined as

$$\beta(s) = \sum_{n=0}^{\infty} \frac{(-1)^n}{(2n+1)^s}, \quad \text{for} \quad Re \quad s = \sigma > 0.$$

This series can also be factorised as an *Euler's product*, leading to the general idea of the Dirichlet character and giving a tool to obtain the results involving the distribution of the prime numbers in arithmetic progressions:

$$\beta(s) = \prod_{p \equiv 1 \bmod 4} \frac{1}{1 - p^{-s}} \prod_{p \equiv 3 \bmod 4} \frac{1}{1 + p^{-s}}.$$

The Dirichlet beta function is a homomorphic function on the whole complex plane due to the functional equation

$$\beta(1-s) = \left(\frac{\pi}{2}\right)^{-s} \sin\left(\frac{\pi}{2}s\right) \Gamma(s)\beta(s),$$

extending the beta function to the left-hand side of the complex plane.

The values of the Dirichlet beta function at odd positive integers are connected with the Euler numbers. The values at negative integer points are rational numbers, also connected with the Euler sequence.

5.6.4. It turns out that values of two number-theoretic functions, the Riemann zeta function $\zeta(s)$ and the Dirichlet beta function $\beta(s)$, at some integer values of argument s can be obtained using the Euler and Bernoulli numbers.

- *For an integer n, it holds that*

$$\zeta(2n) = (-1)^{n+1}(2\pi)^{2n}\frac{B_{2n}}{2(2n)!}, \quad n \geq 1; \quad \zeta(-n) = -\frac{B_{n+1}}{n+1},$$

$$n \geq 1; \quad \zeta(0) = -\frac{1}{2};$$

$$\beta(2n+1) = (-1)^n \left(\frac{\pi}{2}\right)^{2n+1} \frac{E_{2n}}{2(2n)!}, \quad n \geq 0; \quad \beta(-n) = \frac{1}{2}E_n,$$
$$n \geq 1; \quad \beta(0) = \frac{1}{2}.$$

In particular, it follows that

$$\zeta(2) = \frac{\pi^2}{6}; \quad \zeta(4) = \frac{4}{3\cdot 5!}\pi^4; \quad \zeta(6) = \frac{16}{3\cdot 7!}\pi^6;$$
$$\zeta(8) = \frac{192}{5\cdot 9!}\pi^8; \quad \zeta(10) = \frac{5\cdot 2^8}{3\cdot 11!}\pi^{10};$$
$$\beta(1) = \frac{\pi}{4}; \quad \beta(3) = \frac{1}{2^4\cdot 2!}\pi^3; \quad \beta(5) = \frac{5}{2^6\cdot 4!}\pi^5;$$
$$\beta(7) = \frac{61}{2^8\cdot 6!}\pi^7; \quad \beta(9) = \frac{1385}{2^{10}\cdot 8!}\pi^9.$$

□ The results for negative values of the argument hold from the functional equations for the functions $\zeta(s)$ and $\beta(s)$ if the results for positive integer values of the argument are known.

In turn, the results for positive values of the argument can be obtained using the Fourier expansions for the Euler and Bernoulli polynomials:

$$B_n(x) = -2n!\sum_{k=1}^{\infty} \frac{\cos\left(2k\pi x - \frac{n\pi}{2}\right)}{(2k\pi)^n};$$
$$\frac{(-1)^n}{4(2n)!}\pi^{2n+1}E_{2n}(x) = \sum_{k=0}^{\infty} \frac{\sin((2k+1)\pi x)}{(2k+1)^{2n+1}}.$$

In fact,

$$B_{2n} = B_{2n}(0) = -2(2n)!\sum_{k=1}^{\infty} \frac{\cos n\pi}{(2k\pi)^{2n}} = \frac{(-1)^{n+1}2(2n)!}{(2\pi)^{2n}}\sum_{k=1}^{\infty} \frac{1}{k^{2n}}$$
$$= \frac{(-1)^{n+1}2(2n)!}{(2\pi)^{2n}}\zeta(2n);$$

this equality gives the first result of the statement. Similarly,

$$\frac{(-1)^n}{4(2n)!}\pi^{2n+1}\frac{E_{2n}}{2^{2n}} = \frac{(-1)^n}{4(2n)!}\pi^{2n+1}E_{2n}\left(\frac{1}{2}\right) = \sum_{k=0}^{\infty}\frac{\sin\left((2k+1)\frac{\pi}{2}\right)}{(2k+1)^{2n+1}}$$

$$= \sum_{k=0}^{\infty}\frac{(-1)^k}{(2k+1)^{2n+1}} = \beta(2n+1);$$

this equality gives the first result of the statement. □

5.6.5. However, we would like to consider the idea of Euler's original proof. Thus, let us use the decomposition of $\sin x$ into an infinite product ([Fich01]):

$$\sin x = x\left(1-\frac{x^2}{\pi^2}\right)\left(1-\frac{x^2}{4\pi^2}\right)\cdot\dots\cdot\left(1-\frac{x^2}{m^2\pi^2}\right)\cdot\dots\,.$$

By applying logarithm to both parts of the equality, we get that

$$\log\sin x = \log x + \log\left(1-\frac{x^2}{\pi^2}\right) + \log\left(1-\frac{x^2}{4\pi^2}\right)$$

$$+\dots+\log\left(1-\frac{x^2}{m^2\pi^2}\right)+\dots\,.$$

Using the classical decomposition $\log(1-x) = -\frac{x}{1} - \frac{x^2}{2} - \frac{x^3}{3} - \frac{x^4}{4} - \dots - \frac{x^k}{k} - \dots$, we can represent each summand $\log\left(1-\frac{x^2}{m^2\pi^2}\right)$, $m = 1, 2, 3, \dots$, in the form

$$\log\left(1-\frac{x^2}{m^2\pi^2}\right) = -\frac{x^2}{m^2\pi^2} - \frac{x^4}{2m^4\pi^4} - \frac{x^6}{3m^6\pi^6} - \frac{x^8}{4m^8\pi^8}$$

$$-\dots-\frac{x^{2k}}{km^{2k}}\pi^{2k} - \dots\,.$$

So, it holds that

$$\log\sin x = \log x - \frac{x^2}{\pi^2}\left(\frac{1}{1^2}+\frac{1}{2^2}+\dots+\frac{1}{m^2}+\dots\right)$$

$$-\frac{x^4}{2\pi^4}\left(\frac{1}{1^4}+\frac{1}{2^4}+\dots+\frac{1}{m^4}+\dots\right)$$

$$
\begin{aligned}
&- \frac{x^6}{3\pi^6}\left(\frac{1}{1^6}+\frac{1}{2^6}+\cdots+\frac{1}{m^6}+\cdots\right)\cdots \\
&- \frac{x^{2k}}{2\pi^{2k}}\left(\frac{1}{1^{2k}}+\frac{1}{2^{2k}}+\cdots+\frac{1}{m^{2k}}+\cdots\right)-\cdots.
\end{aligned}
$$

As, by definition, $\frac{1}{1^{2k}}+\frac{1}{2^{2k}}+\cdots+\frac{1}{m^{2k}}+\cdots=\zeta(2k)$, we can rewrite the last equality in the form

$$
\begin{aligned}
\log\sin x = \log x &- \frac{x^2}{\pi^2}\zeta(2)-\frac{x^4}{2\pi^4}\zeta(4)-\frac{x^6}{3\pi^6}\zeta(6) \\
&-\cdots-\frac{x^{2k}}{k\pi^{2k}}\zeta(2k)-\cdots.
\end{aligned}
$$

As $(\log\sin x)' = \frac{\cos x}{\sin x} = \cot x$, the differentiation of both sides of the last equality gives the following decomposition of $\cot x$:

$$
\begin{aligned}
\cot x = \frac{1}{x} &- \frac{2\zeta(2)}{\pi^2}x-\frac{2\zeta(4)}{\pi^4}x^3-\frac{2\zeta(6)}{\pi^6}x^5-\frac{2\zeta(8)}{\pi^8}x^7 \\
&-\cdots-\frac{2\zeta(2k)}{\pi^{2k}}x^{2k-1}+\cdots.
\end{aligned}
$$

Using the classical decomposition

$$
\begin{aligned}
\cot x = \frac{1}{x} &- \frac{2^2 B_2}{2!}x+\frac{2^4 B_4}{4!}x^3-\frac{2^6 B_6}{6!}x^5 \\
&+\cdots+(-1)^k\frac{2^{2k} B_{2k}}{(2k)!}x^{2k-1}-\cdots
\end{aligned}
$$

and comparing the coefficients at x^{2k-1} in both representations, we get that $(-1)^k \frac{B_{2k}2^{2k}}{(2k)!} = -\frac{2\zeta(2k)}{\pi^{2k}}$. Therefore, it holds that

$$
B_{2k} = (-1)^{k-1}\frac{2(2k)!}{(2\pi)^{2k}}\zeta(2k).
$$

The obtained representation of B_{2k} is called *Euler's representation*:

$$
B_{2k} = (-1)^{k-1}\frac{2(2k)!}{(2\pi)^{2k}}\left(\frac{1}{1^{2k}}+\frac{1}{2^{2k}}+\frac{1}{3^{2k}}+\frac{1}{4^{2k}}+\cdots+\frac{1}{m^{2k}}+\cdots\right).
$$

In this case, knowing the value of B_{2k}, we can obtain the sum of the corresponding series:

$$\frac{1}{1^{2k}}+\frac{1}{2^{2k}}+\frac{1}{3^{2k}}+\frac{1}{4^{2k}}+\cdots+\frac{1}{m^{2k}}+\cdots=(-1)^{k-1}\frac{(2\pi)^{2k}}{2\,(2k)!}B_{2k}.$$

In other words, we can find the value of the Riemann zeta function at the point $2k$:

$$\zeta\,(2k)=(-1)^{k-1}\,(2\pi)^{2k}2\,(2k)!B_{2k}.$$

If $k=1$, we get that

$$\frac{1}{1^2}+\frac{1}{2^2}+\frac{1}{3^2}+\frac{1}{4^2}+\cdots+\frac{1}{m^2}+\cdots=\frac{(2\pi)^2}{2\cdot 2!}B_2=\pi^2\cdot\frac{1}{6}=\frac{\pi^2}{6}.$$

For $k=2$, it holds that

$$\frac{1}{1^4}+\frac{1}{2^4}+\frac{1}{3^4}+\frac{1}{4^4}+\cdots+\frac{1}{m^4}+\cdots=-\frac{(2\pi)^4}{2\cdot 4!}B_4=\frac{\pi^4}{3}\cdot\frac{1}{30}=\frac{\pi^2}{90},\qquad \text{etc.}$$

Exercises

1. Using the Euler numbers E_{2n}, find the following infinite sums: $\frac{1}{1^3}-\frac{1}{2^3}++\cdots+\frac{(-1)^{m+1}}{(2m+1)^3}+\cdots$; $\frac{1}{1^5}-\frac{1}{2^5}+\cdots+\frac{(-1)^{m+1}1}{(2m+1)^5}+\cdots$; $\frac{cos\pi}{3^5}+\frac{\cos 2pi}{5^5}+\cdots+\frac{\cos m\pi}{(2m+1)^5}+\cdots$.
2. Using the Bernoulli numbers B_{2n}, find the following infinite sums: $\frac{1}{1^6}+\frac{1}{2^6}+\cdots+\frac{1}{m^6}+\cdots$; $\frac{1}{1^8}+\frac{1}{2^8}+\cdots+\frac{1}{m^8}+\cdots$; $\frac{5}{3^2}+\frac{5}{4^2}+\cdots+\frac{5}{m^2}+\cdots$.
3. Find the values $\zeta(-n)$, $n=1,2,3,4$. Find the values $\beta(-n)$, $n=1,2,3,4$.
4. Using the functional equation for the Riemann zeta function $\zeta(s)$ and the formulas for $\zeta(2m)$, $m=1,2,3,4,\ldots$, prove the formulas for $\zeta(1-2m), m=1,2,3,4,\ldots$. Using the functional equation for the Dirichlet beta function $\beta(s)$ and the formulas for $\beta(2m+1)$, $m=0,1,2,3,\ldots$, prove the formulas for $\beta(-2m), m=1,2,3,4,\ldots$.
5. Prove the infiniteness of the set of prime numbers using Euler's product for $\zeta(s)$. Prove the infiniteness of the set of primes of the form $4k-1$ using Euler's product for $\beta(s)$.

5.7. Other Number Constructions Related to Euler

Euler's convenient numbers

5.7.1. In number theory, *Euler's convenient numbers* (or *idoneal numbers*, *suitable numbers*) are the positive integers D such that any integer having the only proper representation by the binary quadratic form $x^2 \pm Dy^2$ (i.e., a representation in which x is relatively prime to Dy^2) is a prime power or twice a prime power.

There are 65 convenient numbers, as found by Euler and Carl Friedrich Gauss: 1, 2, 3, 4, 5, 6, 7, 8, 9, 10, 12, 13, 15, 16, 18, 21, 22, 24, 25, 28, 30, 33, 37, 40, 42, 45, 48, 57, 58, 60, 70, 72, 78, 85, 88, 93, 102, 105, 112, 120, 130, 133, 165, 168, 177, 190, 210, 232, 240, 253, 273, 280, 312, 330, 345, 357, 385, 408, 462, 520, 760, 840, 1320, 1365, and 1848 (sequence A000926 in the OEIS).

Results derived by Peter J. Weinberger, 1973, imply that at most two other convenient numbers exist and that the above list above is complete if the generalised Riemann hypothesis holds.

So, the following question is an open mathematical problem:

- *are there 65, 66, or 67 convenient numbers?*

5.7.2. In order to specify the consideration, let us give several basic definitions.

In number theory, an *integral binary quadratic form* (or simply *quadratic form*) is a quadratic homogeneous polynomial in two variables:

$$ax^2 + bxy + cy^2, \ a, b, c \in \mathbb{Z}.$$

The *discriminant* D of a given quadratic form $ax^2 + bxy + cy^2$ is defined as

$$D = b^2 - 4ac.$$

A form of discriminant D is called *definite* if $D < 0$, *degenerate* if $D \geq 0$ is a perfect square, and *indefinite* if $D > 0$ is not a perfect square.

A definite form $ax^2+bxy+cy^2$ is called *positive definite* if $D<0$, $a>0$ and is called *negative definite* if $D<0, a<0$.

The *content* of a binary quadratic form $ax^2+bxy+cy^2$ is defined as the greatest common divisor $gcd(a,b,c)$ of its coefficients a, b, and c. A form is called *primitive* if its content is equal to 1, that is, if its coefficients are coprime: $gcd(a,b,c)=1$.

For example, the form x^2+y^2 is a primitive positive definite quadratic form with discriminant $D=-4$. On the other hand, the form $2x^2-10xy+4y^2$ is an indefinite quadratic form with content 2 and discriminant 68.

Not every integer can arise as a discriminant of an integral binary quadratic form. An integer D is called a *fundamental discriminant* if and only if it meets one of the following conditions:

- $D\equiv 1(mod\ 4)$ and is square-free;
- $D=4m$, where $m\equiv 2,3(mod\ 4)$ and is square-free.

These conditions ensure that every fundamental discriminant corresponds uniquely to a specific type of quadratic form.

The first 10 positive fundamental discriminants are $1,5,8,12,13,17,21,24,28,29,\ldots$ (sequence A003658 in the OEIS). The first 10 negative fundamental discriminants are -3, -4, -7, -8, -11, $-15,-19,-20,-23,-24,\ldots$ (sequence A003657 in the OEIS).

There exists a fundamental connection between the theory of quadratic forms and the theory of quadratic fields: an integer D is a fundamental discriminant if and only if $D=1$, or D is the discriminant of a quadratic field. For each fundamental discriminant $D\neq 1$, there exists an unique (up to isomorphism) quadratic field with D as its discriminant.

5.7.3. Two forms $ax^2+bxy+cy^2$ and $du^2+euv+fv^2$ are called *equivalent* if there exist integers α,β,γ and δ, $\alpha\delta-\beta\gamma=1$ such that $a(\alpha u+\beta v)^2+b(\alpha u+\beta v)(\gamma u+\delta v)+c(\gamma u+\delta v)^2=du^2+euv+fv^2$, i.e., if the linear transformation

$$x=\alpha u+\beta v,$$
$$y=\gamma u+\delta v,$$

translates the form $ax^2+bxy+cy^2$ into the form $du^2+euv+fv^2$.

For example, the form $5z^2 + 6zt + 2t^2$ is equivalent to the form $x^2 + y^2$ since there exists $\alpha_1 = 2$, $\beta_1 = 1$, $\gamma_1 = 1$, $\delta_1 = 1$, $\alpha_1\delta_1 - \beta_1\gamma_1 = 1$, such that the linear transformation

$$\begin{aligned} x &= 2u + v, \\ y &= u + v, \end{aligned}$$

translates the form $5z^2 + 6zt + 2t^2$ into the form $x^2 + y^2$.

Similarly, the binary form x^2+y^2 is equivalent to the binary form $29u^2 + 34uv + 10v^2$ since there exist $\alpha_2 = 2$, $\beta_2 = 1$, $\gamma_2 = 5$, $\delta_2 = 3$, and $\alpha_2\delta_2 - \beta_2\gamma_2 = 1$ such that the linear transformation

$$\begin{aligned} x &= 2u + v, \\ y &= 5u + 3v, \end{aligned}$$

translates the form $x^2 + y^2$ into the form $29u^2 + 34uv + 10v^2$.

We can easy prove that, for instance, the form x^2+y^2 is equivalent to the form $5z^2 + 6zt + 2t^2$; it is enough to use the linear transformation, defined by the matrix $\left(\begin{smallmatrix}1 & -1\\ -1 & 2\end{smallmatrix}\right)$, which is the inverse of the matrix $\left(\begin{smallmatrix}2 & 1\\ 1 & 1\end{smallmatrix}\right)$. Similarly, we can check that the forms $5z^2+6zt+2t^2$ and $29u^2 + 34uv + 10v^2$ are equivalent using the linear transformation defined by the matrix $\left(\begin{smallmatrix}9 & 5\\ 7 & 4\end{smallmatrix}\right)$, which is the product of the matrices $\left(\begin{smallmatrix}2 & 1\\ 1 & 1\end{smallmatrix}\right)$ and $\left(\begin{smallmatrix}2 & 1\\ 5 & 3\end{smallmatrix}\right)$. At last, the identical linear transformation, defined by the matrix $\left(\begin{smallmatrix}1 & 0\\ 0 & 1\end{smallmatrix}\right)$, translates any binary form into itself.

So, we have shown, that the above equivalence conditions define an equivalence relation on the set of integral quadratic forms. It follows that the quadratic forms are partitioned into equivalence classes, called *classes of quadratic forms.* A class invariant can mean either a function defined on equivalence classes of forms or a property shared by all forms in the same class.

The most important invariant is the discriminant D: all equivalent quadratic forms have the same discriminant. Moreover, the number $h(D)$ of equivalence classes of binary quadratic forms, having the discriminant D, is finite.

Gauss's class number problem consists of the determination, for a given m, of a complete list of fundamental binary quadratic-form discriminants D, such that the class number is given by $h(D) = m$.

In his monumental treatise *Disquisitiones Arithmeticae*, Gauss conjectured that:

- *the class number* $h(-D)$ *of an imaginary quadratic field with binary quadratic-form discriminant* $-D$, $D > 0$, *tends to infinity if* D *tends to infinity.*

A proof was finally given by Hans Arnold Heilbronn (1934) and Carl Ludwig Siegel (1936).

5.7.4. An integer N is said to be *represented* by the quadratic form $ax^2 + bxy + cy^2$ if there exist integers x_0 and y_0 such that $N = ax_0^2 + bx_0y_0 + cy_0^2$. The representation $N = ax_0^2 + bx_0y_0 + cy_0^2$ is called *proper* if $gcd(x_0, y_0) = 1$.

For example, the number 5 has the only representation $5 = 2^2+1^2$ as a sum of two squares (and this representation is proper); the number 25 has two such representations $25 = 5^2+0^2$ and $25 = 4^2+3^2$ (and only the second representation is proper); the number 3 has no representations as a sum of two squares; the number 9 has the only one such representation $9 = 3^2 + 0^2$, and it is non-proper.

The problem of representing integers by sums of two squares was considered in the 3rd century by Diophantus. In the 17th century, inspired by Diophantus's *Arithmetica*, Pierre de Fermat made several observations about representations by specific quadratic forms including that which is now known as *Fermat's theorem on sums of two squares*. Euler provided the first proofs of Fermat's observations and added some new conjectures about representations by specific forms without proofs.

Following Euler, we speak about representations of positive integers by primitive quadratic forms $ax^2 + by^2$, $a, b \geq 1$, $gcd(a, b) = 1$. All such forms are positive definite with the discriminant $D = -4ab$.

We consider without proofs several famous statements connected with such representations in order to give a practical illustration to the notion of *convenient numbers*.

- *If a prime* p *can be represented by the form* $ax^2 + by^2$, *then this representation is unique; moreover, it is proper.*

Therefore, if a positive integer n has two representations by the form ax^2+by^2, then it is composite. For example, all numbers n^4+4, $n \geq 2$, are composite, as they have two representations as a sum of two squares:

$$n^4 + 4 = \left(n^2\right)^2 + 2^2 = \left(n^2 - 2\right)^2 + (2n)^2 .$$

However, a composite number can have only one representation by a given quadratic dorm or have no such representations.

- *Any prime number $p \equiv 1(mod\ 4)$ can be represented as a sum of two squares (Fermat's theorem on sums of two squares).*

It means that any prime number $p \equiv 1(mod\ 4)$ has a representation by the form $x^2 + y^2$. Using the previous statement, we can be sure that this representation is unique and proper.

On the other hand, it is obvious that a prime number $p \equiv -1(mod\ 4)$ cannot be represented as a sum of two squares: as any square x^2 is congruent to 0 or to 1 modulo 4, the sum $x^2 + y^2$ of two squares is congruent to 0, 1, or 2 modulo 4, and $x^2+y^2 \not\equiv -1(mod\ 4)$.

- *For a prime number $p > 2$ and a positive integer k, if the natural power p^k of p has a proper representation as a sum of two squares, then such a representation is unique.*

Note that the number p^k can have also non-proper representations. For example, the only decomposition $9 = 3^2 + 0^2$ of the number $9 = 3^2$, $p = 3, k = 2$, is non-proper. The number $25 = 5^2$, $p = 5, k = 2$, has two decompositions: $25 = 4^2 + 3^2 = 5^2 + 0^2$, but the only first one is proper.

A similar result can be obtained for all primes $p \equiv 1(mod\ 4)$.

- *For a prime number $p \equiv 1(mod\ 4)$ and a positive integer k, the natural power p^k of p has the unique proper representation as a sum of two squares.*

As for primes $p \equiv -1(mod\ 4)$, their natural powers p^k have no proper representations by the form $x^2 + y^2$.

It turns out that any odd positive integer n having a proper representation by the form x^2+y^2 has no prime divisors $p \equiv -1(mod\ 4)$.

Moreover, if two odd coprime positive integers $n > 1$ and $m > 1$ have proper representations by the form $x^2 + y^2$, then their product mn has at least two proper representations by the form $x^2 + y^2$.

From these considerations, we obtain the following important result:

- *An odd positive integer $n > 1$ has the unique proper representation as a sum of two squares if and only if n is a natural power of a prime number p congruent to 1 modulo 4:*
$$n = p^k, p \in P, p \equiv 1 (mod\ 4), k \in \mathbb{N}.$$

The above result allows us to state that the quadratic form x^2+y^2 possesses the following property: *if the odd number n has an unique proper representation by the form $x^2 + y^2$, then n is a natural power of a prime number.*

We call a positive integer d a *convenient number* if the quadratic form $x^2 + dy^2$ possesses the same property: *if the odd number n has a unique proper representation by the form $x^2 + dy^2$, then n is a natural power of a prime number.*

So, we can state that the number $d = 1$ is a convenient number.

On the other hand, it is easy to check that the number 11 is not a convenient number: for example, a composite number 15 has the unique representation $15 = 2^2 + 11 \cdot 1^2$ by the form $x^2 + 11y^2$, and this representation is proper.

Adding to the consideration the only even prime 2, which can always be represented as
$$2 = 1^2 + d \cdot 0^2,$$
and that this decomposition is a unique and proper representation of 2 by the form $x^2 + dy^2$, we obtain the definition of convenient numbers, which was stated in the beginning of this section.

Lucky numbers of Euler

5.7.5. A *lucky number of Euler* is a positive integer number n such that the polynomial
$$x^2 - x + n$$
gives prime numbers for all $x = 1, 2, \ldots, n - 1$.

When x is equal to n, the value cannot be a prime since $n^2 - n + n = n^2$ is divisible by n. Moreover, the number n itself should be prime since, for a composite number $n = n_1 n_2$, $1 < n_1 \leq n_2 < n$, the value $x = n_1$ gives the composite number $n_1^2 - n_1 + n_1 n_2 = n_1(n_1 - 1 + n_2)$. Since the polynomial $x^2 - x + n$ can be written as $x(x-1)+n$, the integers $x = -n+2, -n+3, \ldots, 0$ produce the same set of numbers as $1 \leq x < n$; therefore, taking $x = -n+2, \ldots, -1, 0, 1, \ldots, n-1$, we obtain only primes, each repeating twice.

It is easy to see that the numbers $n = 2, 3, 5, 11, 17$ are lucky Euler numbers. In particular, for $n = 5$, we get the primes $1^1 - 1 = 5$, $2^2 - 2 + 5 = 7$, $3^2 - 3 + 5 = 11$, and $4^2 - 4 + 5 = 17$. On the other hand, the numbers $n = 7, 13, 19, 23, 31$, and 37 are not lucky Euler numbers, as they have no prime pair $n+2$, and the numbers $2^2 - 2 + n = n + 2$ are composite. As for 29, the number $3^2 - 3 + 29 = 35$ is composite.

Euler, in 1772, introduced the polynomial $x^2 + x + 41$, which produces prime numbers for all integer values of x from 0 to 39. The quadratic polynomial $x^2 - x + 41$ (Adrien-Marie Legendre, 1798) gives the same prime numbers for $x = 1, \ldots, 40$ and gives composite numbers for 41 and 42.

It can be proven (see [LeLi83] and [Weis25]) that:

- *only six lucky numbers of Euler exist.*

Namely, they are 2, 3, 5, 11, 17, and 41 (sequence A014556 in the OEIS).

5.7.6. The polynomials $x^2 - x + p$, considered above, are all members of the larger set of prime-generating polynomials, closely related to the classic number-theoretic problem of searching formulas for primes.

A *formula for primes* is a formula generating the prime numbers exactly and without exception. No easily computable formula is known. A great deal is known about what, more precisely, such a "formula" can and cannot be.

In fact, to obtain the prime numbers, it is natural to search for functions, defined for all positive integers n, which are computable in practice and produce some or all prime numbers. For example, one of the following conditions should be satisfied:

1. $f(n) = p_n$ for all $n \geq 1$;
2. $f(n)$ is always a prime number, and if $m \neq n$, then $f(n) \neq f(m)$;
3. the set of prime numbers is equal to the set of positive values of the function $f(n)$.

The simplest candidate for the construction of some convenient formula for these purposes could be a polynomial with integer coefficients. However, it was proven by Christian Goldbach in 1752 that:

- *no non-constant polynomial $f(x)$ with integer coefficients can give a prime for all integers values of x.*

□ Suppose such a polynomial, $f(x) = a_n x^n + a_{n-1}x^{n-1} + \cdots + a_1 x + a_0$, exists. Then, $f(1)$ would evaluate to a prime p, i.e., it holds that $f(1) \equiv 0(mod\ p)$. But for any integer k, we have that

$$\begin{aligned} f(1+kp) &= a_n(1+kp)^n + a_{n-1}(1+kp)^{n-1} + \cdots + a_1(1+kp) + a_0 \\ &\equiv f(1) \equiv 0(mod\ p). \end{aligned}$$

Therefore, the number $f(1+kp)$ cannot also be prime (as it would be divisible by p) unless it is p itself, but the only way $f(1+kp) = f(1)$ for all k is if the polynomial is constant. □

Legendre showed (see [Nage51], [HaWr79], and [Weis25]) that:

- *there is no rational algebraic function which always gives primes.*

Using algebraic number theory further, one can show an even stronger result:

- *no non-constant polynomial $f(n)$ exists that evaluates to a prime number for almost all integers n.*

It was also shown by Irving Reiner, in 1943 ([Rein43]), that:

- *if f_i, g_i, $i = 1, \ldots, n$, are polynomials with positive coefficients, then $\sum_{i=1}^{n} f_i(x)^{g_i(x)}$ cannot give only primes for $x \in \mathbb{N}$.*

5.7.7. However, there exist polynomials which can give many prime numbers. As was mentioned before, the *Euler polynomial*

$f(x) = x^2+x+41$ gives a prime for all non-negative integers less than 40. The primes for $n = 0, 1, 2, 3, \ldots, 39$ are 41, 43, 47, 53, 61, 71, 83, 97, 113, 131, ... (sequence A005846 in the OEIS). For $n = 40$, it produces a square number, 1681, which is equal to 41^2, the smallest composite number for this formula. In fact, if 41 divides n, it divides $f(n)$, too.

Prime numbers, generated by Euler's prime-generating polynomial $x^2 + x + 41$, are known as the *Euler primes.* (These primes are distinct from prime Euler numbers, which are known as Euler number primes.) The first few Euler primes occur for $x = 0, 1, 2, \ldots$ $42, 43, 45, 46, 47, 48, 50, 51, \ldots$ (sequence A056561 in the OEIS). As of 2025, the largest known Euler prime is 1523844527 ... 6354845061, which has 398,204 decimal digits.

The quadratic polynomial $x^2 - x + 41$ gives the same prime numbers for $x = 1, \ldots, 40$ and gives composite numbers for 41 and 42.

The polynomial $g(x) = x^2 - 79x + 1601$ gives primes for $x = 0, \ldots, 79$. The primes 1601, 1523, 1447, 1373, 1301, 1231, 1163, 1097, 1033, 971, ... (sequence A060566 in the OEIS), generated by this polynomial for $x = 0, \cdots, 79$, correspond to the 40 primes given by the above formula, taken twice each, due to the identity $x^2 - 79x + 1601 = (x - 40)^2 + (x - 40) + 41$ ([HaWr79]).

Other examples of low-order polynomials that generate only primes for the first few nonnegative values are as follows:

- $2x^2 + 11$, giving primes 11, 13, 19, 29, 43, 61, 83, 109, 139, 173, 211 (see sequence A050265 in the OEIS) for all $x = 0, 1, 2, \ldots, 10$;
- $2x^2 + 29$ (Legendre, 1798), giving primes 29, 31, 37, 47, 61, 79, 101, 127, 157, 191, ... (see sequence A007641 in the OEIS) for all $x = 0, 1, 2, \ldots, 28$;
- $x^3 + x^2 + 17$, giving primes 17, 19, 29, 53, 97, 167, 269, 409, 593, 827, ... (see sequence A050266 in the OEIS) for all $x = 0, 1, 2, \ldots, 10$, etc.

Particularly "poor" polynomials are $x^6 + 1091$, which does not give primes for $x = 1, 2, \ldots, 3905$, but which gives primes for $x = 0, 3906, 4620, 5166, 5376, 5460, 8190, 13020, 13986, 14490, \ldots$ (sequence A119276 in the OEIS), and $x^{12} + 488669$, which does not give primes for $x = 0, 1, \ldots, 616980$, but which gives primes for $x = 616980$, 764400, 933660, 1051050, 1730820, 1758120, 2467920, 3093090, 3164070, $3461640, \ldots$ (sequence A122131 in the OEIS). For more information, see [Weis25] and [Wiki25].

5.7.8. Based on *Dirichlet's theorem on the existence of primes in a given arithmetic progression* (Johann Peter Gustav Lejeune Dirichlet, 1837), we can state that a linear polynomial function $f(x) = ax + b$ produces infinitely many primes as long as a and b are relatively prime, though no such function will assume prime values for all values of n. In particular, there are infinitely many primes of the form $4x-1$, $4x+1$, $6x-1$, $6x+1$.

It is not known, however, whether there exists a polynomial of degree greater than one that assumes an infinite number of values that are prime. In particular, it is not known whether there are infinitely many primes of the form $ax^2 + bx + c$ and, especially, of the form $x^2 + 1$.

However, it was proven by Dirichlet, that, for integers a, b, and c, such that $gcd(a, 2b, c) = 1$ and $b^2 = ac$, the polynomial in two variables $ax^2 + bxy + cy^2$ represents infinitely many primes. In particular, there are infinitely many primes of the form $x^2 + y^2$ and of the form $x^2 + y^2 + 1$.

5.7.9. Lucky numbers of Euler have deep algebraic meaning. It turns out that they are closely connected with the so-called *Heegner numbers*. In fact, if p lists all lucky numbers, then $4p-1$ lists all *Heegner numbers* of such form, except the trivial case 3.

The *Heegner numbers* are positive integers n such that the imaginary quadratic field $Q(\sqrt{-n})$ features unique factorisation into primes. These are the square-free numbers n, for which if some positive integer N can be written in the form $\left(\frac{x}{2}\right)^2 + n\left(\frac{y}{2}\right)^2$ for some integers x and y, then every prime factor of N which occurs in the factorisation of N in an odd power can also be written in the form $\left(\frac{a}{2}\right)^2 + n\left(\frac{b}{2}\right)^2$ for integers a and b.

The case of $n = 1$ gives the ring $(\mathbb{Z}[i], +, \cdot)$ of *Gaussian integers*, $\mathbb{Z}[i] = \{x + y\sqrt{-1} \mid x, y \in \mathbb{Z}\}$. The case of $n = 2$ gives the ring $(\mathbb{Z}[\sqrt{-2}], +, \cdot)$, $\mathbb{Z}[\sqrt{-2}] = \{x + y\sqrt{-2} \mid x, y \in \mathbb{Z}\}$. Other cases satisfy the condition $n \equiv 3(mod\ 4)$ and correspond to the rings of numbers of the form $\frac{x}{2} + \frac{y}{2}\sqrt{-n}$, with integers x and y of the same parity.

A Heegner number n corresponds to a binary quadratic form discriminant $D = -n$, which has class number $h(-n)$ equal to 1, except for the Heegner numbers 1 and 2, which correspond to $D = -4$ and $D = -8$, respectively.

The determination of such numbers is a particular case of $m = 1$ of *Gauss's class number problem*: for a given m, find a complete list of fundamental binary quadratic-form discriminants $-D$ and $D > 0$ such that their class number is given by $h(-D) = m$.

According to the *Baker–Stark–Heegner theorem* (see [Heeg52]),

- *there are precisely nine Heegner numbers.*

Namely, they are 1, 2, 3, 7, 11, 19, 43, 67, and 163 (sequence A003173 in the OEIS).

This result was conjectured by Gauss and proved with minor flaws by Kurt Heegner in 1952. Alan Baker and Harold Stark independently proved the result in 1966, and Stark further indicated that the gap in Heegner's proof was minor.

The lucky numbers of Euler are those numbers p such that, for $n = 4p - 1$, the imaginary quadratic field $Q(\sqrt{-n})$ has class number 1. As there are only nine numbers n, such that $h(-n) = 1$, and of these only 7, 11, 19, 43, 67, and 163 (except the trivial case 3) are of the required form, the only lucky numbers of Euler are 2, 3, 5, 11, 17, and 41, and there does not exist a better prime-generating polynomial of Euler's form.

5.7.10. Euler also considered polynomials of the form $2x^2 + p$ with primes p. He showed that they gives primes for all $x = 0, 1, 2, \ldots, p-1$ if and only if the ring $Q(\sqrt{-2p})$ has class number 2, which allows for $p = 3, 5, 11$, and 29. Baker (1971, [Bake71]) and Stark (1971, [Star71]) showed that there are no such fields for $p > 29$.

Similar results have been found for polynomials of the form $px^2 + px + n$ (Michael D. Hendy, [Hend74]).

Euler's number

5.7.11. *Euler's number* e is a mathematical constant and is the base of the natural logarithm and the exponential function. The numerical value of e is

$$e = 2.71828182845904523536028747135266249775\,7\ldots$$

(sequence A001113 in the OEIS).

The number e is of great importance in mathematics, alongside the numbers $0, 1, \pi$, and i. All five constants appear in one of the formulations of *Euler's identity*,

$$e^{i\pi} + 1 = 0,$$

and play important and recurring roles across mathematics.

The first references to the constant e were published in 1618 in the appendix table of a work on logarithms by John Napier. However, this did not contain the constant itself but simply a list of logarithms to the base e.

The constant itself was introduced by Jacob Bernoulli in 1683 for solving the problem of continuous compounding of interest. In his solution, the constant e occurs as the limit

$$\lim_{n\to\infty}\left(1+\frac{1}{n}\right)^n,$$

where n represents the number of intervals in a year on which the compound interest is evaluated (for example, $n = 12$ for monthly compounding).

Euler began using the letter e for the constant in 1727 or 1728 in an unpublished paper on explosive forces in cannons and in a letter to Christian Goldbach on 25 November 1731. The first appearance of e in a printed publication was in Euler's *Mechanica* (1736).

Euler proved that e is the sum of the infinite series

$$e = \sum_{n=0}^{\infty}\frac{1}{n!} = \frac{1}{0!}+\frac{1}{1!}+\frac{1}{2!}+\frac{1}{3!}+\frac{1}{4!}+\cdots,$$

where $n!$ is the factorial of n. The equivalence of the two characterisations, using the limit and the infinite series, can be proved via the binomial theorem.

5.7.12. The principal motivation for introducing the number e is to perform differential and integral calculus with exponential functions and logarithms. A general exponential function $y = a^x$ has a derivative, given by the limit

$$(a^x)' = \lim_{\alpha \to 0} \frac{a^{x+\alpha} - a^x}{\alpha} = \lim_{\alpha \to 0} \frac{a^x a^\alpha - a^x}{\alpha} = a^x \cdot \lim_{\alpha \to 0} \frac{a^\alpha - 1}{\alpha}.$$

The limit $\lim_{\alpha \to 0} \frac{a^\alpha - 1}{\alpha}$ is independent of the variable x. Its value turns out to be the logarithm of a to base e. Thus, when the value of a is set to e, this limit is equal to 1, and so one arrives at the following simple identity:

$$(e^x)' = e^x.$$

Consequently, the exponential function with base e is particularly suited for doing calculus. Choosing e (as opposed to some other number) as the base of the exponential function makes calculations involving the derivatives much simpler.

Another motivation comes from considering the derivative of the base-a logarithm $\log_a x$ for $x > 0$:

$$(\log_a x)' = \lim_{\alpha \to 0} \frac{\log_a(x + \alpha) - \log_a x}{\alpha} = \lim_{\alpha \to 0} \frac{\log_a \left(1 + \frac{\alpha}{x}\right)}{x \cdot \frac{\alpha}{x}}$$

$$= \frac{1}{x} \log_a \left(\lim_{t \to 0} (1 + t)^{\frac{1}{t}} \right) = \frac{1}{x} \log_a e.$$

As the base-e logarithm of e is equal to 1, we get that

$$(\log_e x)' = \frac{1}{x}.$$

The logarithm with this special base is called the *natural logarithm*; it behaves well under differentiation since there is no undetermined limit to carry out the calculations.

Thus, there are two ways of selecting such a special number e. One way is to set the derivative of the exponential function a^x equal to a^x and then solve for a. The other way is to set the derivative of the base-a logarithm to $\frac{1}{x}$ and then solve for a. In each case, one arrives at a convenient choice of base a for doing calculus. It turns out that these two give the same solution: the number e.

5.7.13. The Taylor–Maclaurin series for the exponential function can be deduced from the facts that the exponential function is its own derivative and that it equals 1 when evaluated at 0:

$$e^x = \sum_{n=0}^{\infty} \frac{x^n}{n!}.$$

Setting $x = 1$ recovers the definition of e as the sum of an infinite series.

Because this series is convergent for every complex value of x, it is commonly used to extend the definition of e^x to the complex numbers. Thus, the Taylor series for $\sin x$ and $\cos x$ allow us to derive *Euler's formula*:

$$e^{ix} = \cos x + i \sin x,$$

which holds for any complex x. The special case of $x = \pi$ gives *Euler's identity*

$$e^{i\pi} + 1 = 0,$$

which is considered an exemplar of mathematical beauty, as it shows a profound connection between the most fundamental numbers in mathematics, i, π, e, 1, and 0, as well as involving the fundamental operations of equality, addition, multiplication, and exponentiation.

Furthermore, using the laws of exponentiation, we get for any integer n *de Moivre's formula*:

$$(\cos x + i \sin x)^n = \left(e^{ix}\right)^n = e^{inx} = \cos nx + i \sin nx.$$

The expressions for $\cos x$ and $\sin x$ in terms of the exponential function can be deduced from the Taylor series for e^{ix} and e^{-ix}; they have the following form:

$$\cos x = \frac{e^{ix} + e^{-ix}}{2}, \quad \sin x = \frac{e^{ix} - e^{-ix}}{2i}.$$

5.7.14. It is well known that the number e is irrational and, moreover, transcendental. Euler, in 1737, proved that e is irrational by showing that it has an infinite, simple continued fractions decomposition $e = [2, 1, 2, 1, 1, 4, 1, 1, 6, \ldots]$:

$$e = 2 + \cfrac{1}{1 + \cfrac{1}{2 + \cfrac{1}{1 + \cfrac{1}{1 + \cfrac{1}{4 + \cfrac{1}{1 + \cfrac{1}{1 + \ddots}}}}}}}.$$

Joseph Liouville, in 1844, proved that e does not satisfy any quadratic equation with integral coefficients (i.e., if it is algebraic, it must be algebraic of degree greater than 2). At last, Charles Hermite proved that e is transcendental in 1873. It was the first number to be proved transcendental without having been specifically constructed for this purpose. However, e is the "least" transcendental possible with the *irrationality measure* $\mu(e) = 2$. The number e is one of only a few transcendental numbers for which the exact irrationality exponent μ is known.

However, it is still not known whether $\pi + e$ or $\frac{\pi}{e}$ is irrational.

5.7.15. There are numerous applications of the constant e in mathematics. Some of the most well-known applications are listed as follows:

- The number e is the limit (the *force of interest*), as n goes to infinity, in the following question pertaining to compound interest: if an account starts with \$1.00 and pays 100 percent interest per year, what happens if the interest is computed and credited n times per year? Compounding once per year ($n = 1$) yields \$2, compounding weekly ($n = 52$) yields \$2.692596..., and compounding daily ($n = 365$) yields \$2.714567... .
- The number e has applications in probability theory. Suppose that a gambler plays a slot machine that pays out with a probability of one in n and plays it n times. The probability of winning k times

out of n trials is $\binom{n}{k}\left(\frac{1}{n}\right)^k\left(1-\frac{1}{n}\right)^{n-k}$. As n increases, the probability that the gambler will lose all n bets is equal to $\left(1-\frac{1}{n}\right)^n$ and approaches $\frac{1}{e}$.

- Another application of e, also discovered in part by Jacob Bernoulli along with Pierre Remond de Montmort, is in the problem of *derangements*, also known as the *hat check problem*: n guests are invited to a party, and at the door, the guests all check their hats with the butler, who in turn places the hats into n boxes, each labelled with the name of one guest. But the butler has not asked the identities of the guests, and so puts the hats into boxes selected at random. The problem of de Montmort is to find the probability that none of the hats gets put into the correct box. This probability p_n is represented by the formula

$$p_n = 1 - \frac{1}{1!} + \frac{1}{2!} - \frac{1}{3!} + \cdots + \frac{(-1)^n}{n!} = \sum_{k=0}^{n} \frac{(-1)^k}{k!}.$$

As n tends to infinity, p_n approaches $\frac{1}{e}$. Furthermore, for a given positive integer n, the number of ways the hats can be placed into the boxes so that none of the hats are in the correct box is $\left[\frac{n!}{e}\right]$, i.e., the value $\frac{n!}{e}$, rounded to the nearest integer.

- The number e is the solution to *Steiner's problem* (1850) of finding the maximum of the function $f(x) = x^{\frac{1}{x}}$. The maximum value of $x^{\frac{1}{x}}$ occurs at $x = e$.
- The number e occurs naturally in connection with many problems involving asymptotics. An example is *Stirling's formula* for the asymptotics of the factorial function, in which both the numbers e and π appear:

$$n! \sim \sqrt{2\pi n}\left(\frac{n}{e}\right)^n.$$

As a consequence, we can consider e as the following limit:

$$e = \lim_{n\to\infty} \frac{n}{\sqrt[n]{n!}}.$$

Euler–Marcheroni constant

5.7.16. The *Euler–Mascheroni constant* (or *Euler's constant*) γ is a mathematical constant, defined as the limiting difference between the harmonic series and the natural logarithm:

$$\gamma = \lim_{n\to\infty}\left(\sum_{k=1}^{n}\frac{1}{k} - \log n\right).$$

The numerical value of the Euler–Mascheroni constant is

$$\gamma = 0.57721566490153286060651209008 24\ldots$$

(sequence A001620 in the OEIS).

The constant first appeared in the 1734 paper *Progressionibus Harmonicis Observationes* by Euler, where he described it as "worthy of serious consideration" and initially calculated the constant's value to six decimal places. In 1781, he calculated it to 16 decimal places. The Italian mathematician Lorenzo Mascheroni, in 1790, attempted to calculate the constant up to 32 decimal places. Other computations were done by Johann von Soldner in 1809. The notation γ appears nowhere in the writings of either Euler or Mascheroni and was chosen at a later time, perhaps because of the connection to the gamma function. In particular, the German mathematician Carl Anton Bretschneider used the notation γ in 1835, and Augustus De Morgan used it in a textbook published in parts from 1836 to 1842. Euler's constant was also studied by the Indian mathematician Srinivasa Ramanujan, who published a paper on it in 1917. David Hilbert mentioned the irrationality of γ as an unsolved problem that seems "unapproachable".

5.7.17. The number γ has not been proved to be algebraic or transcendental. In fact, it is not even known whether γ is irrational. The ubiquity of γ in mathematics and the fact that γ has been called the third most important mathematical constant after π and e makes the irrationality of γ a major open question in mathematics.

The simple continued fraction expansion of Euler's constant is given by

$$[0, 1, 1, 2, 1, 2, 1, 4, 3, 13, 5, 1, 1, 8, 1, 2, 4, 1, 1, 40, \ldots]$$

(sequence A002852 in the OEIS), i.e., we have the decomposition

$$\gamma = 0 + \cfrac{1}{1 + \cfrac{1}{1 + \cfrac{1}{2 + \cfrac{1}{1 + \cfrac{1}{2 + \cfrac{1}{1 + \cfrac{1}{4 + \ldots}}}}}}}$$

which has no apparent pattern. It is known to have at least 16,695,000,000 terms, and it has infinitely many terms if and only if γ is irrational.

Using a continued fraction analysis, Papanikolaou showed in 1997 that if γ is rational, its denominator must be greater than 10^{244663}. If e^{γ} is rational, then its denominator must be greater than 10^{15000} (see [Wiki25]).

A good simple approximation of γ can be given by the reciprocal of the square root of 3:

$$\frac{1}{\sqrt{3}} = 0 + \cfrac{1}{1 + \cfrac{1}{1 + \cfrac{1}{2 + \cfrac{1}{1 + \cfrac{1}{2 + \cfrac{1}{1 + \cfrac{1}{2 + \ldots}}}}}}}$$

5.7.18. Euler's constant appears frequently in mathematics, especially in number theory and analysis. Examples include the following:

- The Weierstrass product formula for the gamma function:

$$\Gamma(z) = \frac{e^{-\gamma z}}{z} \prod_{n=1}^{\infty} \left(1 + \frac{z}{n}\right)^{-1} e^{z/n}.$$

- The asymptotic expansion of the gamma function:

$$\Gamma\left(\frac{1}{x}\right) \sim x - \gamma.$$

- The Laurent series expansion for the Riemann zeta function:

$$\zeta(s) = \frac{1}{s-1} + \sum_{n=0}^{\infty} \frac{\gamma_n}{n!}(1-s)^n,$$

where the constants γ_n are called the *Stieltjes constants* and can be defined by the limit

$$\gamma_n = \lim_{m\to\infty} \left(\left(\sum_{k=1}^{m} \frac{(\log k)^n}{k} \right) - \frac{(\log m)^{n+1}}{n+1} \right);$$

the constant γ_0 is the Euler–Mascheroni constant γ.

- The asymptotic of the mean of the divisor function $d(n) = \sum_{d|n} 1$:

$$d(1) + d(2) + \cdots + d(n) = n \log n + (2\gamma - 1)n + O(\sqrt{n}).$$

- The *Lenstra–Pomerance–Wagstaff conjecture*: there are infinitely many Mersenne primes, and more precisely, the number of Mersenne primes less than x is asymptotically approximated by

$$e^{\gamma} \cdot \log_2 \log_2 x.$$

5.7.19. The following list gives the most important examples of the representations of the constant γ itself.

- The constant γ is related to the derivative of the gamma function when it is evaluated at 1:

$$-\gamma = \Gamma'(1).$$

- The constant γ can be expressed as an infinite sum whose terms involve the Riemann zeta function evaluated at positive integers:

$$\gamma = \sum_{m=2}^{\infty} (-1)^m \frac{\zeta(m)}{m} = \log \frac{4}{\pi} + \sum_{m=2}^{\infty} (-1)^m \frac{\zeta(m)}{2^{m-1} m}.$$

- The constant γ is equal to the values of a number of definite integrals, including

$$\gamma = -\int_0^{\infty} e^{-x} \log x \, dx, \quad \gamma = -\int_0^1 \log\left(\log \frac{1}{x}\right) dx,$$

$$\gamma = 2\int_0^{\infty} \frac{e^{-x^2} - e^{-x}}{x} \, dx,$$

as well as Catalan's integral (1875):

$$\gamma = \int_0^1 \left(\frac{1}{1+x} \sum_{n=1}^{\infty} x^{2^n - 1} \right) dx.$$

5.7.20. The constant e^{γ} is also important in number theory. Its numerical value is

$$e^{\gamma} = 1.78107241799019798523650410310717954916964521430343\ldots$$

(sequence A073004 in the OEIS).

e^{γ} represents the following limit, concerning all primes up to the nth prime number p_n:

$$e^{\gamma} = \lim_{n\to\infty} \frac{1}{\log p_n} \prod_{i=1}^{n} \frac{p_i}{p_i - 1}.$$

This restates the third *Mertens' theorem*:

$$\lim_{n\to\infty} \log n \prod_{p \le n} \left(1 - \frac{1}{p}\right) = e^{-\gamma} = 0.561459483566885\ldots.$$

Other infinite products relating to e^{γ} include

$$\frac{e^{1+\frac{\gamma}{2}}}{\sqrt{2\pi}} = \prod_{n=1}^{\infty} e^{-1+\frac{1}{2n}} \left(1+\frac{1}{n}\right)^n, \quad \text{and}$$

$$\frac{e^{3+2\gamma}}{2\pi} = \prod_{n=1}^{\infty} e^{-2+\frac{2}{n}} \left(1+\frac{2}{n}\right)^n.$$

Exercises

1. Check that the polynomial $3x^2 + 3x + 23$ gives prime numbers for $x = 0, 1, 2, \ldots, 21$; the polynomial $x^4 + 29x^2 + 101$ gives prime numbers for $x = 0, 1, 2, \ldots, 19$; the polynomial $3n^2+39n+37$ gives prime numbers for $x = 0, 1, 2, \ldots, 18$; the polynomial $n^2 + n + 17$ gives prime numbers for $x = 0, 1, 2, \ldots, 15$ (sequence A007635 in the OEIS); and the polynomial $36x^2 - 810x + 2753$ gives prime numbers for $x - 0, 1, 2, \ldots 44$ (sequence A050268 in the OEIS).
2. Check that the polynomial $2x^2 + p$, $p \in P, p \leq 29$, gives prime numbers for all $x = 0, 1, 2, \ldots, p - 1$ and only for $p = 3, 5, 11$, and 29.
3. Prove that numbers $d = 14, 17, 19, 20, 26, 27, 29$ are non-convenient numbers in terms of the representations of positive integers by the binary quadratic form x^2+dy^2. Find several numbers represented by the binary quadratic forms $x^2 - y^2$, $x^2 + 2y^2$, and $x^2 + 3y^2$; find several binary quadratic forms represented the numbers 5, 25, 50; the numbers 3, 9, 27; and the numbers 15, 45, 135.
4. Prove that any odd prime number p can be uniquely represented by the form $x^2 - y^2$ and that this representation is proper; prove that any odd composite number has at least two representations of the form $x^2 - y^2$.
5. Prove that $\left(1+\frac{1}{x}\right)^x < e < \left(1+\frac{1}{x}\right)^{x+1}$ for all positive x; prove that $e^x \geq x + 1$ for all real x, with equality if and only if $x = 0$.

6. Prove that $\sum_{n\leq x}\frac{1}{n}-\log x=O(1)$; $\sum_{n\leq x}\frac{1}{n}=\log x+\gamma+O(x^{-1})$.
7. Prove that $\sum_{n\leq x}d(n)=x\log x+(2\gamma-1)x+O(\sqrt{x})$.

References

[AbSt72], [Ande03], [Andr79], [Apos86], [BeSl95], [BiBa70], [BLM20], [Buch09], [BuGr94], [ChMo75], [CoGu96], [Comt74], [Deza17], [Deza21], [Deza24], [DeMo10], [Dick05], [Entr66], [Eule36], [Eule48], [Eule49], [Eule55], [FlSe09], [Gaus01], [Geno52], [GKP94], [HaWr79], [Kara83], [KaVo92], [Khin6403], [Knut97], [Knut76], [KoKo77], [KPP94], [Lege79], [Male11], [Mazu10], [Ore48], [Riem59], [Ries94], [Rior80], [Sier64], [SlPl95], [Sloa25], [Stan97], [Step01], [Stru87], [Vell08], [Weis25], [Wiki25].

Chapter 6

Zoo of Numbers

In this chapter, we collect some remarkable individual special numbers related to the Eulerian numbers (see [Abra40], [Boro85], [BoBa03], [CoGu96], [DeDe12], [Deza17], [Deza18], [Deza21], [Deza23], [Deza24], [Dick05], [Gard61], [Gard88], [GKP94], [LcLi83], [Line86], [Litz63], [Mada79], [Malc86], [OLBC10], [Plat80], [Prim24], [RaTo57], [Ribe96], [Rybn82], [Rybn85], [Sier64], [SlPl95], [Sloa25], [Weis25], [Well86], [Wiki25], etc.).

- **1:** empty product (empty sum); $0! = 1$; $\binom{0}{0} = 1$; $E(0,0) = 1$; $E^2(0,0) = 1$; $E^m(0,0) = 1$; $E_0 = 1$; $S(0,0) = 1$; $s(0,0) = 1$; $B(0) = 1$.
- **2:** the only even prime; the value $E^2(2,1)$, the first non-trivial Eulerian number of the second order; the first non-trivial Eulerian Catalan number; the first non-trivial Catalan number; the first non-trivial Lah number; the value $B(2)$; the value $2!$.
- **3:** the first odd prime; the value $N(3,2)$, the first non-trivial Narayana number; the value $S(3,2)$, the first non-trivial Stirling number of the second kind; the first non-trivial value in the sequence of ordered Bell numbers, $OB(2)$; the first Fermat prime, F_0; the first Mersenne prime, M_2.
- **4:** the value $E(3,1)$, the first non-trivial element of Euler's triangle, the first non-trivial Eulerian central number; the first composite element in Pascal's triangle.

- **5:** the value E_4; the value $B(4)$; the second centred square number; the only prime, digit in which a perfect square can end; the only prime whose square is composed of only prime digits.
- **6:** the value $E^2(3,2)$; the first perfect number; the first composite Lah number; the first composite element in Narayana's triangle; the first composite element in the triangle of Stirling numbers of the second kind; the first composite element in the triangle of unsigned Stirling numbers of the first kind; the first composite factorial, 3!; the first non-trivial octagonal number; the only mean between a pair of twin primes which is triangular, $\frac{5+7}{2}$; the largest known number n such that there are n integers for which all pairwise sums are perfect squares.
- **7:** the second Mersenne prime, M_3; the only prime p such that $p+1$ is a perfect cube; the only prime equal to the difference between the product and the sum of the two previous primes; the largest (besides 4 and 5) known solution of *Brocard's problem*: to find integers n such that $n!+1$ is a perfect square; the largest known prime that is not the sum of a triangular number, a square, and a cube, all of them being positive.
- **8:** the value $E^2(3,1)$; the first composite Eulerian number of the second order; the second cubic number; the first non-trivial hyperoctahedral number: the largest Fibonacci number of the form $p+1$ or $p-1$ for a prime p; the largest composite number such that all its proper divisors plus 1 are primes.
- **9:** the second centred cube number; the only known composite number n such that both 2^n+n^2 and 2^n-n^2 are primes (namely, the primes 431 and 593).
- **10:** the first two-dimensional Narayana number; the first non-trivial four-dimensional hyperoctahedron number; the only known integer of the form $n!m!$, excluding $n = k! = m+1$ $(10! = 7!6!)$.
- **11:** the value $E(4,1) = E(4,2)$; the first prime Eulerian number of the first order; the smallest prime repunit; the largest integer that cannot be expressed as a sum of at least two distinct primes; the largest number which is not expressible as a sum of two composite numbers.

- **12:** the first composite Bell number; the smallest two-digit Lah number; the only number n such that $n \pm 1$, $\frac{n}{2} \pm 1$ and $\frac{n}{3} \pm 1$ are all primes.
- **15:** the smallest composite Bell number, $B(4)$; the smallest Mersenne composite, M_4.
- **17:** the only prime that is the average $(\frac{13+21}{2})$ of two consecutive Fibonacci numbers; the only prime that is the sum $(17 = 2+3+5+7)$ of four consecutive primes; the only prime of the form $p^q + q^p$, where p and q are primes: $17 = 2^3 + 3^2$; the only number n having n partitions into prime parts; the largest – if Goldbach's conjecture is true – integer which is not the sum of three distinct primes; the smallest prime whose sum of the digits is a cubic number; the only known prime that is equal to the sum of digits of its cube: $17^3 = 4913$ and $4+9+1+3 = 17$.
- **21:** the value M_5; the first non-trivial octagonal number.
- **22:** the value $E^2(4,1) = 2E(4,1)$; the first two-digit Eulerian number of the second order; the maximum number of regions into which five intersecting circles divide a plane.
- **23:** the smallest odd prime that is not a twin prime; the largest integer n such that no factor of a binomial coefficient $\binom{n}{k}$ is a perfect square; the largest prime, besides $2, 3, 5, 7$, and 11, which is uniquely expressible as a sum of at most four squares.
- **24:** the value $E^2(4,3) = 4!$; the first two-digit composite Stirling number of the first kind, $s(5,1)$; the sum of the twin primes 11 and 13; the only integer $n > 1$ such that $\sum_{i=1}^{n} i^2$ is a perfect square: $\sum_{i=1}^{24} i^2 = 70^2$.
- **25:** the first non-trivial Stirling number of the second kind, $S(5,3)$, which is a square number; the fourth centred square number; the only perfect square of the form $k^3 - 2$.
- **26:** the value of $E(5,1) = \frac{1}{2}E^2(5,1)$; the largest integer n such that the segment $[m, m+100]$ contains n primes (it happens only for $m = 2$).
- **28:** the second perfect number; the only perfect number of the form $n^k + m^k$ with $k > 1$: $28 = 3^3 + 1^3$.
- **29:** the smallest prime equal to the sum of three consecutive squares: $29 = 2^2 + 3^2 + 4^2$; the smallest multi-digit prime which on

adding its reverse gives a perfect square: $29 + 92 = 11^2$; the only *non-titanic prime* (i.e., with less than 1,000 decimal digits) of the form $p^p + 2$.

- **31:** the third Mersenne prime, M_5; the smallest two-digit prime Stirling number of the second kind, $S(6, 2)$; the smallest prime that can be represented as a sum of two triangular numbers in two different ways: $31 = 21 + 10 = 28 + 3$; the smallest prime that can be represented as a sum of two triangular numbers with prime indices; there are only 31 numbers that cannot be expressed as a sum of distinct squares; $3+5+7+11+\cdots+89 = 31^2$, i.e., a sum of the first 31 odd primes is a square of prime.
- **33:** the smallest odd repdigit that is not a prime number; the largest integer which is not a sum of distinct triangular numbers.
- **36:** the first non-trivial square Lah number, $L(4, 2)$; the smallest perfect square expressible as a sum of four consecutive primes, which are also two couples of prime twins: $36 = 5 + 7 + 11 + 13$; the smallest triangular number whose sum of divisors, as well as the sum of its proper divisors, is also a triangular number.
- **41:** the smallest prime whose cube can be written as a sum of three cubes in two ways: $41^3 = 40^3 + 17^3 + 2^3 = 33^3 + 32^3 + 6^3$; a sum of two consecutive squares: $4^2 + 5^2$.
- **42:** the smallest number n such that n^2 is the mean of cubed twin primes, $42^2 = \frac{11^3+13^3}{2}$.
- **51:** the product of the distinct Fermat primes 3 and 17; a regular polygon with 51 sides is constructible with a compass and a straightedge.
- **52:** the value $E^2(5, 1) = 2E(5, 1)$; the value $B(5)$; a vertically symmetrical number; an untouchable number since it is never the sum of proper divisors of any number.
- **53:** the largest known integer that can be expressed as a sum of three non-negative triangular numbers in exactly one way.
- **57:** the value of $E(6, 1) = \frac{1}{2}E^2(6, 1)$; a Leyland number, i.e., a number of the form $x^y + y^x$: $57 = 2^5 + 5^2$.
- **58:** the value of $E^2(4, 2)$; a semiprime; the sum of the first seven consecutive prime numbers: $2 + 3 + 5 + 7 + 11 + 13 + 17 = 58$.

- **61:** the value $|E_6|$; the value of the sixth centred square number; the value of the fifth centred hexagonal number.
- **65:** the only number which gives a square of prime on adding as well as subtracting its reverse from it: $65 + 56 = 11^2$, $65 - 56 = 3^2$; the only number which is the difference $(3^4 - 2^4)$ between two biquadratic numbers with prime indices; $(65!)^2 + 1$ is a prime.
- **66:** the value of $E(5, 2)$; the third central Eulerian number; a semiperfect number, as it is equal to the sum of some of its proper divisors: $66 = 22 + 33 + 11$.
- **67h:** the smallest multi-digit prime whose square 4489 and cube 300763 consist of different digits.
- **83:** the largest known prime that can be expressed as a sum of three positive triangular numbers in exactly one way; the only prime equal to a sum of squares of odd primes: $83 = 3^2 + 5^2 + 7^2$; the only prime of the form $p^4 + 2$, where p is a prime.
- **89:** the smallest positive integer whose square 7921 and cube 704969 are likewise primes upon reversal.
- **90:** the value $S(6, 3)$; expressible as the sum of distinct non-zero squares in six ways – more than any smaller number.
- **100:** the smallest perfect square whose summation of the differences between itself and each of its digits, where each difference is raised to the power of the corresponding digit, is equal to a prime: $101 = (100 - 1)^1 + (100 - 0)^0 + (100 - 0)^0$ is a prime.
- **109:** the smallest number (coincidentally prime) that has more distinct digits than its square 11881.
- **113:** the smallest prime which is a sum of three biquadratic numbers with prime indices: $113 = 2^4 + 2^4 + 3^4$; the value of the eighth centred square number.
- **114:** the value of $E^2(6, 1) = 2E(6, 1)$; the sum of the first four hyperfactorials: $114 = 0^0 + 1^1 + 1^1 \cdot 2^2 + 1^1 \cdot 2^2 \cdot 3^3$.
- **120:** the value of $E(7, 1)$; the value of $E^2(5, 4) = 5!$; the 15th triangular number, as well as the sum of the first eight triangular numbers, making it also a tetrahedral number; the smallest number to appear six times in Pascal's triangle (as all triangular and tetragonal numbers appear in it); since 15 is also triangular,

120 is a doubly triangular number; 120 is divisible by the first five triangular numbers and the first four tetrahedral numbers.

- **121:** the only perfect square of the form $1 + p + p^2 + p^3 + p^4$, $p \in P$: $121 = 1 + 3 + 3^3 + 3^3 + 3^4$; the only perfect square (besides 4) of the form $n^3 - 4$ ($121 = 5^3 - 4$).
- **127:** the fourth Mersenne prime, M_7; the smallest three-digit prime Stirling number of the second kind, $S(8, 2)$; the exponent for 12th Mersenne prime $M_{127} = 2^{127} - 1$ (it is the largest prime ever discovered by hand calculations, as well as the largest known double Mersenne prime); the smallest prime that can be written as the sum of the first two or more odd primes ($127 = 3 + 5 + 7 + 11 + 13 + 17 + 19 + 23 + 29$).
- **132:** the 99th composite number; the number of irreducible trees with 15 vertices.
- **144:** the largest – besides 0 and 1 – perfect square which is a Fibonacci number; the sum of a twin prime pair $(71; 73)$.
- **149:** the only known prime in the concatenate square sequence.
- **173:** the largest known prime whose square 29929 and cube 5177717 consist of different digits.
- **203:** the value $B(6)$; 203 different triangles can be made from three rods with integer lengths of at most 12.
- **211:** the largest known prime that cannot be written as a sum of a prime and a positive triangular number.
- **239:** the largest integer (besides 23) which is not a sum of fewer than nine cubic numbers: $239 = 2 \cdot 4^3 + 4 \cdot 3^3 + 3 \cdot 1^3$.
- **257:** the fourth Fermat prime, F_3; the largest prime in a sequence of 15 primes of the form $2n + 17$, where n runs through the first 15 triangular numbers.
- **289:** the square of the sum of the first four primes: $289 = (2+3+5+7)^2$.
- **323:** a semiprime, $323 = 17 \cdot 19$; the sum of nine consecutive primes: $19 + \cdots + 53$; the sum of 13 consecutive primes: $5 + \cdots + 47$.
- **343:** the only cubic number (besides 1), 7^3, such that the sum of its divisors is a perfect square: $1 + 7 + 7^2 + 7^3 = 20^2$.

- **367:** the largest number (in fact, a prime) whose square 134689 has strictly increasing digits.
- **400:** the only known square of the form $1 + k + k^2 + k^3$, where $k \in \mathbb{N} \backslash \{1\}$ (in fact, $k = 7$).
- **407:** the largest integer (besides $1, 153, 370$ and 371) which is the sum of the cubes of their decimal digits.
- **463:** the smallest multi-digit prime such that both a sum of the digits and a product of the digits of its square remain squares.
- **496:** the third perfect number; the perfect Stirling number of the second kind, $S(32, 31)$; the smallest triangular number such that the sum of the cubes of its digits is prime: $4^3 + 9^3 + 6^3 = 1009$.
- **541:** the 100th prime number, the 10th hexagonal star (i.e., the Star of David) number.
- **576:** the only known perfect square that can be represented as a difference between a squared sum of consecutive primes and the sum of their squares: $576 = 24^2 = (2 + 3 + 5 + 7 + 11)^2 - (2^2 + 3^2 + 5^2 + 7^2 + 11^2)$ – it is the only such case for all primes up to $2 \cdot 10^9$.
- **613:** a prime which presents a mathematical enigma in the story *Number of the End* by Jason Earls: *bring the first digit back to get* 136*, it is triangular, now bring the first digit of that back to get* 361*, it is a square*; the square 375769 of 613 is the largest known perfect square that divides a number of the form $n! + 1$, which happens when $n = 229$, another prime.
- **631:** a prime which is the reverse concatenation of the first three triangular numbers.
- **691:** the only known prime which is a square when turned upside down, 169, and another square when reversed, 196; moreover, 169 and 196 are the smallest consecutive squares using the same digits.
- **701:** the smallest prime whose square 491401 contains all of square digits only; it is equal to $5^4 + 4^3 + 3^2 + 2^1 + 1^0$.
- **720:** the value $E^2 6, 5) = 6!$; a Harshad number in every base from binary to decimal: in particular, 720 is divisible by the sum $0 + 2 + 7 = 9$ of its decimal digits.

- **727:** the first prime whose square (528529) can be represented as the concatenation of two consecutive numbers.
- **773:** replacing each digit of the prime 773 with its square and its cube results in two new primes: 49499 and 34334327, respectively.
- **786:** the largest known number n such that the binomial coefficient $\binom{2n}{n}$ is not divisible by the square of an odd prime.
- **877:** the value $B(7)$; a Chen prime, as $879 = 3 \cdot 293$ is a semiprime; a prime index prime: $877 = p_{151}$.
- **900:** the smallest perfect square which is a sum of different primes by using all the 10 digits: $900 = 503 + 241 + 89 + 67 = 509 + 283 + 61 + 47$.
- **1385:** the value of E_8; the first composite Euler number.
- **1729:** the *Hardy–Ramanujan number*, the first nontrivial *taxicab number* expressed as the sum of two cubic numbers in two different ways: $1729 = 1^3 + 12^3 = 9^3 + 10^3$; the 10th centred cube number.
- **2047:** the first Mersenne composite with a prime index, M_{11}.
- **2416:** the fourth central Eulerian number, $E(7, 3)$.
- **65537:** the biggest known Fermat prime, F_4.
- **4294967297:** the smallest Fermat composite (in fact, semiprime), F_5.
- **$\mathbf{10^{100}}$:** one *googol*; in decimal notation, it is written as the digit 1 followed by one hundred zeroes; a googol is approximately 70!; 1 googol $\approx 2^{332.19280949}$.
- **1000...0007=$\mathbf{10^{999} + 7}$:** the smallest *titanic prime*, i.e., a prime of at least 1,000 decimal digits.
- **$\mathbf{2^{4253}-1}$:** the smallest Mersenne titanic prime, M_{4253}; the 19th Mersenne prime; it has 1,281 digits.
- **1000...00033603= $\mathbf{10^{9999}+33603}$:** the smallest *gigantic prime*, i.e., a prime number with at least 10,000 decimal digits; it has exactly 10,000 digits.
- **$\mathbf{2^{44497}-1}$:** the smallest Mersenne giganic prime, M_{44497}; the 27th Mersenne prime; it has 13,395 digits.
- **$\mathbf{191273 \cdot 2^{3321908}-1}$:** the smallest known *megaprime*, i.e., a prime with at least one million decimal digits; it has exactly 1,000,000 digits.

- $\mathbf{2^{6972593}{-}1}$: the smallest Mersenne megaprime, $M_{6972593}$; the 38th Mersenne prime; it has 2,098,960 digits.
- $\mathbf{2^{82589933}{-}1}$: the largest known Mersenne prime, $M_{82589933}$; the 51st known Mersenne prime; it has 24,862,048 digits.
- $\mathbf{10^{10^{100}}} = \mathbf{10^{googol}}$: one *googolplex*; written out in ordinary decimal notation, it is 1 followed by 10^{100} zeroes, that is, a unity followed by a googol zeroes.

Chapter 7

Mini Dictionary

In this chapter, a mini dictionary, i.e., a list of all special numbers related to the Eulerian numbers, is presented (see [Abra40], [Boro85], [CoGu96], [DeDe12], [Deza17], [Deza18], [Deza21], [Deza23], [Deza24], [Dick05], [Dorr65], [Goul85], [GGL95Û], [Hons91], [Line86], [Plat80], [Prim24], [SlPl95], [Stan15], [Weis25], [Wiki25], etc.).

- $S_d(n,k)$ – *associated* (in fact, *d-associated*) *Stirling numbers of the second kind*; the number of partitions of an n-set into k subsets, such that each of these subsets contains at least d elements; $S_1(n,k) = S(n,k)$ are the ordinary *Stirling numbers of the second kind*; $S_2(n,k)$ are the *Ward numbers*.
- $|s_d(n,k)|$ – *associated* (in fact, *unsigned d-associated*) *Stirling numbers of the first kind*; the number of partitions of an n-set into k cycles such that each of these cycles contains at least d elements; $|s_1(n,k)| = |s(n,k)|$ are the ordinary *unsigned Stirling numbers of the first kind*; $|s_2(n,k)|$ gives the number of derangements of n elements with k cycles.
- $BN(n,k) = \frac{k}{2n+k}\binom{2n+k}{n}$ – *ballot numbers*; $C_n = BN(n,1)$.
- $B(n) = \sum_{k=0}^{n-1}\binom{n-1}{k}B(k)$, with $B(0) = 1$ – *Bell numbers*: $B(n)$ is the number of partitions of an n-set into non-empty parts.
- $B_n = -\frac{1}{n+1}\sum_{k=1}^{n}\binom{n+1}{k+1}B_{n-k}$ with $B_0 = 0$ – *Bernoulli numbers*.

- $\binom{n}{m}$, $n = 0, 1, 2, \ldots, m = 0, 1, \ldots, n$ – *binomial coefficients*; they form *Pascal's triangle* – a number triangle whose sides are formed by 1 and with any internal entry obtained by adding the two entries diagonally above it.
- $a_n = \int_0^1 x^{\underline{n}} dx$, $x^{\underline{n}} = x(x-1)\cdots(x-n+1)$ – *Cauchy numbers of the first type.*
- $b_n = \int_0^1 x^{\overline{n}} dx$, $x^{\overline{n}} = x(x+1)\cdots(x+n-1)$ – *Cauchy numbers of the second type.*
- $C_n = \frac{1}{n+1}\binom{2n}{n}$, $n = 0, 1, 2, \ldots$ – *Catalan numbers.*
- $CS_m(n) = \frac{mn^2-mn+2}{2}$, $n = 1, 2, 3, \ldots$ – *centred m-gonal numbers.*
- $CS_4(n) = n^2 - (n-1)^2$, $n = 1, 2, 3, \ldots$ – *centred square numbers.*
- $\overline{C}(n) = n^3 - (n-1)^3$, $n = 1, 2, 3, \ldots$ – *centred cube numbers* (or *centred cubic numbers*); $\overline{C}^k(n) = n^k - (n-1)^k$, $n = 1, 2, 3, \ldots$ – *centred k-dimensional hypercube numbers* (or *centred k-hypercube numbers, centred regular k-orthotopic numbers*).
- $E(2n+1, n)$ – *central Eulerian numbers of the first order*; $E(2n+1, n)$ is the number of permutations of the set $\{1, 2, \ldots, 2n+1\}$ which have n ascents.
- $\widetilde{B}(n)$ – *complementary Bell numbers* (or *Uppuluri–Carpenter numbers*): $\widetilde{B}(n) = S(n,0) - S(n,1) + S(n,2) - \cdots + (-1)^n S(n,n)$, where $S(n,k)$ are the Stirling numbers of the second kind.
- $C(n) = n^3$, $n = 1, 2, 3, \ldots$ – *cubic numbers* (or *cube numbers*).
- $D(n,k)$ – *Delannoy numbers*: $D(m,n) = D(m-1,n) + D(m-1, n-1) + D(m, n-1)$, $D(0,0) = 1$, $n, k = 0, 1, 2, \ldots$; $D(n) = D(n,n)$ – *central Delannoy numbers.*
- $n!!$ – *double factorials* (or *semifactorials*): $n!! = n(n-2)(n-4)\cdots$.
- $E_{n,k}$, $n = 0, 1, 2, \ldots, k = 0, 1, 2, \ldots, n$ – *Entringer numbers*; the number of reverse alternating (down-up) permutations of the set $\{1, 2, \ldots, n, n+1\}$ with the first term $k+1$.
- E_n, $n = 0, 1, 2, \ldots$ – *Euler numbers*: $\frac{2}{e^x+e^{-x}} = \sum_{n=0}^{\infty} \frac{E_n}{n!} \cdot x^n$.
- $5, 61, 23489580527043108252017828576198947741, \ldots$ – *Euler number primes*, i.e., prime numbers $|E_n|$.
- $41, 43, 47, 53, 61, \ldots$ – *Euler primes*; prime numbers generated using Euler's prime-generating polynomial $x^2 + x + 41$.

- $1, 2, 3, \ldots, 1848$ – *Euler's convenient numbers* (or *idoneal numbers, suitable numbers*); the positive integers D such that any integer having the only proper representation by the binary quadratic form $x^2 \pm Dy^2$ is a prime power or twice a prime power.
- $e = 2.718281828\ldots$ – *Euler's number*; the base of the natural logarithm and the exponential function.
- $E(n, k, \rho)$, $0 \leq \rho \leq 1$, $n = 0, 1, 2, \ldots, k = 0, 1, 2, \ldots, n$ – *Euler–Frobenius numbers*; $E(n, k, 0) = E(n, k-1)$ and $E(n, k, 1) = E(n, k)$, where $E(n, k)$ are the Eulerian numbers of the first order; $B(n, k) = 2^n E(n, k, \frac{1}{2})$ – *Eulerian numbers of type B*; $D(n, k, N) = N^n E(n, k, \frac{1}{N})$.
- $\gamma = \lim_{n\to\infty} \left(\sum_{k=1}^{n} \frac{1}{k} - \log n\right)$ – *Euler–Mascheroni constant.*
- A_n, $n = 0, 1, 2, \ldots$ – *Euler zigzag numbers* (or *up-down numbers*): numbers of alternating up-down permutations of the set $\{1, 2, \ldots, n\}$; the number Z_n of all alternating (up-down and down-up) permutations of $\{1, 2, \ldots, n\}$ is $2A_n$, $n \geq 2$; A_{2k+1} – *tangent numbers* (or *zag numbers*); A_{2n} – *secant numbers* (or *zig numbers*).
- EC_n, $n = 0, 1, 2, \ldots$ – *Eulerian Catalan numbers*: $EC_n = \frac{E(2n+1,n)}{n+1}$, where $E(2n+1, n)$ is the nth central Eulerian number of the first order.
- $E(n, k)$, $n = 0, 1, 2, \ldots, k = 0, 1, 2, \ldots, n$ – *Eulerian numbers of the first order*; the number of permutations of the set $\{1, 2, \ldots, n\}$ which have k ascents.
- $B(n, k) = 2^n E(n, k, \frac{1}{2})$, $n = 0, 1, 2, \ldots, k = 0, 1, 2, \ldots, n$ – *Eulerian numbers of type B.*
- $E^2(n, k)$, $n = 0, 1, 2, \ldots, k = 0, 1, 2, \ldots, n$ – *Eulerian numbers of the second order*; the number of Stirling permutations of the set $\{1, 1, 2, 2, \ldots, n, n\}$ which have k ascents.
- $E^m(n, k)$, $n = 0, 1, 2, \ldots, k = 0, 1, 2, \ldots, n$ – *Eulerian numbers of the mth kind*; the number of m-Stirling permutations of the set $\{1, \ldots, 1, 2, \ldots, 2, \ldots, n, \ldots, n\}$ which have k ascents.
- $n! = 1 \cdot 2 \cdot \cdots \cdot n$, $n = 1, 2, 3, \ldots$; $0!=1$ – *factorial numbers* (or *factorials*).
- $n^{\underline{k}}$ – *falling powers* (or *falling factorials*): $n^{\underline{k}} n(n-1)\cdots(n-k+1) = \frac{n!}{(n-k)!}$.

- $F_n = 2^{2^n} + 1$, $n = 0, 1, 2, \ldots$ – *Fermat numbers*.
- $u_{n+2} = u_{n+1} + u_n$, $u_1 = u_2 = 1$ – *Fibonacci numbers*.
- $OB(n)$ – *Fubini numbers* (or *ordered Bell numbers*): $OB(n) = \sum_{k=0}^{n} k!S(n,k)$, where $S(n,k)$ are the Stirling numbers of the second kind.
- $A_m(p,r) = \frac{r}{mp+r}\binom{mp+r}{m}$ – *Fuss–Catalan numbers* (or *two-parameter Fuss–Catalan numbers*, *Raney numbers*).
- $C_{n,k} = \frac{1}{(k-1)n+1}\binom{kn}{n}$ – *k-Catalan numbers*; $C_n = C_{n,2}$.
- $B_n^{(x)}$, $n = 0, 1, 2, \ldots, x \in \mathbb{R}$ – *generalised Bernoulli numbers*; $\sum_{n=0}^{\infty} B_n^{(x)} \frac{z^n}{n!} = \left(\frac{z}{e^z-1}\right)^x$.
- $H_n^{(m)}$ – *generalised harmonic numbers*: $H_n^{(m)} = \sum_{k=1}^{n} \frac{1}{k^m}$, $m \in \mathbb{R}$, $n = 1, 2, 3, \ldots$.
- $|s(n,k)|_\alpha$, $\alpha \in \mathbb{N}$ – *generalised Stirling numbers of the first kind* (or *generalised α-factorial coefficients*): $|s(n,k)|_\alpha = (\alpha n + 1 - 2\alpha)|s(n-1,k)|_\alpha + |s(n-1,k-1)|_\alpha + \delta_{n,0}\delta_{k,0}$.
- $S(n,m,k)$, $n, m, k \in \mathbb{N}$ – *generalised Stirling numbers of the second kind*: $S(n,m,k)$ is the divided by $k!$ number of ways to place n different elements into m different boxes, such that none of the k fixed boxes are empty; $S(n,k,k) = S(n,k)$.
- G_n – *Genocchi numbers*: $G_n = 2(1-2^n)B_n$, where B_n is a Bernoulli number, $n = 0, 1, 2, 3, \ldots$.
- $p \in P$, $p > 10^{9999}$ – *gigantic primes*: prime numbers with at least 10,000 decimal digits.
- H_n – *harmonic numbers*: $H_n = \sum_{k=1}^{n} \frac{1}{k}$, $n = 1, 2, 3, \ldots$; $H_n^{(m)}$ – *generalised harmonic numbers*: $H_n^{(m)} = \sum_{k=1}^{n} \frac{1}{k^m}$, $m \in \mathbb{R}$, $n = 1, 2, 3, \ldots$.
- 1, 2, 3, 7, 11, 19, 43, 67, and 163 – *Heegner numbers*; positive integers n such that the imaginary quadratic field $Q(\sqrt{-n})$ has unique factorisation into primes.
- $hf(n) = 1^1 \cdot 2^2 \cdot \cdots \cdot n^n$, $n \in \mathbb{N}$ – *hyperfactorials*.
- $HO(n) = HO^4(n) = \frac{n^2(n^2+2)}{3}$ – *hyperoctahedral numbers* (or *four-dimensional hyperoctahedron numbers*, *hexadecahoron numbers*, *4-cross-polytope numbers*, *4-orthoplex numbers*, *16-cell numbers*, *β^4-numbers*, *$\{3,3,4\}$ numbers*).

- $S_3^4(n) = \frac{n(n+1)(n+2)(n+3)}{24}$ – *hypertetrahedral numbers* (or *triangulo-triangular numbers*, *pentatope numbers*).
- $T(n)$, $n \in \mathbb{N}$ – *involution numbers* (or *telephone numbers*): $T(n)$ is the number of involutions in the symmetric group S_n of all permutations of a given n-set.
- $1, 2, 4, 6, 8, 12, 16, 24, 32, 36, \dots$ – *Jordan Pólya numbers*: positive integers which can be represented as products of factorials, allowing repetitions.
- $C_{n,k,d} = \frac{d}{(k-1)n+d}\binom{kn+d-1}{n}$ – (k,d)*-Catalan numbers*; $C_{n,k,1} = C_{n,k} = d_n^k$ – k*-Catalan numbers* (or *Fuss numbers*); $C_n = C_{n,2,1}$ – Catalan numbers.
- $C^k(n) = n^k$, $n = 1, 2, 3, \dots$ – k*-dimensional hypercube numbers* (or k*-hypercube numbers*, or *regular* k*-orthotopic numbers*), $k \in \mathbb{N}$, $k > 1$.
- $HO^k(n) = \sum_{j=0}^{k-1}(-1)^j\binom{k-1}{j}2^{k-j-1}S_3^{k-j}(n)$ – k*-dimensional hyperoctahedron numbers*, $k \in \mathbb{N}$, $k > 1$.
- $S_3^k(n) = \frac{n(n+1)(n+2)\cdots(n+(k-1))}{k!}$ – k*-dimensional hypertetrahedron numbers* (or k*-dimensional simplicial numbers*).
- $L(n,k)$ – *Lah numbers*; the number of partitions of an n-set into k totally ordered parts; $L(n,k) = C_{n-1}^{k-1}\frac{n!}{k!}$, $n = 0, 1, 2, \dots, k = 0, 1, 2, \dots, n$.
- $x^y + y^x$, $x, y \in \mathbb{Z}, x, y > 1$ – *Leyland numbers*.
- $SH_0 = 1$, $SH_n = \sum_{k=0}^{n}\frac{\binom{n}{k}\binom{n+k}{k}}{2k+2}$, $n \geq 1$ – *little Schröder numbers* (or *Schröder–Hipparchus numbers*, *super-Catalan numbers*, *Hipparchus numbers*).
- $L_{m,n} = \frac{2m+1}{m+n+1}\binom{2n}{m+n}$, $n \geq m \geq 0$ – *Lobb numbers*.
- $L_n = L_{n-1} + L_{n-2}$, $L_0 = 2$, $L_1 = 1$ – *Lucas numbers*.
- 2, 3, 5, 11, 17 and 41 – *Lucky numbers of Euler*; positive integers n such that the polynomial $x^2 - x + n$ gives prime numbers for all $x = 1, 2, \dots, n-1$.
- $[x^n]\prod_{m=1}^{n}\left(\sum_{i=0}^{m-1}x^i\right)$ – *Mahonian numbers*; the number of permutations of n elements with k inversions.
- $p \in P$, $p > 10^{999999}$ – *megaprimes*: prime numbers with at least one million decimal digits.

- $M_n = 2^n - 1$, $n = 1, 2, 3, \ldots$ – *Mersenne numbers.*
- $S_m(n) = \frac{n((m-2)n-m+4)}{2}$, $n = 1, 2, 3, \ldots$ – *m-gonal numbers,* $m \in \mathbb{N}$.
- $n!_\alpha$, $\alpha \in \mathbb{N}$ – *multifactorials* (or *α-factorials*): $n!_\alpha = n(n-\alpha)(n-2\alpha)\cdots$; $n!_1 = n!$; $n!_2 = n!!$.
- $\binom{n}{k_1,\, k_2,\, \ldots,\, k_m}$ – *multinomial coefficients:* $\binom{n}{k_1,\, k_2,\, \ldots,\, k_m} = \frac{n!}{k_1!k_2!\cdots k_m!}$, $k_1, k_2, \ldots, k_m$ – non-negative integers such that $k_1 + k_2 + \cdots + k_m = n$; $\binom{n}{k,l}$ – *threenomial coefficients.*
- $N(n,k)$ – *Narayana numbers:* $N(n,k) = \frac{1}{n} \cdot \binom{n}{k} \cdot \binom{n}{k-1}$, $n = 0, 1, 2, \ldots, k = 0, 1, 2, \ldots, n$.
- $N_m(n,k)$ – *Narayana numbers of order m:* $N_m(n,k) = \frac{m+1}{n+1} \cdot \binom{n+1}{k+1} \cdot \binom{n-m-1}{k}$, $n = 0, 1, 2, \ldots, k = 0, 1, 2, \ldots, n$; $N(m,k) = N_0(m,k)$.
- $OB(n) = \sum_{k=0}^{n} k!S(n,k)$ – *ordered Bell numbers* (or *Fubini numbers*).
- $OS(n,k) = k!S(n,k)$, $n = 0, 1, 2, \ldots, k = 0, 1, 2, \ldots, n$ – *ordered Stirling numbers of the second kind.*
- $2^{n-1}(2^n - 1)$, $n \in \mathbb{N}$, where $2^n - 1 \in P$ – *perfect numbers* (in fact, even perfect numbers): positive integers equal to the sum of their proper divisors.
- $n\#$, $n \in \mathbb{N}$ – *primorials:* $n\#$ is the product of prime numbers less than or equal to n.
- $n(n+1)$, $n = 1, 2, 3, \ldots$ – *pronic numbers.*
- $O(n) = \frac{n(2n^2+1)}{3}$, $n = 1, 2, 3, \ldots$ – *octahedral numbers.*
- $S^d(n,k)$ – *reduced* (in fact, *d-reduced*) *Stirling numbers of the second kind:* $S^d(n,k)$ is the number of partitions of the set $\{1, 2, \ldots, n\}$ into k non-empty subsets so that the distance between any two elements in each subset is not less than d; $S^1(n,k) = S(n,k)$.
- $A_{n,k,m}$, $n = 0, 1, 2, \ldots, k, m = 0, 1, 2, \ldots, n$ – *refined Eulerian numbers*; the number of permutations of the set $\{1, 2, \ldots, n\}$ with k descents, ending with the element m.
- $C^k(n) = n^k$, $n = 1, 2, 3, \ldots$ – *regular k-orthotopic numbers* (or *k-hypercube numbers, k-dimensional hypercube numbers*).

- $RD(n) = 4n^3 - 6n^2 + 4n - 1$, $n = 1, 2, 3, \ldots$ – *rhombic dodecahedral numbers*.
- $R(n) = (n-1)(2R(n-1) + 3R(n-2))/(n+1)$, $R(0) = 1$, $R(1) = 0$ – *Riordan numbers* (or *Motzkin sums*, *Motzkin summands*, *ring numbers*).
- $n^{\overline{k}}$ – *rising powers* (or *rising factorials*): $n^{\overline{k}} = n(n+1)\cdots(n+k-1) = \frac{(n+k-1)!}{(n-1)!}$.
- $R_n = 2^n \cdot 1 \cdot 3 \cdot 5 \cdot \cdots \cdot (2n-1) = 2^n(2n-1)!!$, $n \in \mathbb{N}$ – *Rodriguez numbers*.
- $S_n = \sum_{k=0}^{n} \frac{\binom{n}{k}\binom{n+k}{k}}{k+1}$, $n \geq 0$ – *Schröder numbers* (or *large Schröder numbers*, *big Schröder numbers*).
- A_{2n}, $n = 0, 1, 2, \ldots$ – *secant numbers* (or *zig numbers*): numbers of alternating up-down permutations of the set $\{1, 2, \ldots, 2n\}$; $A_{2n} = |E_{2n}|$.
- $S_4(n) = n^2$, $n = 1, 2, 3, \ldots$ – *square numbers*.
- $|s(n,k)|$, $n = 0, 1, 2, \ldots$, $k = 0, 1, 2, \ldots, n$ – *unsigned Stirling numbers of the first kind*; the number of partitions of an n-set into k cycles.
- $s(n,k)$, $n = 0, 1, 2, \ldots$, $k = 0, 1, 2, \ldots, n$ – *Stirling numbers of the first kind*; $s(n,k) = (-1)^{n+k}|s(n,k)|$.
- $S(n,k)$ $n = 1, 2, 3, \ldots$, $k = 1, 2, 3, \ldots, n$ – *Stirling numbers of the second kind*; the number of partitions of an n-set into k parts.
- $!n = [\frac{n!}{e}]$, $n \in \mathbb{N}$ – *subfactorials* (or *derangement numbers*, *de Montmort numbers*).
- $C^S(n,m) = \frac{(2m)!(2n)!}{(m+n)!m!n!}$, $m, n \in \mathbb{Z}, m \geq 0, n \geq 0$ – *super Catalan numbers*.
- $sf(n) = 1! \cdot 2! \cdot \cdots \cdot n!$, $n \in \mathbb{N}$ – *superfactorials*.
- $E^s(n,k)$ – *symmetric Eulerian numbers of the first order*; $E^s(n,k) = E(n+k+1,k)$.
- A_{2n+1}, $n = 0, 1, 2, \ldots$ – *tangent numbers* (or *zag numbers*): numbers of alternating up-down permutations of the set $\{1, 2, \ldots, 2n+1\}$.

- $T(n)$, $n \in \mathbb{N}$ – *telephone numbers* (or *involution numbers*): the number of ways n people can be connected by person-to-person telephone calls.
- $S_3^3(n) = \frac{n(n+1)(n+2)}{6}$, $n = 1, 2, 3, \ldots$ – *tetrahedral numbers*.
- $p \in P$, $p > 10^{999}$ – *titanic primes*: prime numbers with at least 1,000 decimal digits.
- $\binom{n}{k,l}$ – *threenomial coefficients*: $\binom{n}{k,l} = \frac{n!}{(n-k-l)!k!l!}$, $0 \le l + k \le n$.
- $S_3(n) = \frac{n(n+1)}{2}$, $n = 1, 2, 3, \ldots$ – *triangular numbers*.
- $t_{n+3} = t_{n+2} + t_{n+1} + t_n$, $t_1 = 0, t_2 = t_3 = 1$ – *Tribonacci numbers*.
- $\widetilde{B}(n)$ – *Uppuluri–Carpenter numbers* (or *complementary Bell numbers*): $\widetilde{B}(n) = S(n,0) - S(n,1) + S(n,2) - \cdots + (-1)^n S(n,n)$, where $S(n,k)$ are the Stirling numbers of the second kind.
- $|s_2(n,k)|$ – *Ward numbers* (or *2-associated Stirling numbers of the second kind*): $|s_2(n,k)|$ is the number of partitions of an n-set into k subsets such that each of these subsets contains at least two elements.

Chapter 8

Exercises

In this chapter, we present some interesting problems, concerning the Eulerian and Euler numbers, as well as their generalisations and relatives, and give (sketches of) their solutions.

Problems Connected with Eulerian Numbers

1. Prove that the number of permutations π of the set $\{1, 2, \ldots, n\}$ with k excedances $i : \pi_i > i$, $1 \leq i \leq n$, as well as the number of permutations π of the set $\{1, 2, \ldots, .n\}$ with $k+1$ weak excedances $i : \pi_i \geq i$, $1 \leq i \leq n$, is equal to the Eulerian number $E(n, k)$.
2. For a given positive integer m, give a bijective proof of Worpitzky's identity:

$$m^n = E(n,0)\binom{m}{n} + E(n,1)\binom{m+1}{n} + E(n,2)\binom{m+2}{n} + \cdots + E(n,n-1)\binom{m+(n-1)}{n}.$$

3. For a given positive integer $g \geq 2$, choose n random k-digit integers in base g, with all $k \leq m$. Add these integers using the usual addition algorithm. For every $0 \leq j \leq g-1$, let $f(j)$ be the number of times that we carry the digit j in the addition process.

Prove that, as $n \to \infty$, the expected value of $\frac{f(j)}{n}$ approaches the value $\frac{E(n,j)}{n!}$, where $E(n,j)$ is an Eulerian number.

4. Let a_n be the number of possible ways to write the *Fubini formula* for a summation of integration of order n (see [Comt74]). Prove that $a_n = \sum_{k=0}^{n} E(n,k)2^k$.
5. Prove that $\int_n^{n+1} x^{k-1}dx = \frac{1}{k}\sum_{j=0}^{k-1} E(k,j)\binom{n+j}{k-1}$.
6. Prove that the sequence $E(n,k)$, for fixed n, is *unimodal*, that is, it has a single maximum, which is achieved at one point or at two consecutive points. Give other examples of unimodal sequences.
7. For a given permutation π of the set $\{1,2,\ldots,n\}$, define its *descent set* $D(\pi)$ as

$$D(\pi) = \{i = 1,2,\ldots,n-1 \mid \pi_i > \pi_{i+1}\}.$$

For any $S \subset \{1,2,\ldots,n-1\}$, let $\alpha_n(S)$ be the number of permutations π whose descent set is contained in S. Prove that, for $S = \{s_1,\ldots,s_k\}$, it holds that

$$\alpha_n(S) = \binom{n}{s_1, s_2 - s_1, s_3 - s_2, \ldots, n - s_k}.$$

8. Prove the Carlitz identity, $\frac{A_n(t)}{(1-t)^{n+1}} = \sum_{k\geq 0}(k+1)^n t^k$, using the properties of the function $s_n(t) = (1-t)^{n+1}\sum_{k\geq 0}(k+1)^n t^k$.
9. Give a combinatorial proof of the Carlitz identity $\frac{A_n(x)}{(1-x)^{n+1}} = \sum_{k\geq 0}(k+1)^n x^k$.
10. Let $A_n(x)$ be the nth Eulerian polynomial. Give a combinatorial proof that the value of the polynomial $2xA_n(x)$ at $x = 2$ is equal to the number of ordered partitions (i.e., partitions whose blocks are linearly ordered) of an n-element set.
11. Prove that $A_n(x) = \sum_\pi x^{des(\pi)}$, where π goes over all permutations of the set $\{1,2,\ldots,n\}$ and $des(\pi)$ is the number of descents $i : \pi_i > \pi_{i+1}$, $1 \leq i \leq n-1$, in a given permutation π. Prove that

$$A_n(x) = \sum_\pi x^{1+des(\pi)}(1+x)^{n-1-2des(\pi)},$$

where π goes over all permutations of the set $\{1,2,\ldots,n\}$ with no *proper double descents* i: $\pi_i > \pi_{i+1} > \pi_{i+2}$, $1 \leq i \leq n-2$.

12. For a given permutation π of the set $\{1, 2, \ldots, n\}$, let $cdes(\pi)$ be the number of *cyclic descents* of π, i.e., the number of ordinary descents plus one if $\pi_n > \pi_1$. Prove that
$$\sum_{\pi \in S_n} x^{cdes \pi} = nxA_{n-1}(x).$$

13. For a given permutation π of the set $\{1, 2, \ldots, n\}$, define the *major index* $maj(\pi)$ as the sum of descents of π: $maj(\pi) = \sum_{i:\pi_i > \pi_{i+1}} i$. Let $inv(\pi)$ is the number of inversions of π, i.e., the number of pairs $1 \leq i < j \leq n$ such that $\pi_i > \pi_j$. Prove that, for any k, the following holds:
$$|\{\pi \in S_n \mid inv(\pi) = k\}| = |\{\pi \in S_n \mid maj(\pi) = k\}|.$$
Show that, in terms of generating functions, the result takes the form
$$\sum_{\pi \in S_n} x^{inv(\pi)} = \sum_{\pi \in S_n} x^{maj(\pi)}.$$

14. Prove that the absolute value of the Eulerian polynomial $A_{2n+1}(x)$ at the point -1 is equal to the $(2n+1)$th *tangent number* A_{2n+1}, where $\tan z = \sum_{k=0}^{\infty} A_{2k+1} \frac{z^{2k+1}}{(2k+1)!}$. More exactly,
$$A_{2n+1}(-1) = (-1)^n A_{2n+1}.$$

15. Find an expression for the exponential generating function $A(t, z) = \sum_{n \geq 0} A_n(t) \frac{z^n}{n!}$ using the recurrent relation $A_n(x) = A_{n-1}(x) + x \sum_{i=0}^{n-2} \binom{n-1}{i} A_i(x) A_{n-1-i}(x)$ for the Eulerian polynomials.
16. Prove that
$$\sum_{i=1}^{m} i^n t^i = \sum_{l=1}^{n} (-1)^{n+l} \binom{n}{l} \frac{t^{m+1} A_{n-l}(t)}{(t-1)^{n-l+1}} m^l + (-1)^n \frac{t(t^m - 1)}{(t-1)^{n+1}} A_n(t).$$

17. Prove that
$$Li_{-n}(z) = z(1-z)^{-n-1} \sum_{k=0}^{n} E(n, k) z^k = \frac{z A_n(x)}{(1-z)^{n+1}},$$

where $\mathrm{Li}_s(z) = \sum_{k=1}^{\infty} \frac{z^k}{k^s} = z + \frac{z^2}{2^s} + \frac{z^3}{3^s} + \cdots, |z| < 1$, is the *polylogarithm function.*

18. Define the *Jordatz function* $J_k(n)$ as the following double sequence:

$$J_k(n) = n^k \prod_{p|n} (1 - p^{-k}),\ p \in P.$$

Check that it is a generalisation of the Euler totient function $\phi(n)$. For any integer $k \geq 1$, show that $J_k(n)$ is equal to the number of $(k+1)$-tuples $(x_1, x_2, \ldots, x_k, n)$ of integers $x_i \in \{1, 2, \ldots, n\}$, $i = 1, 2, \ldots, k$, whose greatest common divisor is equal to 1. Show that $\sum_{d|n} J_k(d) = n^k$. Obtain the *Lambert generating function*

$$\sum_{n=1}^{\infty} J_k(n) \frac{x^n}{1 - x^n} = \frac{x A_k(x)}{(1 - x)^{k+1}}.$$

19. Prove that the generating function of the sequence of *square numbers* n^2, $n = 1, 2, 3, \ldots$, has the form $\frac{1+x}{(1-x)^3} = \frac{A_2(x)}{(1-x)^3}$. Prove that the generating function of the sequence of *cubic numbers* n^3, $n = 1, 2, 3, \ldots$, has the form $\frac{1+4x+x^2}{(1-x)^4} = \frac{A_3(x)}{(1-x)^4}$. Prove that the generating function of the sequence of *k-dimensional hypercube numbers* (or *k-hypercube numbers, regular k-orthotopic numbers*) n^k, $n = 1, 2, 3, \ldots$, has the form $\frac{A_k(x)}{(1-x)^{k+1}}$.
20. Prove that the generating function of the sequence of *centred square numbers* $2n^2 - 2n + 1 = n^2 + (n-1)^2$, $n = 1, 2, 3, \ldots$, has the form $\frac{(1+x)^2}{(1-x)^3} = \frac{(1+x)A_2(x)}{(1-x)^3}$. Prove that the generating function of the sequence of *centred cubic numbers* $n^3 + (n-1)^3$, $n = 1, 2, 3, \ldots$, has the form $\frac{(1+x)(1+4x+x^2)}{(1-x)^4} = \frac{(1+x)A_3(x)}{(1-x)^4}$. Prove that the generating function of the sequence of *k-dimensional centred hypercube numbers* (or *centred k-hypercube numbers, centred regular k-orthotopic numbers*) $n^k + (n-1)^k$, $n = 1, 2, 3, \ldots$, has the form $\frac{(1+x)A_k(x)}{(1-x)^{k+1}}$.
21. Call a generating function an *Eulerian generating function* if it has the form $\frac{g(x)A_k(x)}{(1-x)^{k+1}}$ for some $k \geq 0$ and some polynomial $g(x)$. Check that the generating functions of the sequence

of k-hypercube numbers and centred k-hypercube numbers are Eulerian generating functions. Give other examples of Eulerian generating functions.

22. Prove that

$$A_n(x) = \cfrac{1}{1 - t - \cfrac{x\,t^2}{1 - (2+x)\,t - \cfrac{4x\,t^2}{1 - (3+2x)\,t - \cfrac{9x\,t^2}{1 - (4+3x)\,t - \cfrac{16x\,t^2}{\ddots}}}}}$$

23. Prove that the polynomial $xU_n(x) = \sum_{k=0}^{n} E^2(n,k)x^{k+1}$ is the enumerator of *trapezoidal words* with n elements by number of distinct elements. Here, a *trapezoidal word* of length n is a string of length n such that the its ith element can take the values $1, 2 \ldots, 2i-1$.

Problems Connected with Relatives of Eulerian Numbers

1. For a fixed ρ, $0 \le \rho \le 1$, define the nth *Euler–Frobenius polynomial* $P_{n,\rho}(x)$ by the equality

$$\frac{P_{n,\rho}(x)}{(1-x)^{n+1}} = \sum_{k=0}^{\infty} (k+\rho)^n x^k.$$

Find the first four Euler–Frobenius polynomials $P_{n,\rho}(x)$, $n = 0, 1, 2, 3$. Prove that there exists the recurrent relation

$$P_{n,\rho}(x) = (nx + \rho(1-x))P_{n-1,\rho}(x) + x(1-x)P'_{n-1,\rho}(x),$$

$$P_{0,\rho}(x) \equiv 1.$$

Check that the special cases $\rho = 0, 1$ yield the Eulerian polynomials.

2. Let $Q_n = [0,1]^n$ be an n-dimensional unit cube. For a given $s \in \mathbb{R}$, let Q_n^s be the slice $Q_n^s = \{x \in Q_n \mid s - 1 \le \sum_{i=1}^{n} x_i \le s\}$. Prove that the volume of $Q_n^{k+\rho}$ is $\frac{A_{n,k,\rho}}{n!}$, where $A_{n,k,\rho}$ are *Euler–Frobenius numbers*: $P_{n,\rho}(x) = \sum_{k=0}^{n} A_{n,k,\rho} x^k$, $\rho \in [0,1]$.

3. Define a *refined Eulerian number* $A_{d,k,j}$ as the number of permutations of the set $\{1, 2, \ldots, n\}$ with k descents, which end with the element j. Prove that the recurrent relations for these numbers have the form

$$A_{d+1,k,d-j+1} = (k+1)A_{d,k,d-j} + (d-k)A_{d,k-1,d-j},$$
$$A_{d+1,k,d-j+1} = kA_{d,k,d-j+1} + (d-k+1)A_{d,k-1,d-j+1}.$$

Check that the closed formula for $A_{d,k,j}$ is

$$A_{d,k,j} = \sum_{i=0}^{k} \binom{f}{i} (-1)^i (k-1)^{d-j} (k-i+1)^{j-1}.$$

4. The *indexed permutation* of length d with index-set $\{0, 1, \ldots, n-1\}$ is an ordinary permutation in the symmetric group S_d, where each letter is assigned an integer between 0 and $n-1$. The set of all such indexed permutations is denoted by S_d^n. Let $D(d, n, k)$ be the number of indexed permutations in S_d^n with k descents. Prove that the numbers $D(d, n, k)$ satisfy the three-term recurrent relation

$$D(d, n, k) = (nk+1)D(d-1, n, k) + (n(d-k) + (n-1))D(d-1, n, k-1).$$

Check that there exists the following closed formula:

$$D(d, n, k) = \sum_{i=0}^{k} \binom{d+1}{i} (-1)^i (n(k-i)+1)^d.$$

5. Consider the set of all *m-Stirling permutations* of the nth order, i.e., all permutations of the multiset $Q_n^m = \{1, \ldots, 1, 2, \ldots, 2, \ldots, n, \ldots, n\}$, with the multiplicity of m for each integer, such that, if $i < j < k$, and $\pi_i = \pi_k$, then $\pi_j > \pi_i$. Define the *Eulerian numbers of the mth order* $E^m(n, k)$ as the number of such permutations π with exactly k ascents i: $\pi_i < \pi_{i+1}$, $i = 1, 2, \ldots, mn-1$. Prove that $E^1(n, k) = E(n, k)$ are the ordinary Eulerian numbers and, for $m = 2$, we get the Eulerian

numbers of the second order. Find the first several Eulerian numbers of the mth order, $m = 3, 4, 5, 6$. Prove that there exists the recurrent relation

$$E^m(n,k) = (k+1)E^m(n-1,k) \\ + (mn-k-m+1)E^m(n-1,k-1)$$

with the initial conditions $E^m(0,0) = 1$, $E^m(n,0) = 1$, $n \geq 1$, and $E^m(n,k) = 0, k \geq n \geq 1$. Construct the first 10 rows of the triangles of Eulerian numbers of the mth kind, $m = 3, 4, 5, 6$.

6. For $n = 0, 1, 2, 3, 4$ and $k = 0, 1, 2, \ldots, n$, check that for the Eulerian numbers of the mth order, it holds that

$$E^m(n,k) = \sum_{t_1+t_2+\cdots+t_{k+1}=n-k-1, t_i \geq 0} 1^{t_1} 2^{t_2} \\ \cdots (k+1)^{t_{k+1}} (mt_1+m)(m(y_1+t_2)+2(m-1)+1) \\ \cdots (m(t_1+\cdots+t_k)+k)m-1)+1).$$

7. Prove that the sum of the elements of the nth row of the triangle of Eulerian numbers of the $(m-1)$th order is equal to the last non-zero element of the nth row of the triangle of Eulerian numbers of the mth order:

$$\sum_{k=0}^{n-1} E^{m-1}(n,k) = E^m(n,n-1).$$

8. Define the nth *Eulerian polynomial of the mth order* as $A_n^m(x) = \sum_{k=0}^{n} E^m(n,k)x^k$. Construct several such polynomials for $m = 3, 4, 5, 6$. Prove that there exists the recurrent relation $A_{n+1}^m(x) = (1+mnx)A_n^m(x) + x(1-x)(A_n^m(x))'$.
9. Using the Eulerian polynomials of the mth order, prove that the sum of the elements of the $(n+1)$th row of the triangle of Eulerian numbers of the mth order is $1+mn$ times the sum of the elements of the nth row of this triangle: $\sum_{k=0}^{n} E^m(n+1,k) = (1+mn)\sum_{k=0}^{n-1} E^m(n,k)$. Using this fact, prove that the sum of

the elements of the nth row of the triangle of Eulerian numbers of the mth order is a multifactorial:

$$\sum_{k=0}^{n-1} E^m(n,k) = 1\cdot(m+1)(2m+1)\cdots((n-1)m+1)$$
$$= ((n-1)m+1)!!!\cdots!.$$

10. For a given Eulerian polynomial of the mth order $A_n^m(x)$, let

$$\alpha_n^m(x) = \frac{A_n^m(x)}{(1-x)^{mn-m+1}}, \quad \text{and} \quad \beta_n^m(x) = \frac{xA_n^m(x)}{(1-x)^{mn+1}}.$$

Check that $\alpha_n^1(x) = \frac{A_n(x)}{(1-x)^{n+1}}$, $\beta_n^1(x) = \frac{xA_n(x)}{(1-x)^{n+1}}$, where $A_n(x)$ is the classical Eulerian polynomial. Prove that

$$\left(\frac{x}{(1-x)^{m-1}}\alpha_n^m(x)\right)' = \alpha_{n+1}^m(x),$$
$$x\left(\beta_n^m(x)\right)' = (1-x)^{m-1}\beta_{n+1}^m(x).$$

11. Let $\beta_n^m(x) = \sum_{k\geq 0} f_{m,n}(k)x^k$. Prove that $f_{m,n}(k) = \sum_{j=0}^{k-1} E^m(n,j)\binom{mn+k-j-1}{mn}$.
12. Let $\beta_n^m(x) = \sum_{k\geq 0} f_{m,n}(k)x^k$. Prove that $E^m(n,k) = \sum_{j=1}^{k+1}(-1)^{k-j+1}\binom{mn+1}{k-j+1}f_{m,n}(j)$.
13. Prove that the two-dimensional exponential generating function for the *Entringer numbers* $E_{m+n,[m,n]}$ has the form $\frac{\cos x+\sin x}{\cos(x+y)}$, i.e., it holds that $\sum_{m\geq 0}\sum_{n\geq 0} E_{m+n,[m,n]}\frac{x^m y^n}{m!n!} = \frac{\cos x+\sin x}{\cos(x+y)}$.
14. An *even tree* is a rooted tree such that every vertex has an even number of children. Prove that the number $g(2n+1)$ of increasing even trees with $2n+1$ vertices is equal to $A_{2n} = |E_{2n}|$.
15. Prove that the tangent number A_{2n+1} is divisible by 2^n.
16. Prove that the Euler numbers E_{2n} can be given by the following integrals:

$$(-1)^n E_{2n} = \int_0^\infty \frac{t^{2n}}{\cosh\frac{\pi t}{2}}\,dt = \left(\frac{2}{\pi}\right)^{2n+1}\int_0^\infty \frac{x^{2n}}{\cosh x}\,dx$$
$$= \left(\frac{2}{\pi}\right)^{2n}\int_0^1 \log^{2n}\left(\tan\frac{\pi t}{4}\right)dt$$

$$= \left(\frac{2}{\pi}\right)^{2n+1} \int_0^{\pi/2} \log^{2n}\left(\tan\frac{x}{2}\right) dx$$

$$= \frac{2^{2n+3}}{\pi^{2n+2}} \int_0^{\pi/2} x \log^{2n}(\tan x)\, dx$$

$$= \left(\frac{2}{\pi}\right)^{2n+2} \int_0^{\pi} \frac{x}{2} \log^{2n}\left(\tan\frac{x}{2}\right) dx.$$

17. Prove that

$$\sum_{n=0}^{\infty}(-1)^{n+1}\left(\sum_{k=0}^{2n+1}\binom{2m-1}{2n+1-k}E(-2m,k)\right)$$
$$= (-1)^m \left(\frac{\pi}{2}\right)^{2m+1} \frac{E_{2m}}{2(2m)!}.$$

18. Define the nth *telephone number* (or *involution number*) $T(n)$, $n = 0, 1, 2, \ldots$, as the number of ways n people can be connected by person-to-person telephone calls. Check that this number can be defined as the number of permutations on n elements that are involutions. Check that the first few values of the sequence $T(n)$, $n = 0, 1, 2, \ldots$, are $1, 1, 2, 4, 10, 26, 76, 232, 764, 2620, 9496, \ldots$ (sequence A000085 in the OEIS). Prove that the telephone numbers satisfy the recurrence relation $T(n) = T(n-1) + (n-1)T(n-2)$.
19. Prove that $T(n) = \sum_{k=0}^{\lfloor n/2 \rfloor} \binom{n}{2k}(2k-1)!! = \sum_{k=0}^{\lfloor n/2 \rfloor} \frac{n!}{2^k (n-2k)!k!}$.
20. Prove that, for a fixed positive integer n, the sequence $\binom{n}{k}$, $k = 0, 1, 2, \ldots, n$, is *unimodal*, that is, it has a single maximum, which is achieved for an even n at one point, $k_n = \lfloor \frac{n}{2} \rfloor = \frac{n}{2}$, and for an odd n at two consecutive points, $k_n = \lfloor \frac{n}{2} \rfloor = \frac{n-1}{2}$ and $k_n + 1 = \lfloor \frac{n}{2} \rfloor + 1 = \frac{n+1}{2}$.
21. Prove that, as $k \to \infty$, the following holds: $\binom{z}{k} \sim \frac{(-1)^k}{\Gamma(-z)k^{z+1}}$; $\binom{z+k}{k} \sim \frac{e^{z(H_k - \gamma)}}{\Gamma(z+1)}$. Here, H_k is the kth harmonic number, $\Gamma(z)$ is the Euler gamma function, and $\gamma = 0.5772215\ldots$ is the Euler–Mascheroni constant.
22. Check that the sequence of central binomial coefficients $\binom{2n}{n} = \frac{(2n)!}{(n!)^2}$, $n \geq 0$, starts with the elements $1, 2, 6, 20, 70, 252, 924,$

3432, 12870, 48620, ... (sequence A000984 in the OEIS). Get for central binomial coefficient $\binom{2n}{n}$ the following estimations:

(a) $\frac{4^n}{2n+1} \leq \binom{2n}{n} \leq 4^n$, $n \geq 1$;
(b) $\frac{4^n}{\sqrt{4n}} \leq \binom{2n}{n} \leq \frac{4^n}{\sqrt{3n+1}}$, $n \geq 1$;
(c) $\binom{2n}{n} = \frac{4^n}{\sqrt{\pi n}}\left(1 - \frac{c_n}{n}\right)$, where $\frac{1}{9} < c_n < \frac{1}{8}$, and $n \geq 1$.

Problems Connected with Number Triangles

1. Construct the first 10 rows of the *triangle of Stirling numbers of the second kind*:

$$S(n,k) = kS(n-1,k) + S(n-1,k-1),$$

with the initial conditions $S(0,0) = 1$ and $S(n,0) = 0$ for $n > 0$, $S(n,k) = 0$ for $k > n > 0$. Prove several properties of the constructed triangle.

2. Construct the first 10 rows of the *triangle of unsigned Stirling numbers of the first kind*:

$$|s(n,k)| = (n-1)|s(n-1,k)| + |s(n-1,k-1)|,$$

with the initial conditions $|s(0,0) = 1|$ and $|s(n,0)| = 0$ for $n > 0$, $|s(n,k)| = 0$ for $k > n > 0$. Prove several properties of the constructed triangle.

3. Construct the first 10 rows of the *factorial triangle*, which is a number triangle in which the rightmost element of the nth row, $n \geq 0$, is equal to 1, the leftmost element of the nth row of the triangle is $n!$, and any internal element of the nth row is obtained as the sum of the element located in the previous row on the left and the element increased by n times located in the previous row on the right. Check that the elements of the factorial triangle $e(n,m)$, where n is the row's number and m is the number of the element in the row, satisfy the following recurrent relation: $e(n+1,m) = e(n,m-1) + (n+1)e(n,m)$. Find the initial conditions.

4. Prove that the nth row of the factorial triangle gives the coefficients of the expansion of the polynomial $(1+x)(2+x)\cdot\dots\cdot(n+x)$.
5. Prove that the sum of the elements of the nth row of the factorial triangle is equal to $(n+1)!$: $\sum_{m=0}^{n} e(n,m) = (n+1)!$.
6. Construct the first 10 rows of *Lucas's triangle*, each element L_n^k of which is determined using Pascal's rule, $L_n^k = L_{n-1}^k + L_{n-1}^{k-1}$, under the initial conditions $L_0^0 = 2$, $L_n^0 = 1$, and $L_n^n = 2$ for all positive integers n. Prove that Lucas's triangle consists of the coefficients of the expansion of the polynomial $(x+2y)(x+y)^n$.
7. Verify that the sums of the elements of the ascending diagonals of Lucas's triangle give the sequence of *Lucas numbers*, which are defined recursively: $L_n = L_{n-1} + L_{n-2}$, and $L_0 = 2, L_1 = 1$.
8. Prove that there is the following relationship between the elements L_n^k of Lucas's triangle and the binomial coefficients: $L_n^k = \binom{n}{k} + \binom{n-1}{k-1}$, i.e., *the nth row of Lucas's triangle can be obtained by adding the nth and* $(n-1)$*th rows of Pascal's triangle.*
9. The following figure shows the first few rows of the *Tribonacci triangle*:

1					
1	1				
1	3	1			
1	5	5	1		
1	7	13	7	1	
1	9	25	25	9	1

   Formulate the recurrent rule for the construction of the triangle. Check that the sums of the elements located in the ascending diagonals form the sequence of Tribonacci numbers $1, 1, 2, 4, 7, 13, 24, 44, 81, 149, \ldots$ (sequence A000073 in the OEIS).
10. Construct the first 10 rows of the number triangle with the following recurrent rule: $T(0,0) = 1$, $T(n,j) = T(n-1,j) + T(n-2,j) + T(n-3,j)$ (every interior number can be obtained by adding the three previous numbers on its vertical), and $T(n,n) = T(n,0)$. Find in this triangle the Tribonacci numbers. Prove the corresponding statement.

11. Construct the first 10 rows of *Losanitsch's triangle*, named in honour of the Serbian chemist Sima Losanitsch. It refers to a number triangle in which the right- and left-hand sides consist of unities and each internal element is equal to the sum of the two numbers that are directly above it, except for the elements located in an odd position k in an even row with the number $2t$. For their construction, one should subtract from the sum of the two numbers located directly above this element, the element located at the position with the number $\frac{k-1}{2}$ in the $(t-1)$th row of Pascal's triangle.
12. Check that Losanitsch's triangle is symmetrical with respect to the principal diagonal. Check that the first right (left) descending diagonal of Losanitsch's triangle consist of the consecutive positive integers $1, 2, 3, \ldots$, each repeated twice.
13. Prove that the second right (left) descending diagonal of Losanitsch's triangle is formed from alternating the *square numbers* n^2 and the *pronic numbers* $n(n+1)$.
14. Prove that the sum of the elements of the nth row of Losanitsch's triangle is $2^{n-2} + 2^{\lfloor n/2 \rfloor - 1}$.
15. Check that the sum of the elements of the ascending diagonal of Losanitsch's triangle is either $\frac{u_{2n-1}+u_{n+1}}{2}$ or $\frac{u_{2n}+u_n}{2}$, where u_n is the nth Fibonacci number.
16. The *Bell triangle* (or *Aitken's array*, *Peirce triangle*) is a number triangle obtained by beginning the first row with the number one and beginning subsequent rows with last number of the previous row. Rows are filled out by adding the number in the previous column to the number above it. Construct the first 10 rows of this triangle. Find the Bell numbers in this triangle. Give a combinatorial interpretation for the elements of the constructed triangle.
17. *Clark's triangle* is a number triangle created by setting the vertex equal to zero, filling one diagonal with unities, the other diagonal with multiples of an integer α, and filling in the remaining entries the result of summing the elements on either side from one row above. Construct the first five rows of Clark's triangle with $\alpha = 1, 2, 3, 4, 5, 6$. Obtain the recurrent relation for the

entries $T^{\alpha}_{m,n}$, $n = 0, 1, 2, \ldots$, $m = 0, 1, 2, \ldots, n$, of the triangle. Prove that $T^{\alpha}_{m,n} = \alpha\binom{m}{n+1} + \binom{m-1}{n-1}$.

18. The *Leibniz harmonic triangle* is a triangular arrangement of unit fractions in which the sides consist of the reciprocals of the row's numbers and each internal element is the sum of the two elements below it. Construct the first five rows of the triangle. Prove that for the elements $L_{(n,k)}$, $n = 1, 2, 3, \ldots$, $k = 1, 2, 3, \ldots, n$, of the triangle, the following recurrent relation holds: $L_{(n,1)} = \frac{1}{n}$, and $L_{(n,k)} = L_{(n-1,k-1)} - L_{(n,k-1)}$. Prove that $L_{(n,k)} = \frac{1}{n\binom{n-1}{k-1}} = \frac{1}{k\binom{n}{k}}$.
19. Define a *generalised Pascal's triangle* (more exactly, an *(α, β, L, R)-Pascal's triangle*) as a number triangle constructed using the following rule: if $GT(n, k)$ is the kth element of the nth row of the triangle, then $GT(n, k) = \alpha GT(n-1, k-1) + \beta GT(n-1, k)$, while $GT(0, n) = L(n)$ and $GT(n, 0) = R(n)$. Prove that the classical Pascal's triangle can be obtained for $\alpha = \beta = 1$, $L(n) = R(n) = 1$. Give several examples of generalised Pascal's triangles. Is it possible to obtain a closed formula for $GT(n, k)$?
20. The *Bernoulli triangle* is a number triangle composed of the partial sums of binomial coefficients: $BT_{nk} = \sum_{i=0}^{k} \binom{n}{i}$. Construct the first five rows of this triangle. Prove that $BT_{nn} = 2^n$; $BT_{n,n-1} = 2^n - 1$.
21. A *Magog triangle* of order n is a number triangle of order n with entries from 1 to n such that the entries are nondecreasing across its rows and down its columns, and all entries in column k, $k = 1, 2, \ldots, n$, are less than or equal to k. Construct several Magog triangles. Which properties of Magog triangles can we prove?
22. A *monotone triangle* (or *strict Gelfand pattern*, *gog triangle*) of order n is a number triangle with n numbers along each side and the base containing entries between 1 and n, such that there is strict increase across rows and weak increase diagonally up or down to the right. Construct monotone triangles of order n, $n = 1, 2, 3, 4, 5$. Which properties of monotone triangles can we prove?

Solutions to Problems Connected with Eulerian Numbers

1. See [Stan97]. For more details and the full description of the fundamental bijection used in the classical proof, see also Chapter 2.
2. Interpret the left-hand side m^n as counting the set of all integer vectors $(x_1, x_2, \ldots, x_n)$, with $0 \leq x_i \leq m-1$, and try to group these according to the permutations of n by rearranging the sequence in weakly increasing order. See [Pete15].
3. For instance, if we add 71801, 80914, and 62688 in base 10, then $f(0) = 1$ and $f(1) = f(2) = 2$. The probability that we carry j in a certain column is roughly the probability that if $i_1, \ldots, i_n$ are random integers in the interval $[0, g-1]$, then $gj \leq i_1 + \cdots + i_n < g(j+1)$. See Chapter 3. For more details, see [Holt97] and [Stan97].
4. It is easy to check that $a_1 = 1$, $a_2 = 3$, $a_3 = 13$, $a_4 = 75$, $a_5 = 541$, $a_6 = 46683$, $a_7 = 47293$, $a_8 = 545835$, $a_9 = 7087261$, and $a_{10} = 102247563$.

   In fact, it holds that $a_n = \sum_{k=1}^{n} k!S(n,k)$ and $\sum_{n=0}^{\infty} a_n \frac{x^n}{n!} = \frac{1}{2-e^x}$.
5. Using Worpinsky's identity, we can get that

$$\begin{aligned}\int_n^{n+1} x^{k-1}dx &= \frac{(n+1)^k - n^l}{k}\\ &= \frac{1}{k}\sum_{j=0}^{k-1} E(k,j)\left(\binom{n+1+j}{k} - \binom{n+j}{k}\right)\\ &= \frac{1}{k}\sum_{j=0}^{k-1} E(k,j)\binom{n+j}{k-1}.\end{aligned}$$

   The left-hand side is the area of the region under $f(x) = x^{k-1}$ for $n \leq x \leq n+1$, while the right-hand side is the average area of rectangles with dimensions $E(k,j)$ and $\binom{n+j}{k-1}$. For details and some generalisations, see [Tsao11].

6. Use for the proof the fact that the polynomials $A_n(x)$ form a *Sturm sequence*, that is, $A_n(x)$ has n real roots (≤ 0), separated by the roots of $A_{n-1}(x)$. See [Comt74].
7. In order to obtain a permutation $\pi = (\pi_1, \pi_2, \ldots, \pi_n)$ satisfying $D(\pi) \subseteq S$, first choose $\pi_1 < \pi_2 < \cdots < \pi_{s_1}$ in $\binom{n}{s_1}$ ways. Then, choose $\pi_{s_1+1} < \pi_{s_1+2} < \cdots < \pi_{s_2}$ in $\binom{n-s_1}{s_2-s_1}$ ways, and so on. So, it holds that
$$\alpha_n(S) = \binom{b}{s_1}\binom{n-s_1}{s_2-s_1}\cdots\binom{n-s_k}{n-s_k}$$
$$= \binom{n}{s_1, s_2-s_1, s_3-s_2, \ldots, n-s_k}.$$
8. For $n \geq 0$, define the function $s_n(t) = (1-t)^{n+1}\sum_{k\geq 0}(k+1)^n t^k$. Then, $s_0 = 1$, and it is easy to check that $s_{n+1}(t) = (1+nt)$ $s_n(t) + t(1-t)s'_n(t)$. Thus, $s_n(t) = A_n(t)$, and we have that $A_n(t) = (1-t)^{n+1}\sum_{k\geq 0}(k+1)^n t^k$, and hence, $\frac{A_n(t)}{(1-t)^{n+1}} - \sum_{k\geq 0}(k+1)^n t^k$.
9. Use the method of "balls in boxes" (see [Pete15]). Obviously, the number $(k+1)^n$ is the number of ways to place n distinct balls into $k+1$ boxes. Try to partition the ways to put the balls into boxes according to the permutations of n, and show that for a fixed permutation $\pi \in S_n$, the generating function for the arrangements of ball in boxes that correspond to π is $\frac{z^{des(w)}}{(1-x)^{n+1}}$.
10. We should use the decomposition $A_n(x) = \sum_{k=0}^{n-1}(n-k)!$ $S(n, n-k)(x-1)^k$. The value $(n-k)!S(n, n-k)$ is the number of ordered partitions of an n-set into $n-k$ blocks.
11. The first result is obvious. Moreover, it holds that $A_n(x) = \sum_\pi x^{exc(\pi)}$, where π goes over all permutations of the set $\{1, 2, \ldots, n\}$ and $exc(\pi)$ is the number of excedances $i : \pi_i > i$, $1 \leq i \leq n$, in a given permutation π. For the second result, see [FoSh70]. See also [Stan97].
12. For example, $cdes(31524) = 3$. See [Pete15].
13. See Chapter 2. Note that $\sum_{\pi\in S_n} x^{inv(\pi)} = (1+x)(1+x+x^2)\cdots(1+x+x^2+\cdots+x^{n-1})$.

14. In fact, $A_1(-1) = 1 = A_1$, $A_3(-1) = 1 - 4 + 1 = -2 = -A_3$, $A_5(-1) = 1 - 26 + 66 - 26 + 1 = 16 = A_5$, etc. The general result follows from the connections of the tangent numbers with the Bernoulli numbers. See Chapters 3 and 4.
15. In fact, it holds that

$$\begin{aligned} A'(t,z) &= \sum_{n\geq 1} A_n(t)\frac{z^{n-1}}{(n-1)!} \\ &= \sum_{n\geq 1} A_{n-1}(t)\frac{z^{n-1}}{(n-1)!} \\ &\quad + \sum_{n\geq 1} t \sum_{i=0}^{n-2} \binom{n-1}{i} A_i(t)A_{n-1-i}(t)\frac{z^{n-1}}{(n-1)!} \\ &= A(t,z) + t\sum_{n\geq 1} t \sum_{i=0}^{n} A_i(t)\frac{z^i}{i!}A_{n-1-i}(t)\frac{z^{n-1-i}}{(n-1-i)!} \\ &= A(t,z) + tA(t,z)(A(t,z)-1). \end{aligned}$$

So, we get the differential equation $f'(z) = tf^2(z) + (1-t)f(z)$ with the initial condition $f(0) = 1$. It gives us the classical result for $A(t,z)$.
16. See [Wiki25].
17. The result can be obtained using the Eulerian fractions $\alpha_n(z) = \sum_{k=0}^{\infty}(k+1)^n x^k = \frac{A_n(x)}{(1-x)^{n+1}}$. In particular, one has

$$\begin{aligned} Li_0(z) &= z\alpha_0(z) = \frac{z}{1-z}; \\ Li_{-1}(z) &= z\alpha_1(z) = z\left(\frac{1}{1-z}\right)' = \frac{z}{(1-z)^2}; \\ Li_{-2}(z) &= z\alpha_2(z) = z(z\alpha_1(z))' = \frac{z(1+z)}{(1-z)^3}; \\ Li_{-3}(z) &= z\alpha_3(z) = z(z\alpha_2(z))' = \frac{z(1+4z+z^2)}{(1-z)^4}. \end{aligned}$$

18. It is easy to check that $\phi(n) = J_1(n)$. For more information, see [Comt74].

19. Use the decomposition $\sum_{n=0}^{\infty}(n+1)^k x^n = \frac{A_k(x)}{(1-x)^{k+1}}$. However, we can use for the proof the recurrent relation for $C^k(n) = n^k$:

$$\begin{aligned} C^k(n+1) &= (n+1)^k \\ &= C^k(n) + \binom{k}{1}C^{k-1}(n) + \binom{k}{2}C^{k-2}(n) \\ &\quad + \cdots + \binom{k}{k-1}C^1(n) + 1, \quad C^k(1) = 1. \end{aligned}$$

For more details, see [DeDe12].

20. It follows from the simple fact that the generating function $g_k(x)$ of the sequence of *centred k-hypercube numbers* is $(1+x)f_k(x)$, where $f_k(x)$ is the generating function of the sequence of *k-hypercube numbers*. For the definition of centred figurate numbers and other useful information, see [DeDe12].

21. In fact, the function $g(x) \equiv 1$ gives the generating function of the regular orthotopic numbers, while the function $g(x) = 1 + x$ gives the generating function of the centred orthotopic numbers. Many elementary classes of sequences have an Eulerian generating function. For example, if $g(x) = 1 - x$, we get, for $k = 0$, the sequence $1, 0, 0, 0, 0, 0, 0, 0, 0, 0, \ldots$ (sequence A000007 in the OEIS), i.e., the sequence 0^n, $n = 0, 1, 2, \ldots$. For $k = 1$, we get the sequence $1, 1, 1, 1, 1, 1, 1, 1, 1, 1, \ldots$ (sequence A000012 in the OEIS), i.e., a sequence with all 1's. For $k = 2$, we get the sequence $1, 3, 5, 7, 9, 11, 13, 15, 17, 19, \ldots$ (sequence A005408 in the OEIS) of odd numbers. For $k = 3$, we have the sequence $1, 7, 19, 37, 61, 91, 127, 169, 217, 271, \ldots$ (sequence A003215 in the OEIS) of central hexagonal numbers. For $k = 5$, we get the sequence $1, 15, 65, 175, 369, 671, 1105, 1695, 2465, 3439, \ldots$ (sequence A005917 in the OEIS) of *rhombic dodecahedral numbers*. If $g(x) = 1 - x^2$ and, for example, $k = 4$, we get the sequence $0, 16, 80, 240, 544, 1040, 1776, 2800, 4160, 5904, \ldots$ (sequence A008511 in the OEIS), which counts the number of points on the surface of a four-dimensional cube. See [DeDe12].

22. See [Wiki25].

23. For $n = 2$, we have the words 11; 12, and 13. Their enumerator is $xU_2(x) = x + 2x^2$. For $n = 3$, the trapezoidal words,

classified by the number of distinct elements, are 111; 112, 113, 114, 115, 121,122, 131, 133; 123, 124, 125, 132, 134, 135. Thus, their enumerator is $xU_3(x) = x + 8x^2 + 6x^3$. See [Rior76].

Solutions to Problems Connected with Relatives of Eulerian Numbers

1. It is obvious that $P_{n,\rho}(x)$ is a polynomial in x of degree at most n. If we write $P_{n,\rho}(x) = \sum_{k=0}^{n} A_{n,k,\rho} x^k$, we obtain the Euler–Frobenius numbers $A_{n,k,\rho}$ (see Chapter 3). We usually regard ρ as a fixed parameter, but $P_{n,\rho}(x)$ can also be considered as a polynomial in ρ. See [Jans13].
2. The case of $\rho = 0$ (or $\rho = 1$), when the volumes are given by the Eulerian numbers, is well known; see Chapter 3. In general, we should use the results of the previous problem and similar probabilistic considerations. See [Jans13].
3. See [WaXu10].
4. The numbers $D(d, n, k)$ are the coefficients of the so-called *descent polynomials*, denoted by $D_d^n(t)$, which are defined as $D_d^n(t) = \sum_{k=0}^{d} D(d, n, k) t^k$. For details, see [WaXu10].
5. In fact, the triangle of numbers $E^m(n, k)$, $n \geq 0$, $1 \leq k \leq n$, starts with the elements 1, 1, 0, 1, m, 0, 1, $4m$, $m(2m-1)$ 0 .... In particular, for $E^5(n, k)$, $0 \leq n \leq 4$, $0 \leq k \leq n$, we have the following triangle:

$n\backslash k$	0	1	2	3	4
0	1				
1	1	0			
2	1	5	0		
3	1	20	45	0	
4	1	55	415	585	0

For more details, see Chapter 4.

6. The proof is similar to those of the corresponding properties for $E(n,k)$ and $E^2(n,k)$. See Chapters 3 and 4. For additional information, see [He23]. Substituting $k = n-1$ and $k = 1$ into the formula, we get that

$$E^m(n, n-1) = m(2(m-1)+1)\cdots((n-1)(m-1)+1),$$
$$E^m(n,1) = m(2^n - n - 1).$$

7. For the proof, we should simply compare two results concerning multifactorials. However, there also exists a combinatorial proof. For instance, we know that $E^2(2,0) = 1$ counts the permutation in Q_2^2 with no ascents, i.e., $(2,2,1,1)$, and $E^2(2,1) = 2$ is the number of permutations with a single ascent: $(1,1,2,2)$, and $(2,1,1,2)$. To prove that $E^2(2,0) + E^2(2,1) = 3 = E^3(2,1)$, we may insert 12 at the beginning of $(2,2,1,1)$, as well as in each ascent of $(1,1,2,2)$ and $(2,1,1,2)$, in order to obtain three permutations in Q_3^2 with a single ascent: $(1,2,2,2,1,1)$, $(1,1,1,2,2,2)$, and $(2,1,1,1,2,2)$, which are exactly three permutations counted by $E^3(2,1)$.
8. It follows from the definition of $A_n^m(x)$ and the recurrent relation for the numbers $E^m(n,k)$.
9. In fact, using the recurrent relation for $A_n^m(x)$, we get that $A_{n+1}^m(1) = (1+mn)A_n^m(1)$. So, we have that

$$\sum_{k=0}^{n} E^m(n+1,k) = A_{n+1}^m(1) = (1+mn)A_n^m(1)$$
$$= (1+mn)\sum_{k=0}^{n-1} E^m(n,k);$$

therefore, it holds that

$$\sum_{k=0}^{n-1} E^m(n,k)$$
$$= ((n-1)m+1)\sum_{k=0}^{n-2} E^m(n-1,k)$$

$$= ((n-1)m+1)((n-2)m+1)\sum_{k=0}^{n-3} E^m(n-2,k) = \cdots$$

$$\cdots = ((n-1)m+1)((n-2)m+1)\cdots(m+1)E^m(1,0)$$

$$= ((n-1)m+1)((n-2)m+1)\cdots(m+1)\cdot 1$$

$$= ((n-1)m+1)!!!\cdots!.$$

10. In fact, the functions $\alpha_n^m(x)$ and $\beta_n^m(x)$ can be considered the m-order analogue of the Eulerian fractions $\alpha_n(x) = \frac{A_n(x)}{(1-x)^{n+1}}$, having the property $(x\alpha_n(x))' = \alpha_{n+1}(x)$. See Chapters 3 and 4.
11. For $m = 1$, we have that $f_{1,n}(k) = (k+1)^n$. For $m = 2$, we have that $f_{2,n}(k) = S(n+k,k)$. See [He23].
12. One should use two representations of $xA_n^m(x)$. On the one hand, $xA_n^m(x) = \sum_{k=0}^{n} E^m(n,k)t^{k+1}$. On the other hand, $xA_n^m(x) = (1-x)^{mn+1}\sum_{l\geq 0} f_{m,n}(l)x^l$. Using the binomial theorem, we can state that

$$(1-x)^{mn+1} = \sum_{j=0}^{mn+1} \binom{mn+1}{j}(-1)^j x^j.$$

After simple transformations, we get the result by comparing the coefficients at t^{k+1} in the two decompositions of $xA_n^m(x)$.
13. Rotate the triangle of Entringer numbers and change the sign of $E_{m,n}$ when $m+n \equiv 1,2(mod\ 4)$. See [Stan97].
14. Use the exponential generating function $\sum_{n\geq 1} \frac{g(n)x^n}{n!}$. For further information and a bijective proof, see [Stan97] and [KPP94].
15. Use the Genocchi numbers $1, 1, 3, 17, 155, 2073, 38227, 929569, 28820619, 1109652905, \ldots$ (sequence A110501 in the OEIS). See [Comt74].
16. See, for example, [Wiki25].
17. See [BHM23].
18. One way to explain this recurrence is to partition the $T(n)$ connection patterns of the n subscribers to a telephone system into patterns in which the first person is not calling anyone else and

patterns in which the first person is making a call. There are $T(n-1)$ connection patterns in which the first person is disconnected, explaining the first term of the recurrence. If the first person is connected to someone, there are $n-1$ choices for that person and $T(n-2)$ patterns of connection for the remaining $n-2$ people, explaining the second term of the recurrence.

19. In each term of the first sum, k gives the number of matched pairs, the binomial coefficient $\binom{n}{2k}$ counts the number of ways of choosing the $2k$ elements to be matched, and the double factorial $(2k-1)!! = \frac{(2k)!}{2^k\,k!}$ is the product of the odd integers up to its argument and counts the number of ways of completely matching the $2k$ selected elements.

    It follows from the summation formula and Stirling's approximation that, asymptotically, $T(n) \sim \left(\frac{n}{e}\right)^{n/2} \frac{e^{\sqrt{n}}}{(4e)^{1/4}}$. Moreover, the exponential generating function of the telephone numbers is $e^{\left(\frac{x^2}{2}+x\right)}$. See [Deza24].
20. See [Jablo01]. See also a similar proof for the sequence $S(n,k)$ in [Deza24].
21. Use the decomposition $(-1)^k\binom{z}{k} = \binom{-z+k-1}{k} = \frac{1}{\Gamma(-z)}\frac{1}{(k+1)^{z+1}}$ $\prod_{j=k+1} \frac{(1+\frac{1}{j})^{-z-1}}{1-\frac{z+1}{j}}$.
22. (a) These simple bounds follow from the relation $4^n = (1+1)^{2n} = \sum_{k=0}^{2n}\binom{2n}{k}$.

Solutions to Problems Connected with Number Triangles

1. Using the conditions $S(0,0) = 1$, $S(n,0) = 0, n > 0$, and $S(n,k) = 0, k > n$, we can obtain all values of the Stirling numbers of the second kind, i.e., we can construct the triangle of Stirling numbers of the second kind (sequence A008277 in the OEIS):

$n\backslash k$	0	1	2	3	4	5	6	7	8	9
0	1									
1	0	1								
2	0	1	1							
3	0	1	3	1						
4	0	1	7	6	1					
5	0	1	15	25	10	1				
6	0	1	31	90	65	15	1			
7	0	1	63	301	350	140	21	1		
8	0	1	127	966	1701	1050	266	28	1	
9	0	1	255	3025	7770	6951	2646	462	36	1

2. Using the recurrent relation $s(n,k) = -(n-1)s(n-1,k) + s(n-1,k-1)$, we can build the *triangle of Stirling numbers of the first kind* (sequence A048994 in the OEIS):

n/k	0	1	2	3	4	5	6	7	8	9
0	1									
1	0	1								
2	0	−1	1							
3	0	2	−3	1						
4	0	−6	11	−6	1					
5	0	24	−50	35	−10	1				
6	0	−120	274	−225	85	−15	1			
7	0	720	−1764	1624	−735	175	−21	1		
8	0	−5040	13268	−13132	6769	−1960	322	−28	1	
9	0	40320	−109584	118124	−67284	22449	−4536	546	−36	1

Examining the sums of the elements of the nth row of the constructed triangle, we can see that the sum of the elements of the nth row, $n \geq 2$, of the triangle is equal to zero; the alternating sum of the elements of the nth row, $n \geq 2$, of the triangle is equal to $(-1)^n n!$.

3. The following figure shows the first six rows of the triangle:

$$\begin{array}{ccccccccccc} & & & & & 1 & & & & & \\ & & & & 1 & & 1 & & & & \\ & & & 2 & & 3 & & 1 & & & \\ & & 6 & & 11 & & 6 & & 1 & & \\ & 24 & & 50 & & 35 & & 10 & & 1 & \\ 120 & & 274 & & 225 & & 85 & & 15 & & 1 \end{array}$$

4. Let $(1+x)(2+x)(3+x)\cdots(n+x) = E_n^0+E_n^1x+E_n^2x^2+\cdots+E_n^nx^n$. As $E_n^0 = n!$, $E_n^n = 1$, it holds that:

$$\begin{aligned}
&(1+x)(2+x)(3+x)\cdots(n+x)((n+1)+x)\\
&\quad = (E_n^0 + E_n^1x + E_n^2x^2 + \cdots + E_n^nx^n)((n+1)+x)\\
&\quad = (n+1)E_n^0 + (n+1)E_n^1x + (n+1)E_n^2x^2 + \cdots\\
&\qquad + (n+1)E_n^nx^n + E_n^0x + E_n^1x^2 + E_n^2x^3 + \cdots + E_n^nx^{n+1}\\
&\quad = E_n^0(n+1) + (E_n^0 + (n+1)E_n^1)x\\
&\qquad + (E_n^1 + (n+1)E_n^2)x^2 + \cdots + E_n^nx^{n+1}\\
&\quad = E_{n+1}^0 + E_{n+1}^1x + E_{n+1}^2x^2 + \cdots + E_{n+1}^{n+1}x^{n+1}.
\end{aligned}$$

So, $E_{n+1}^0 = (n+1)!$, $E_{n+1}^{n+1} = 1$, and $E_n^{k-1} + (n+1)E_n^k = E_{n+1}^k$, $1 \le k \le n-1$. As we obtain the same recurrent relation with the same initial conditions, it holds that $E_n^k = e(n,k)$.

5. For $n=0$, it holds that $e(0,0) = (0+1)! = 1$. For $n=1$, it holds that $e(1,0)+e(1,1) = (1+1)! = 2$. Let the statement be true for $n=k$: $e(k,0)+e(k,1)+\cdots+e(k,k) = (k+1)!$. Then, for $n=k+1$, it holds that $e(k+1,0)+e(k+1,1)+\cdots+e(k+1,k+1) = (k+2)!$ In fact,

$$\begin{aligned}
&e(k+1,0) + e(k+1,1) + \cdots + e(k+1,k+1)\\
&\quad = (k+1)! + e(k,0) + (k+1)e(k,1)
\end{aligned}$$

$$\begin{aligned}
&+ e(k,1) + (k+1)e(k,2) + \cdots \\
&\cdots + e(k,k-1) + (k+1)e(k,k) \\
&= (k+1)! + (e(k,0) + e(k,1) + \cdots \\
&\cdots + e(k,k-1) + e(k,k)) + (k+1)(e(k,1) + e(k,2) + \cdots \\
&\cdots + e(k,k-1)) + ke(k,k) \\
&= 2(k+1)! + (k+1)((k+1)! - e(k,0) - e(k,k)) + ke(k,k) \\
&= 2(k+1)! + (k+1)((k+1)! - k! - 1) + k \\
&= 2(k+1)! + k(k+1)! = (k+1)!(k+2) = (k+2)!.
\end{aligned}$$

6. The first five rows of Lucas's triangle are given as follows:

2					
1	2				
1	3	2			
1	4	5	2		
1	5	9	7	1	
1	6	14	16	9	2

7. Similar to the original Pascal's triangle giving the Fibonacci numbers as sums of its diagonals, this triangle gives the Lucas numbers $2, 1, 3, 4, 7, 11, 18, 29, 47, 76, \ldots$ (sequence A000032 in the OEIS). The proof is similar.
8. In follows from the construction. As it is true for small values of the parameters, then

$$L_n^k = L_{n-1}^k + L_{n-1}^{k-1} = \binom{n-1}{k} + \binom{n-2}{k-1} + \binom{n-1}{k-1} + \binom{n-2}{k-2}$$
$$= \binom{n}{k} + \binom{n-1}{k-1}.$$

9. It follows from the construction of the triangle ($TT_0^0 = 1$, $TT_n^0 = TT_n^n = 1$, and $TT_n^k = T_{n-1}^k + TT_{n-1}^{k-1} + T_{n-2}^{k-1}$), and the recurrent definition of the Tribonacci numbers: $t_{n+3} = t_{n+2} + t_{n+1} + t_n$ with the initial conditions $t_0 = t_1 = 1, t_2 = 2$.

10. The triangle, often also called the *Tribonacci triangle*, starts with the rows

$$\begin{array}{llllll}
1 & & & & & \\
1 & 1 & & & & \\
2 & 1 & 2 & & & \\
4 & 2 & 2 & 4 & & \\
7 & 4 & 4 & 4 & 7 & \\
13 & 7 & 8 & 8 & 7 & 13
\end{array}$$

(sequence A082793 in the OEIS). By the construction, the Tribonacci numbers form its sides.

11. The first few rows of the triangle are shown as follows (sequence A034851 in the OEIS):

$$\begin{array}{ccccccccccccccc}
 & & & & & & & 1 & & & & & & & \\
 & & & & & & 1 & & 1 & & & & & & \\
 & & & & & 1 & & 1 & & 1 & & & & & \\
 & & & & 1 & & 2 & & 2 & & 1 & & & & \\
 & & & 1 & & 2 & & 4 & & 2 & & 1 & & & \\
 & & 1 & & 3 & & 6 & & 6 & & 3 & & 1 & & \\
 & 1 & & 3 & & 9 & & 10 & & 9 & & 3 & & 1 & \\
1 & & 4 & & 12 & & 19 & & 19 & & 12 & & 4 & & 1
\end{array}$$

12. It follows from the construction.

13. The second pair of diagonals contain the "quarter-squares" (sequence A002620 in the OEIS), or the square numbers and pronic numbers interleaved.

    The next pair of diagonals contain the *alkane numbers* $l(6, n)$ (sequence A005993 in the OEIS). And the next pair of diagonals contain the alkane numbers $l(7, n)$ (sequence A005994 in the OEIS), while the next pair has the alkane numbers $l(8, n)$ (sequence A005995 in the OEIS), then alkane numbers $l(9, n)$ (sequence A018210 in the OEIS), then $l(10, n)$ (sequence A018211 in the OEIS), then $l(11, n)$ (sequence A018212 in the OEIS), then $l(12, n)$ (sequence A018213 in the OEIS), etc.

14. The first few elements are $1, 2, 3, 6, 10, 20, 36, 72, 136, 272, \ldots$ (sequence A005418 in the OEIS).

15. Use the recurrent definition of the Fibonacci numbers.

16. The first few rows of the Bell triangle are

$$\begin{array}{ccccccccc} & & & & 1 & & & & \\ & & & 1 & & 2 & & & \\ & & 2 & & 3 & & 5 & & \\ & 5 & & 7 & & 10 & & 15 & \\ 15 & & 20 & & 27 & & 37 & & 52 \end{array}$$

(sequence A011971 in the OEIS). Numbering the rows of the resulting triangle, starting with the zeroth row, and the elements of a given row from a zero element, we get for the numbers t_n^k, $n = 0, 1, 2, 3, \ldots,$ $k = 0, 1, 2, \ldots, n$, the relations immediately following from the above algorithm:

$$t_0^0 = 1, \quad t_n^0 = t_{n-1}^{n-1}, \quad n \geq 1, \quad t_n^k = t_n^{k-1} + t_{n-1}^{k-1}, \quad 1 \leq k \leq n.$$

It turns out that the number t_n^k is the number of such partitions of the set $\{1, 2, \ldots, n+1\}$ that the element $n+1$ does not fall into the same class as any of the elements $k+1, k+2, \ldots, n$.

17. The following illustration shows Clark's triangle for $n = 6$ (sequence A090850 in the OEIS):

$$\begin{array}{ccccccccc} & & & & 0 & & & & \\ & & & 6 & & 1 & & & \\ & & 12 & & 7 & & 1 & & \\ & 18 & & 19 & & 8 & & 1 & \\ 24 & & 37 & & 27 & & 9 & & 1 \end{array}$$

In this case, $T^6_{m2} = (m-1)^3$ and $T^6_{m3} = \left(\frac{(m-1)(m-2)}{2}\right)^2 = (S_3(m-2))^2$.

18. The first eight rows of the triangle are

$$\begin{array}{ccccccccccccccc} & & & & & & & 1 & & & & & & & \\ & & & & & & \frac{1}{2} & & \frac{1}{2} & & & & & & \\ & & & & & \frac{1}{3} & & \frac{1}{6} & & \frac{1}{3} & & & & & \\ & & & & \frac{1}{4} & & \frac{1}{12} & & \frac{1}{12} & & \frac{1}{4} & & & & \\ & & & \frac{1}{5} & & \frac{1}{20} & & \frac{1}{30} & & \frac{1}{20} & & \frac{1}{5} & & & \\ & & \frac{1}{6} & & \frac{1}{30} & & \frac{1}{60} & & \frac{1}{60} & & \frac{1}{30} & & \frac{1}{6} & & \\ & \frac{1}{7} & & \frac{1}{42} & & \frac{1}{105} & & \frac{1}{140} & & \frac{1}{105} & & \frac{1}{42} & & \frac{1}{7} & \\ \frac{1}{8} & & \frac{1}{56} & & \frac{1}{168} & & \frac{1}{280} & & \frac{1}{280} & & \frac{1}{168} & & \frac{1}{56} & & \frac{1}{8} \\ & & & \vdots & & & & \vdots & & & & \vdots & & & \end{array}$$

(see sequence A003506 in the OEIS).

19. For example, the condition $GT(n,k) = 2GT(n-1,k-1) + T(n-1,k)$, with $GT(n,0) = n, GT(n,n) = n^2$, give the following triangle (sequence A228576 in the OEIS):

$$\begin{array}{ccccccccc}
 & & & & 0 & & & & \\
 & & & 1 & & 1 & & & \\
 & & 2 & & 3 & & 4 & & \\
 & 3 & & 7 & & 10 & & 9 & \\
4 & & 13 & & 24 & & 29 & & 19
\end{array}$$

Glark's triangle gives another example. See also *Riordan arrays*, [Spru94].

20. The first few rows of the triangle are presented as follows:

$$\begin{array}{ccccccccc}
 & & & & 1 & & & & \\
 & & & 1 & & 2 & & & \\
 & & 1 & & 3 & & 4 & & \\
 & 1 & & 4 & & 7 & & 8 & \\
1 & & 5 & & 11 & & 15 & & 16
\end{array}$$

The properties are obvious. So, the triangle contains all the Mersenne numbers.

21. The Magog triangles are in one-to-one correspondence with totally symmetric self-complementary plane partitions (see [Weis25]). An example is

$$\begin{array}{ccccc}
1 & & & & \\
1 & 1 & & & \\
1 & 1 & 1 & & \\
1 & 1 & 1 & 3 & \\
1 & 2 & 3 & 4 & 5
\end{array}$$

22. A monotone triangle of order 5 is presented as follows:

$$\begin{array}{ccccccccc}
 & & & & 4 & & & & \\
 & & & 2 & & 5 & & & \\
 & & 1 & & 4 & & 5 & & \\
 & 1 & & 3 & & 4 & & 5 & \\
1 & & 2 & & 3 & & 4 & & 5
\end{array}$$

There is a bijection between the monotone triangles of order n and the alternating sign matrices of order n (see [Weis25]).

Bibliography

[Abra40] Abramov A.N. *Amazing Numbers*, M.-L.: Childish Publishing House, 1940.

[Abra74] Abramovitch V. Bernoulli numbers, *Kvant*, 1974, 6, 10–14.

[AbSt72] Abramowitz M. and Stegun I.A. *Handbook of Mathematical Functions with Formulas, Graphs, and Mathematical Tables*, New York: Dover Publications, 1972.

[Adam97] Adamchik V. On Stirling numbers and Euler sums, *J. Comput. Appl. Math.*, 1997, 79, 119–130.

[Aign82] Aigner M. *Combinatorial Theory*, Moscow: Mir, 1982.

[Ande03] Anderson D.A. *Discrete Mathematics and Combinatorics*, Moscow: Williams, 2003.

[Andr79] André D. Developpements de séc x et de tang x, *Compt. Acad. Sci.*, 1879, 88, 965–967.

[Andr98] Andrews G.E. *The Theory of Partitions*, Cambridge: Cambridge University Press, 1998.

[Apos86] Apostol T.M. *Introduction to Analytic Number Theory*, New York: Springer-Verlag, 1986.

[Bake71] Baker A. Imaginary quadratic fields with class number two, *Ann. Math.*, 1971, 94, 139–152.

[BaCo87] Ball W.W.R. and Coxeter H.S.M. *Mathematical Recreations and Essays*, 13th ed., New York: Dover, 1987.

[Bell34] Bell E.T. Exponential numbers, *Amer. Math. Monthly*, 1934, 41, 411–419.

[BLM20] Benjamin A.T., Lentfer J. and Martinez T.C. Counting on Euler and Bernoulli numbers identities, 2020, arXiv:2007.12295v1.

[BeSl95] Bernstein M. and Sloane N.J.A. Some canonical sequences of integers, *Linear Algebra Appl.*, 1995, 226–228, 57–72.

[Bhar00] Bhargava M. The factorial function and generalizations, *Am. Math. Monthly*, 2000, 107(9), 783–799.

[BiSi11] Bidkhori H. and Sullivant S. Eulerian-Catalan numbers, 2011, arXiv:1101.1108v1.

[BiBa70] Birkhoff G. and Bartee T.C. *Modern Applied Algebra*, New York: McGraw Hill, 1970.

[Bond93] Bondarenko B.A. *Generalized Pascal's Triangles and Pyramids, Their Fractals, Graphs, and Applications*, San José, CA: Fibonacci Association, 1993.

[Boro85] Boro V. *Live Numbers*, Moscow: Mir, 1985.

[BoBa03] Borwein J. and Bailey D. *Mathematics by Experiment: Plausible Reasoning in the 21-st Century*. Wellesley, MA: A K Peters, 2003.

[Bron01] Bronstein E.M. Generating functions, *Sorosovsky Educ. J.*, 2001, 7(2), 10.

[Brua97] Brualdi R.A. *Introductory Combinatorics*, 4th ed., New York: Elsevier, 1997.

[Buch09] Buchstab A.A. *Number Theory*, Moscow: Nauka, 2009.

[BuGr94] Buhler J. and Graham R. Juggling drops and descents, *Am. Math. Monthly*, 1994, 101, 507–519.

[BHM23] Butzer P., He T.-X. and Markett C. Eulerian fractions and Stirling, Bernoulli and Euler functions with complex order parameters and their impact on the polylogarithm function, *Appl. Anal. Discrete Math.*, 2023, 17(00): 2–2.

[Call09] Callan D. A combinatorial survey of identities for the double factorial, 2009, arXiv:0906.1317.

[Carl60] Carlitz L. Note on Norlund's polynomial $B_n^{(z)}$, *Proc. Am. Math. Soc.*, 1960, 11, 452–455.

[Carl65] Carlitz L. The coefficients in an asymptotic expansion, *Proc. Am. Math. Soc.*, 1965, 16(2), 248–252.

[Carl75] Carlitz L. A combinatorial property of q-Eulerian numbers, *Am. Math. Monthly*, 1975, 82(1), 51–54.

[Carl78] Carlitz L. Stirling pairs, *Rendiconti del Seminario Matematico della Universita di Padova*, 1978, Tome 59, 19–44.

[CKSS72] Carlitz L., Kurtz D.C., Scoville R. and Stackelberg O.P. Asymptotic properties of Eulerian numbers, *Z. Wahrscheinlichkeitstheorie*, 1972, 23, 47–54.

[CaRi53] Carlitz L. and Riordan J. Congruences for Eulerian numbers, *Duke Math. J.*, 1953, 20, 339–344.

[CaSc74] Carlitz L. and Scoville R. Generalized Eulerian numbers: Combinatorial applications, *J. Reine Angew. Math.*, 1974, 265, 110–137.

[Cata44] Catalan E. Note Extraite d'une Lettre Adressée à l'Éditeur, *J. Reine Angew. Math.*, 1844, 27, 192.

[Choi13] Choi J. Note on Stirling polynomilas, *J. Chungcheong Math. Soc.*, 26,(3), 1–9.

[CYY20] Choi H., Yeh Y.-N. and Yoo S. Catalan-like number sequences and Hausdorff moment sequences, *Discrete Math.*, 2020, 343(5), 111808.

[ChMo75] Chorneyko I.Z. and Mohanty S.G. On the enumeration of certain sets of planted trees, *J. Combin. Th., Ser. B*, 1975, 18, 209–221.

[ClKi08] Claesson A. and Kitaev S. Classification of bijections between 321- and 132-avoiding permutations, *Sem. Lothar. Combin.*, 2008, 60, Article B60d.

[Cofm75] Cofman A. *Introduction to Applied Combinatorics*, Moscow: Nauka, 1975.

[Comt74] Comtet L. *Advanced Combinatorics: The Art of Finite and Infinite Expansions*, Netherlands: Reidel, 1974.

[CoGu96] Conway J.H. and Guy R.K. *The Book of Numbers*, New York: Springer-Verlag, 1996.

[Dave99] Davenport H. *The Higher Arithmetic: An Introduction to the Theory of Numbers*, 8th ed., Cambridge: Cambridge University Press, 2008.

[Dede63] Dedekind R. *Essays on the Theory of Numbers*, Cambridge: Cambridge University Press, 1963.

[DeZa80] Dershowitz N. and Zaks S. Enumeration of ordered trees, *Disc. Math.*, 1980, 31, 9–28.

[DeMo10] Deza E. and Model D. *Elements of Discrete Mathematics*, Moscow: URSS, 2010.

[DeDe12] Deza E. and Deza M.M. *Figurate Numbers*, World Scientific, 2012.

[Deza17] Deza E. *Special Positive Integer Numbers*, Moscow: URSS, 2017.

[Deza18] Deza E. *Special Combinatorial Numbers*, Moscow: URSS, 2018.

[Deza21] Deza E. *Mersenne Numbers and Fermat Numbers*, World Scientific, 2021.

[Deza23] Deza E. *Prefect and Amicable Numbers*, World Scientific, 2023.

[Deza24] Deza E. *Stirling Numbers*, World Scientific, 2024.

[Deza25] Deza E. *Catalan Numbers*, World Scientific, 2025.

[Dick05] Dickson L.E. *History of the Theory of Numbers*, New York: Dover, 2005.

[Dorr65] Dörrie H. *100 Great Problems of Elementary Mathematics: Their History and Solutions*, New York: Dover, 1965.

[Entr66] Entringer R.S. A combinatorial interpretation of the Euler and Bernoulli number, *Nieuw Arch. Wisk.*, 1966, 14, pp. 241–246.

[Erus00] Erusalemskii J.M. *Discrete Mathematics*, Moscow: University Book, 2000.

[Eule36] Euler L. Methodus universalis series summandi ulterius promota, *Commentarii academiae scientiarum imperialis Petropolitanae*, 1736, 8, 147–158.

[Eule48] Euler L. *Introductio in Analysin Infinitorum*, Vol. 1, 2, Geneva: Marc Michel Bousquet, 1748.

[Eule49] Euler L. *Remarques sur un beau rapport entre les séries des puissances tant direct que réciproques*, Memoires de l'Académie Royale des Sciences et des Belles-Lettres 17, in Histoire de l'Académie Royale des Sciences et des Belles-Lettres de Berlin 1761, Berlin: Haude et Spener, 1768, pp. 83–106.

[Eule55] Euler L. *Institutiones calculi differentialis cum eius usu in analysi finitorum ac doctrina serierum*, St. Petersburg: Academia imperialis scientiarum Petropolitana, 1755.

[Fich01] Fichtenholtz G.M. *Course of Differential and Integral Calculus* (in three volumes), Moscow: Phtismatlit, 2001.

[FlSe09] Flajolet P. and Sedgewick R. *Analytic Combinatorics*, Cambridge: Cambridge University Press, 2009.

[Foat10] Foata D. Eulerian polynomials: From Euler's time to the present, In: Alladi, K., Klauder, J., Rao, C. (eds.), *The Legacy of Alladi Ramakrishnan in the Mathematical Sciences*, New York: Springer, 2010.

[FoSh70] Foata D. and Schützenberger M.-P. *On the Rook Polynomials of Ferrers Relations,* Colloquia Mathematica Societatis Jano Bolyai, 4, Combinatorial Theory and Its Applications, Vol. 2 (P. Erdős, A. Renyi, and V. Sós, eds.), North-Holland, Amsterdam, 1970, pp. 413–436.

[FoSh70a] Foata D. and Schützenberger M.-P. *Théorie Géométrique des Polynômes Eulériens*, New York: Springer-Verlag, 1970.

[Frob10] Frobenius G. Über die Bernoullischen Zahlen und die Eulerschen Polynome, *Sitzungsberichte der Preußischen Akademie der Wissenschaften*, 1910, 809–847.

[Fu21] Fu A.M. Some identities related to the second-order Eulerian numbers, 2021, arXiv:2104.09316v2.

[Gard61] Gardner M. *Second Book of Mathematical Puzzles and Diversions*, New York: Freeman, 1961.

[Gard88] Gardner M. *Time Travel and Other Mathematical Bewilderments*, New York: Freeman, 1988.

[Gaus01] Gauss C.F. *Disquisitiones Arithmeticae*, Leipzig: Gerhard Fleischer, 1801.

[GaSa92] Gavrilov G.P. and Sapozhenko A.V. *Tasks and Exercises on the Course of Discrete Mathematics*, Moscow: Nauka, 1992.

[GaNe13] Gawronski W. and Neuschel T. Euler-Frobenius numbers, *Integral Transforms Spec. Funct.*, 2013, 24(10), 817–830.

[Gelf98] Gelfand I.M. *Lectures on Linear Algebra*, Dover: Courier Dover Publications, 1998.

[Gelf59] Gelfond A.O. *Calculus of Finite Differences*, Moscow: GIFML, 1959.

[Geno52] Genocchi A. Intorno all'espressione generale de'numeri Bernulliani, *Ann. Sci. Mat. Fis.*, 1852, 3, 395–405.

[Gess92] Gessel I.M. Super ballot numbers, *J. Symbolic Comput.*, 1992, 14, 179–194.

[GeSt78] Gessel I.M. and Stanley R.P. Stirling polynomials, *J. Comb. Theor. Ser. A*, 1978, 24, 24–33.

[Gins28] Ginsburg J. Note on Stirling's numbers, *Am. Math. Monthly*, 1928, 35, 77–80.

[Gins65] Ginsburg J. *Stirling Numbers*, Great Britain: Encyclopedia Britannica, 1965.

[Goul60] Gould H.W. Stirling number representation problems, *Proc. Am. Math. Soc.*, 1960, 11, 447–451.

[Goul85] Gould H.W. *Bell and Catalan Numbers: Research Bibliography of Two Special Number Sequences*, 6th ed., Morgantown, WV: Math Monongliae, 1985.

[GoJa83] Goulden I.P. and Jackson D.M. *Combinatorial Enumeration*, New York: Wiley, 1983.

[GrCh17] Graham R. and Chung F. *The Digraph Drop Polynomial*, 2017, https://api.semanticscholar.org/CorpusID:129940137.

[GKP94] Graham R.L., Knuth D.E. and Patashnik O. *Concrete Mathematics: A Foundation for Computer Science*, 2nd ed., Reading, MA: Addison-Wesley, 1994.

[GGL95Û] Graham R., Grötschel M. and Lovász L. *Handbook of Combinatorics*, New York: Elsevier Science B.V., 1995.

[GuJa90] Gulden J. and Jacson D. *Enumerating Combinatorics*, Moscow: Nauka, 1990.

[Hagg04] Haggarty R. *Discrete Mathematics for Programmers*, Moscow: Technosphera, 2004.

[Hall70] Hall M. *Combinatorics*, Moscow: Mir, 1970.

[Hara03] Harari F. *Graph Theory*, Moscow: Mir, 2003.

[HaPa77] Harari F. and Palmer E. *Graph Enumeration*, Moscow: Mir, 1977.

[HaWr79] Hardy G.H. and Wright E.M. *An Introduction to the Theory of Numbers*, 5th ed., Oxford: Clarendon Press, 1979.

[He23] He T.-X. The mth-order Eulerian numbers, 2023, arXiv:2312.17153v1.

[Heeg52] Heegner K. Diophantische Analysis und Modulfunktionen, *Mathematische Zeitschrift* (in German), 1952, 56(3), 222–253.

[Hend74] Hendy M.D. Prime quadratics associated with complex quadratic fields of class number 2, *Proc. Am. Math. Soc.*, 1974, 43, 253–260.
[Hogg69] Hoggatt V.E.Jr. *The Fibonacci and Lucas Numbers*, Boston, MA: Houghton Mifflin, 1969.
[Holt97] Holte J. Carries, combinatorics and an amazing matrix, *Am. Math. Monthly*, 1997, 104(2), 138-1-49.
[Hons91] Honsberger R. *More Mathematical Morsels*, Washington: Math. Assoc. Amer., 1991.
[HCD19] Hsien-Kuei Hwang H.-K., Chern H.-H. and Duh G.-H. An asymptotic distribution theory for Eulerian recurrences with applications, 2019, arXiv:1807.01412v2.
[Jablo01] Jablonskii S.V. *Introduction to Discrete Mathematics*, Moscow: Higher School, 2001.
[JaWa24] Janson S. and Warren D.S. Better-than-average uniform random variables and Eulerian numbers, or: How many candidates should a voter approve?, 2024, arXiv:2403.02670v1.
[Jans13] Janson S. Euler-Frobenius numbers and rounding, 2013, arXiv:1305.3512v1.
[Jord65] Jordan C. *Calculus of Finite Differences*, 3rd ed., New York: Chelsea, 1965.
[IrRo90] Ireland K. and Rosen M. *A Classical Introduction to Modern Number Theory*, New York: Springer-Verlag, 1990.
[Kara83] Karatsuba A.A. *Basic Analytic Number Theory*, Moscow: Nauka, 1983.
[KaVo92] Karatsuba A.A. and Voronin S.M. *The Riemann Zeta-Function*, Berlin: Walter de Gruyter & Co., 1992.
[Khin6403] Khinchin A.Ja. *Continued Fractions*, Chicago, IL: University of Chicago Press, 1964.
[Klar70] Klarner D.A. Correspondences between plane trees and binary sequences, *J. Comb. Theory*, 1970, 9, 401–411.
[Knot77] Knott G.D. A numbering system for binary trees, *Commun. ACM*, 1977, 20(2), 113–115.
[Knut76] Knuth D.E. *Mathematics and Computer Science. Coping with Finiteness*, Moscow: Nauka, 1976.
[Knut97] Knuth D.E. *The Art of Computer Programming*, Vol. 1: Fundamental Algorithms, 3rd ed., Reading, MA: Addison-Wesley, 1997.
[Kord95] Kordemskii B.A. *Great Lives in Mathematics*, Moscow: Prosvestcheniye, 1995.
[KoKo77] Korn G. and Korn T. *Handbook of Mathematics (for Scientists and Engineers)*, Moscow: Nauka, 1977.
[Kost82] Kostrikin A. *Introduction to Algebra*, Berlin: Springer-Verlag, 1982.

[KLS01] Krízek M., Luca F. and Somer L. 17 *Lectures on Fermat Numbers: From Number Theory to Geometry*, CMS Books in Mathematics, 9, New York: Springer-Verlag, 2001.

[Kudr36] Kudryavtsev V.A. *Summation of Degrees of Positive Integers and Bernoulli Numbers*, Moscow-Leningrad: ONTI, 1936.

[Kurt72] Kurtz D.C. A note on concavity properties of triangular arrays of numbers, *J. Comb. Theory, Ser. A*, 1972, 13, 135–139.

[KuAd88] Kuznetsov O.P. and Adelson-Velsky G.B. *Discrete Mathematics for Engineers*, Moscow: Energoatomizdat, 1988.

[KPP94] Kuznetsov A.G., Pak I.M. and Postnikov A.E. Increasing trees and alternating permutations, *Russian Math. Surveys*, 1994, 49(6), 79–114.

[Lah54] Lah I. A new kind of numbers and its application in the actuarial mathematics, *Boletim do Instituto dos Actuários Portugueses*, 1954, 9, 7–15.

[Land94] Lando S.K. *Lectures on Combinatorics*, Moscow: MCCME, 1994.

[Land02] Lando S.K. *Lectures on Generating Functions*, Moscow: MCCME, 2002.

[Lege79] Legendre A.-M. *Théorie des Nombres*, 4th ed., Paris: A. Blanchard, 1979.

[LeNi92] Lesieur L. and Nicolas J.-L. On the Eulerian Numbers $M_n = \max_{1 \leq k \leq n} A(n, k)$, *Eur. J. Combinatorics*, 1992, 13, 379–399.

[LiNi96] Lidl R. and Niederreiter H. *Finite Fields*, Cambridge: Cambridge University Press, 1996.

[Line86] Lines M.E. *A Number for Your Though*, Bristol: Adam Hilger, 1986.

[LiWi92] Lint J.H. and Wilson R.M. *Course in Combinatorics*, Cambridge: Cambridge University Press, 1992.

[LeLi83] Le Lionnais F. *Les Nombres Remarquables*, Paris: Hermann, 1983.

[Lips88] Lipsky V. *Combinatorics for Programmers*, Moscow: Mir, 1988.

[Litz63] Litzman V. *Fun and Entertaining about Numbers and Figures*, Moscow: Fizmatgiz, 1963.

[MacM08] MacMahon P.A. Second memoir on the compositions of integers, *Philos. Trans.*, 1908, 207, 65–134.

[MacM16] MacMahon P.A. *Combinatory Analysis*, Vols. 1, 2, Cambridge: Cambridge University Press, 1916.

[Mada79] Madachy J.S. *Madachy's Mathematical Recreations*, New York: Dover, 1979.

[Malc86] Malcolm E. *A Number for Your Thoughts: Facts and Speculations about Number from Euclid to the Latest Computers*, Abingdon-on-Thames: CRC Press, 1986.

[Male11] Malenfant J. Finite, closed-form expressions for the partition function and for Euler, Bernoulli, and Stirling numbers, 2011, arXiv:1103.1585v6.
[Mark0] Markushevich A.I. *Return Sequences*, Leningrad: Technical and Theoretical Literature, 1950.
[Mazu10] Mazur D.R. *Combinatorics. A Guided Tour*, USA: The Mathematical Association of America, 2010.
[Mend07] Mendes A. A note on alternating permutations, *Am. Math. Monthly*, 2007, 114(5), 437–440.
[MSY96] Millar J., Sloane N.J.A. and Young N.E. A new operation on sequences: The Boustrouphedon transform, *J. Comb. Theory Ser. A.*, 1996, 76(1), 44–54.
[MSC96] Mitrinovic D.S., Sandor J. and Crstici B. *Handbook on Number Theory*, Dordrecht: Kluwer Acad. Publ., 1996.
[Nage51] Nagell T. *Introduction to Number Theory*, New York: Wiley, 1951.
[OLBC10] Olver F.W.J., Lozier D.M., Boisvert R.F. and Clark C.W. *Handbook of Mathematical Functions*, Cambridge: Cambridge University Press, 2010.
[Ore48] Ore O. *Number Theory and Its History*, New York: Dover Publications, Inc., 1948.
[Ore80] Ore O. *Graph Theory*, Moscow: Nauka, 1980.
[Otte48] Otter R. The number of trees, *Ann. Math.*, Second Series, 1948, 49(3), 583–599.
[Pasc54] Pascal B. *Traité du triangle arithmétique, avec quelques autres petits traitez sur la mesme matière*, Paris: G. Desprez, 1654.
[Pete15] Petersen T.K. *Eulerian Numbers*, New York: Birkhäuser/ Springer, 2015.
[Plat80] Platonov M.L. *Combinatorial Numbers*, Irkutsk: Irkutsk University, 1980.
[Plou92] Plouffe S. 1031 *Generating Functions and Conjectures*, Montréal: Université du Québec à Montréal, 1992.
[Prim24] *The Prime Pages (prime number research, records and resources)*, https://primes.utm.edu/, 10.03.2025.
[Pudw24] Pudwell L. Catalan numbers and permutations, *Math. Mag.*, 2024, 97(3), 279–283.
[Qi11] Qi D. Note: On the second order Eulerian numbers, *Aust. J. Comb.*, 2011, 50, 183–185.
[QiBa17] Qi F. and Guo B.-N. Explicit formulas and recurrence relations for higher order Eulerian polynomials, *Indag. Math.*, 2017, 28(4), 884–891.
[RaTo57] Rademacher H. and Toeplitz O. *The Enjoyment of Mathematics: Selections from Mathematics for the Amateur*, Princeton, NJ: Princeton University Press, 1957.

[Rein43] Reiner I. *Functions not Formulas for Primes, Am. Math. Monthly*, 1943, 50(10), 619–621.

[Ribe96] Ribenboim P. *The New Book of Prime Numbers Records*, New York: Springer-Verlag, 1996.

[Riem59] Riemann B. *Über die Anzahl der Primzahlen unter einer gegebenen Größe*, Berlin: Monatsberichte der Berliner Akademie, 1859.

[Ries94] Riesel H. *Prime Numbers and Computer Methods for Factorization*, 2nd ed., Basel: Birkhouser, 1994.

[Rior76] Riordan J. The blossoming of Schröder's fourth problem, *Acta Math.*, 1976, 137, 1–16.

[Rior79] Riordan J. *Combinatorial Identities*, New York: Wiley, 1979.

[Rior80] Riordan J. *An Introduction to Combinatorial Analysis*, New York: Wiley, 1980.

[RoTe09] Roberts F.S. and Tesman B. *Applied Combinatorics*, Boca Raton, FL: CRC Press, 2009.

[Rose18] Rosen K.H. *Handbook of Discrete and Combinatorial Mathematics*, Boca Raton, FL: CRC Press, 2018.

[Rota64] Rota G.-C. On the foundations of combinatorial theory I. Theory of Möbius functions, *Zeitschrift fur Wahrscheinlichkeitstheorie und Verwandte Gebiete*, 1964, 2(4), 340–368.

[Rybn82] Rybnikov K.A. *Combinatorial Analysis. Tasks and Exercises*, Moscow: Nauka, 1982.

[Rybn85] Rybnikov K.A. *Introduction to Combinatorial Analysis*, Moscow: Moscow State University, 1985.

[Sachk77] Sachkov V.N. *Combinatorial Methods of Discrete Mathematics*, Moscow: Nauka, 1977.

[Sach97] Sachkov V.N. *Probabilistic Methods in Combinatorial Analysis*, Cambridge: Cambridge University Press, 1997.

[Sand78] Sands A.D. On generalized Catalan numbers, *Disc. Math.*, 1978, 21, 218–221.

[Sier64] Sierpiński W. *Elementary Theory of Numbers*, Warszawa: Państwowe Wydawn, Naukowe, 1964.

[SlPl95] Sloane N.J.A. and Plouffe S. *The Encyclopedia of Integer Sequences*, San Diego, CA: Academic Press, 1995.

[Sloa25] Sloane N.J.A. at all. *On-Line Encyclopedia of Integer Sequences*, http://oeis.org/, 25.03.2025.

[Smit84] Smith D.E. *A Source Book in Mathematics*, New York: Dover, 1984.

[Spru94] Sprugnoli R. Riordan arrays and combinatorial sums, *Disc. Math.*, 1994, 132(1–3), 267–290.

[Stan79] Stanley R.P. *Generating functions*, in Studies in Combinatorics (G.-C. Rota, ed.), Vol. 17 in MAA Stafies in Mathematics, New York: Mathematical Association of America, 1979.

[Stan97] Stanley R.P. *Enumerative Combinatorics,* Vols. 1, 2, Cambridge: Cambridge University Press, 1997.

[Stan10] Stanley R.P. A survey of alternating permutations, *Combinatorics and Graphs, Contemporary Mathematics*, Vol. 531, Providence, RI: American Mathematical Society, 2010.

[Stan15] Stanley R.P. *Catalan Numbers*, Cambridge: Cambridge University Press, 2015.

[Stan86] Stanton D. *Constructive Combinatorics*, New York: Springer-Verlag, 1986.

[Star71] Stark H.M. A transcendence theorem for class number problems, *Ann. Math.*, 1971, 94, 153–173.

[Step01] Stepanova L.L. *Selected Chapters in Number Theory,* Moscow: MSPU, 2001.

[Stei94] Steingrímsson E. Permutation statistics of indexed permutations, *Europ. J. Combinatorics*, 1994, 15(2), 187–205.

[Stra16] Strang G. *Linear Algebra and Its Applications*, 5th ed., Cambridge: Wellesley-Cambridge Press, 2016.

[Stru87] Struik D.J. *A Concise History of Mathematics*, Dover: Dover Publications, 1987.

[Tsao11] Tsao H.P. Sums of powers and Eulerian numbers, *Math. Gaz.*, 2011, DOI: 10.1017/S0025557200003247.

[Uspe76] Uspensky V.A. *Pascal's Triangle: Certain Applications of Mechanics to Mathematics*, Moscow: Mir, 1976.

[Vell08] Vella D.C. Explicit formulas for Bernoulli and Euler numbers, *Integers*, 2008, 8(1), Article A01, 7.

[Vile14] Vilenkin N.Ia. *Combinatorics*, Moscow: Academic Press, 2014.

[Voro61] Vorob'ev N.N. *Fibonacci Numbers*, New York: Blaisdell, 1961.

[WaXu10] Wang R. and Xu Y. The asymptotic properties of Eulerian numbers and refined Eulerian numbers: A Spline perspective, 2010, arXiv:1002.0056v1.

[WeQi15] Wei Ch.-F. and Qi F. Several closed expressions for the Euler numbers, *J. Inequal. Appl.*, 2015, 219(2015), DOI 10.1186/s13660-015-0738-9.

[Weis25] Weisstein E.W. *MathWorld — A Wolfram Web Resource*, http://mathworld.wolfram.com/, 25.03.2025.

[Well86] Wells D. *The Penguin Dictionary of Curious and Interesting Numbers*, Middlesex: Penguin Books, 1986.

[Went99] Wentzel E.S. *Probability Theory*, Moscow: Higher School, 1999.

[Wiki25] Wikipedia, the Free Encyclopedia, http://en.wikipedia.org, 25.03.2025.

[Worp83] Worpitzky J. Studien über die Bernoullischen und Eulerschen Zahlen, *Journal für die reine und angewandte Mathematik*, 1883, 94, 203–232.

[XTH12] Xiong T., Tsao H. and Hall J.I. General Eulerian numbers and Eulerian polynomials, 2012, arXiv:1207.0430.

[Zhan98] Zhang W. Some identities involving the Euler and the central factorial numbers, *Fibonacci Quart.*, 1998, 36(4), 154–157.

Index

www.ingramcontent.com/pod-product-compliance
Ingram Content Group UK Ltd.
Pitfield, Milton Keynes, MK11 3LW, UK
UKHW020404150926
8624IPUK00004B/333